ABSTRACT ALGEBRA

An Introduction to Groups, Rings and Fields

ABSTRACT ALGEBRA

An Introduction to Groups, Rings and Fields

Clive Reis

University of Western Ontario, Canada

World Scientific

NEW JERSEY · LONDON · SINGAPORE · BEIJING · SHANGHAI · HONG KONG · TAIPEI · CHENNAI

Published by

World Scientific Publishing Co. Pte. Ltd.

5 Toh Tuck Link, Singapore 596224

USA office: 27 Warren Street, Suite 401-402, Hackensack, NJ 07601

UK office: 57 Shelton Street, Covent Garden, London WC2H 9HE

British Library Cataloguing-in-Publication Data
A catalogue record for this book is available from the British Library.

ABSTRACT ALGEBRA
An Introduction to Groups, Rings and Fields

ISBN-13 978-981-4335-64-5
ISBN-10 981-4335-64-9
ISBN-13 978-981-4340-28-1 (pbk)
ISBN-10 981-4340-28-6 (pbk)

Printed in Singapore.

To my children,
Helen, Rheannon, Jonathan, Karen and Mark.

Preface

This book has evolved over the years from notes I made for courses—ranging from second-year to fourth-year honors—I taught at the University of Western Ontario, Camosun College and the University of Victoria. As a result, it still bears some of the features of its humble origins: many side remarks, colloquial language and an informal style. I have also tried to convey to the reader how to translate into everyday English the somewhat stodgy language of formal definitions so that the ideas take on a more vivid and dynamic aspect.

The first nine chapters (together with the applications in Chapters 11 or 15) cover the basic material traditionally taught at the second-year level, although I have included topics which are often avoided at this stage: cosets, Lagrange's Theorem, the isomorphism theorems for groups and rings and an introduction to finite fields.

I have several reasons for doing this. First, a course which avoids the concept of a coset and consequently, all the ideas which stem from it, conveys very little of the flavour of abstract algebra and, moreover, makes for a very boring course. Second, in my experience, second-year students with some background in linear algebra are quite capable of dealing with the level of abstraction required to tackle these concepts provided they first encounter cosets in the guise of congruence classes in the integers. Time spent computing in the algebraic system consisting of congruence classes under the usual operations of addition and multiplication makes the transition to factor groups much smoother. Moreover, the natural map from the integers to the integers mod n which identifies an integer with its congruence class prepares the student for the notion of a homomorphism. Lastly, by introducing these topics, the student is able to tackle some interesting applications to Number Theory, Combinatorics and Error-Correcting Codes.

Chapters 10, 12, 13, 14 and 16 are more suitable for students at the third- or fourth-year honors level. In Chapter 12 the reader will find such topics as group actions, the class equation of a group, the Sylow theorems and applications to determining the structure of some of the simpler groups. The chapter ends with a discussion of simple groups and a proof that the alternating groups on five or more symbols form one class of simple groups.

Chapter 13 deals with isometries in Euclidean n-space with emphasis on finite Euclidean groups of dimensions 2 and 3. The rotation groups of the Platonic Solids are obtained and are shown to be the only finite Euclidean groups in Euclidean 3-space. A good background in linear algebra is needed.

Polya-Burnside enumeration with applications to counting is the topic of Chapter 14. A theorem due to Polya is proved and applied to counting the number of inequivalent switching circuits.

Applications of groups and polynomials to the construction of group and polynomial codes will be found in Chapters 15 and 16. Chapter 15 can be covered after Chapter 6 provided the students have acquired enough mathematical maturity to be comfortable with the notions of subgroup and homomorphism. Chapter 16, however, requires the material from Chapter 10 and a certain degree of mathematical maturity which comes only after solving a large number of the problems found at the end of each chapter.

To make the book as self-contained as possible, I have included two appendices, one on the more familiar fields of rational, real and complex numbers, the other on the elementary theorems of linear algebra with emphasis on linear transformations. These appendices can either be used in the classroom as abbreviated reviews of the material or as a resource for the student to be used as a reference.

List of Tables

List of Figures

Contents

Chapter 1

Logic and Proofs

1.1 Introduction

Mathematics in the ancient world was born of practical necessity. In Ancient Egypt the yearly flooding of the Nile obliterated boundaries between properties which had to be redrawn once the waters receded, thus giving rise to a system of linear and areal measurement. In other ancient civilizations of the Middle East, increasing trade brought about the evolution of the concept of number to keep a tally of goods-by volume or weight-and of earnings.

Out of these practical needs there grew a considerable body of "facts" and "formulae", some valid and some merely poor approximations. It was not, however, until the classical Greek period (ca 600-300 BC) that mathematics emerged as a deductive discipline wherein the validity of statements and formulae required justification. It was no longer acceptable to assert that a 3-4-5-triangle is right angled (a fact that was known to the Ancient Egyptians); a "proof" was required.

What precisely is a mathematical proof of an assertion? It is a finite sequence of propositions, each deduced from previously established truths and ending with the required assertion. A moment's reflection will lead us to the conclusion that we must ultimately base our sequence of propositions on some fundamental assumptions which we agree to accept without proof; otherwise, we would be involved in an infinite regress by proving proposition A from B, B from C, C from D, etc.

These fundamental assumptions are called *axioms* or *postulates* and serve as a foundation upon which to build the various branches of mathematics. Later on we shall be studying a number of mathematical systems, among them groups and rings, and we shall see that each has its own set of axioms from which we shall deduce the various properties of the respective systems.

Since proofs are at the very heart of modern mathematics, it is essential that we familiarize ourselves with the fundamentals of deductive reasoning. This chapter is therefore devoted to making the reader aware of some of the simple rules of logic.

1.2   Statements, Connectives and Truth Tables

1.2.1   *Definition*

A *statement* is an assertion which is either true or false. "True" and "False" are the *truth values* of the statement which we abbreviate as T and F, respectively.

1.2.2   *Examples*

    (i)   The sea is salty. (Truth value T)
    (ii) The moon is made of blue cheese. (Truth value F)
    (iii) 4 is an even number. (Truth value T)
    (iv) This is a beautiful painting. (Since beauty is subjective and different people may differ in their assessment of the truth value of this assertion, we deem this assertion not to be a statement)

We denote statements by lower case letters. For example we might let p stand for the statement "The sea is salty". We would indicate this by writing: p: the sea is salty.

We string two or more statements together to form a *compound statement* using *logical connectives*. The truth value of the compound statement naturally depends on the truth values of the components comprising the compound statement. We exhibit this dependence by means of a *truth table* as shown below.

The connectives we shall be using are: $\neg$, $\wedge$, $\vee$, $\rightarrow$, $\leftrightarrow$.

1.2.3 *The Connective* ¬ *(Negation)*

This connective is read "not".

1.2.4 *Example*

$\neg p$ is read "not p". Hence if p stands for "the sea is salty", then $\neg p$ stands for "not (the sea is salty)", which in English we would express as "the sea is not salty". If this connective is to reflect the meaning of "not" in ordinary English, then we must assign the truth value F to the negation of a true statement and T to the negation of a false statement. We obtain the table below.

Table 1.1. The truth table for ¬

p	$\neg p$
T	F
F	T

1.2.5 *Connective* ∧ *(Conjunction)*

This corresponds to the English "and". Bearing in mind that in ordinary English, if we assert that it is a sunny day and the temperature is below freezing, then we have told the truth if it is indeed both sunny and freezing. In all other cases we would be guilty of a falsehood.

Table 1.2. The truth table for ∧

p	q	$p \wedge q$
T	T	T
T	F	F
F	T	F
F	F	F

1.2.6 *Example*

p: cobras are poisonous; *q*: 5 is an odd number
$p \wedge q$: cobras are poisonous *and* 5 is an odd number.

1.2.7 *The Connective* ∨ *(Disjunction)*

We translate this by "or" in the inclusive sense. (Disjunction). It has the meaning "either *p* or *q* or both". It is thus clear that only when both *p* and *q* are false will the compound statement be false. Observe that in colloquial English, "or" is often used in the exclusive sense. For example, if I say "For my graduate work, I shall go to Harvard or Yale", clearly, only one of these events will occur. There is a symbol for exclusive "or" but we shall not have occasion to use it since the compound statement $(p \vee q) \wedge \neg (p \wedge q)$ has the same truth table as that for exclusive "or".

Table 1.3. Truth table for ∨

p	*q*	$p \vee q$
T	T	T
T	F	T
F	T	T
F	F	F

1.2.8 *The Connective* → *(Implication, Conditional)*

This is translated in English by "if _____ , then_____ ."

Table 1.4. Truth table for →

p	*q*	$p \to q$
T	T	T
T	F	F
F	T	T
F	F	T

Note. In English, the following are alternatives for "if p then q": p only if q; p is a sufficient condition for q; q is a necessary condition for p.

Here is an example which might help the reader understand how the truth values are arrived at: Let p: you do your homework; q: I shall take you to the movies, and consider under what circumstances you would deem me to have lied. The first two rows of the truth table are clear. In the third row, you do not do your homework but I still take you to the movies. At worst I can be accused of being lenient, but certainly not of lying. Since we must assign a truth value, by default it must be T. As for the last line, neither you nor I have kept the bargain and so I cannot be accused of lying. Therefore, again, by default, the truth value is T.

1.2.9 *The Connective* $\leftrightarrow$ *(Bi-Implication, Biconditional)*

This corresponds in ordinary English to "____ if, and only if,_____". (Bi-implication, biconditional)

Table 1.5. Truth table for $\leftrightarrow$

p	q	$p \leftrightarrow q$
T	T	T
T	F	F
F	T	F
F	F	T

1.2.10 *Example*

Let p: I go to the movies; q: I finish my work. In the statement $p \leftrightarrow q$, "I go to the movies" is synonymous with "I finish my work" in the sense that either action can be deduced from the other: my presence at the movies means that I have finished my homework and, conversely, my completed homework means that I go to the movies. The truth values assigned in the table then become self-evident.

1.3 Relations Between Statements

1.3.1 *Definition*

A statement x *logically implies* a statement y if, whenever x is true, so is y. We write $x \Rightarrow y$.

1.3.2 *Example*

$$(p \leftrightarrow q) \Rightarrow (p \to q).$$

Let x denote the statement $p \leftrightarrow q$ and y the statement $p \to q$. Suppose x is true. Then either both p and q are true or both are false. In both cases y is true.

Note. To check that $x \Rightarrow y$, draw up the truth table of $x \to y$. Every truth value of $x \to y$ should be T. If all the entries in the truth table of a statement are T's, then the statement is called a *tautology*. Thus the statement $(p \leftrightarrow q) \to (p \to q)$ is a tautology. Roughly speaking, if $x \Rightarrow y$ then x conveys more information than y.

1.3.3 *Definition*

A statement x is *logically equivalent* to a statement y if, whenever x is true, so is y, and whenever y is true, so is x. We write $x \Leftrightarrow y$ in this case.

The reader can easily check that $x \Leftrightarrow y$ precisely when $x \leftrightarrow y$ is a tautology.

Below is a list of equivalences which the reader is urged to prove.

1.3.4 *Example*

(i) $\neg(\neg p) \Leftrightarrow p$;

(ii)　$(p \to q) \Leftrightarrow (\neg p \vee q) \Leftrightarrow (\neg q \to \neg p)$
　　$(\neg q \to \neg p$ is the *contrapositive* of $p \to q$.);

(iii)　$\neg(p \wedge q) \Leftrightarrow (\neg p \vee \neg q)$ (DeMorgan's Law);

(iv)　$\neg(p \vee q) \Leftrightarrow (\neg p \wedge \neg q)$ (DeMorgan's Law);

(v)　$p \wedge (q \vee r) \Leftrightarrow (p \wedge q) \vee (p \wedge r)$ ($\wedge$ is distributive over $\vee$);

(vi)　$p \vee (q \wedge r) \Leftrightarrow (p \vee q) \wedge (p \vee r)$ ($\vee$ is distributive over $\wedge$);

(vii)　$p \leftrightarrow q \Leftrightarrow (p \to q) \wedge (q \to p)$.

In later chapters you will often be asked to prove statements of the form $x \leftrightarrow y$. This is done by proving $x \to y$ and then proving *the converse of* $x \to y$, namely, $y \to x$ (see (vii) above)

1.4 Quantifiers

An assertion such as $x^2 - 3x + 2 = 0$ has no truth value as it stands since x is a variable taken from some *universal set* (or *universe of discourse*), in this case, the real numbers, say, and the truth value of the assertion depends on the value of x. Such a statement is called an *open statement*. By prefacing it by a *quantifier*, its truth value can be determined as we shall see.

We shall use two quantifiers, namely, "for all", called the *universal quantifier* and "there exists", called the *existential quantifier*". In English there are many ways of expressing each of these. Here are some examples: for "for all" we have: "for each", "for every", "for any"; for "there exists", we have: "for some", "for at least one".

The symbols used for the universal and existential quantifiers are "$\forall$" and "$\exists$", respectively.

Let us go back to the assertion $x^2 - 3x + 2 = 0$ and preface it by $\forall x \in \mathbf{R}$ where $\mathbf{R}$ is the set of real numbers and "$\in$" is the symbol for membership of an element in a set. (See Chapter 2.) We then have

$$(\forall x \in \mathbf{R})(x^2 - 3x + 2 = 0).$$

This should be read: For all real numbers x, we have $x^2 - 3x + 2 = 0$. Is this statement true or false? Because of the universal quantifier, for it to be true, every real number, when substituted for x in the expression $x^2 - 3x + 2$ would have to yield 0. This is obviously not the case and so the statement above is false.

Consider now the statement:

$$(\exists x \in \mathbf{R})(x^2 - 3x + 2 = 0).$$

In words this states that there is a real number a such that $a^2 - 3a + 2 = 0$, a valid statement since, for example, $a = 1$ yields 0 on substitution.

As another example, consider the open statement "$y = x^2$" with universal set $\mathbf{R}$ as above. This open statement involves two variables, namely, x and y. In order to be able to assign a truth value to this statement, we need to quantify both variables. Here are the possibilities:

(i) $(\forall x \forall y)(y = x^2)$; .
(ii) $(\forall x \exists y)(y = x^2)$;
(iii) $(\exists x \forall y)(y = x^2)$;
(iv) $(\exists x \exists y)(y = x^2)$;
(v) $(\forall y \exists x)(y = x^2)$;
(vi) $(\exists y \forall x)(y = x^2)$.

Note that it is tacitly assumed that the variables come from the universal set $\mathbf{R}$ of real numbers.

We translate each of these six statements into words and establish the truth value of each:

(i)   For all real numbers x and y, $y = x^2$; (F). Take, for example, $y = 1$ and $x = 2$.

(ii) For each real number x there is a real number y such that $y = x^2$; (T).

(iii) There exists a real number x such that the equality holds for all real numbers y; (F).

(iv) There exist real numbers x and y such that the equality holds; (T).

(v)   For every given value y there exists x such that $y = x^2$; (F). $y = -2$ yields no such x.

(vi) There exists a value y such that, for all values of x, we have $y = x^2$; (F).

Suppose that P is some statement involving variables $x, y, z, \ldots$ all taken from a set U. We write $P(a, b, c, \ldots)$ to mean that the statement is true for this particular choice of variables. Thus for example, $(\forall x \forall y \forall z \ldots)(P(x, y, z, \ldots))$ means that the statement is true for all values of the variables.

We now establish rules for the negation of such statements. We restrict ourselves to two variables, though the reader should be able to extend the rules to any number of variables.

We begin with the statement $(\exists x \forall y)(P(x, y))$, which states that there is a value x taken from the universal set U such that the statement $P(x, y)$ is true whatever y we choose from U. The negation would therefore assert that no such x exists. That is, whatever x we choose, we can find a y such that $P(x, y)$ is false. But if $P(x, y)$ is false, then $\neg P(x, y)$ is true. Therefore, for all values of x we can find a y so that $\neg P(x, y)$ is true. Thus

$$\neg[(\exists x \forall y)(P(x, y))] \Leftrightarrow (\forall x \exists y)(\neg P(x, y)).$$

In a similar manner, we can establish the following:

$$\neg[(\forall x \exists y)(P(x, y))] \Leftrightarrow (\exists x \forall y)(\neg P(x, y)).$$

$$\neg[(\forall x \forall y)(P(x, y))] \Leftrightarrow (\exists x \exists y)(\neg P(x, y)).$$

$$\neg[(\exists x \exists y)(P(x, y))] \Leftrightarrow (\forall x \forall y)(\neg P(x, y)).$$

The pattern is now apparent: In negating a statement with quantifiers, each $\forall$ turns into $\exists$ and conversely. In addition, the statement $P(x, y)$ is negated.

1.4.1 *Example*

Negate $(\forall x \in U)(P(x) \to Q(x))$.

Abstract Algebra

Recall that $p \rightarrow q \Leftrightarrow \neg p \vee q$. Hence,

$$\neg[(\forall x \in U)(P(x) \rightarrow Q(x))] \Leftrightarrow (\exists x \in U)(\neg[P(x) \rightarrow Q(x)])$$

$$\Leftrightarrow (\exists x \in U)(\neg[\neg P(x) \vee Q(x)]) \Leftrightarrow (\exists x \in U)(P(x) \wedge \neg Q(x)).$$

In the last equivalence, we have used one of DeMorgan's Laws. ∎

1.4.2 *Example*

Negate $(\forall x \in \mathbf{R})(x^2 > 0 \rightarrow x > 0)$.
The negation is, as we have seen above:

$$(\exists x \in \mathbf{R})[\neg\{\neg(x^2 > 0) \vee (x > 0)\}] \Leftrightarrow (\exists x \in \mathbf{R})[(x^2 > 0) \wedge (x \le 0)].$$

This last statement is true as one can see by taking $x = -2$ and so the original statement is false. ∎

1.5 Methods of Proof

In this course you will often be asked to prove (or disprove) a given statement. If the statement is universal, that is, one which asserts that all members of a given set have a certain property, then providing an example of a member of the set having the property is clearly not sufficient. On the other hand, if you are able to find a member of the set without the given property, then you have proved that the universal statement is false; that is, you have disproved the statement.

To firm up our ideas, let us consider some examples.

1.5.1 *Example*

If n is an odd integer, then $n^2 - 1$ is divisible by 8. Here the universal set is the set $\mathbf{Z}$ of all whole numbers, positive, negative and zero.

Proof. We can restate this symbolically as:

$$(\forall n \in \mathbf{Z})(n \text{ odd} \rightarrow n^2 - 1 \text{ is divisible by } 8).$$

This is a universal statement which asserts something about a whole class of objects.

We test the validity of the statement by substituting. Take $n = 3$. Then $3^2 - 1$ is divisible by 8. We try another couple of values to convince ourselves that the statement is true, say $n = 5$ and $n = 9$ which yield 24 and 80, respectively. We could try many other values of n and convince ourselves of the validity of the statement. However, since there are infinitely many integers, no matter how many values of n we check, there will always be the possibility that for one of the integers we haven't checked, our assertion is false. Therefore, we must construct an argument which, in a finite amount of time, will cover all cases.

Since n is odd, it is of the form $2k + 1$ for some k in $\mathbf{Z}$. Now $n^2 - 1 = (2k + 1)^2 - 1$ and expanding, we get

$$n^2 - 1 = 4k^2 + 4k + 1 - 1 = 4k(k + 1).$$

Since $k, k + 1$ are consecutive integers, exactly one is even. Hence $k(k + 1) = 2t$ for some t in $\mathbf{Z}$ and so

$$n^2 - 1 = 4(2t) = 8t.$$

But $8t$ is divisible by 8 proving that $n^2 - 1$ is divisible by 8. ∎

1.5.2 *Example*

If n is a positive integer, then $2^n > 2n$.

Proof. Letting $\mathbf{Z}^+$ denote the set of all positive integers, this can be written symbolically as

$$(\forall n \in \mathbf{Z}^+)(2^n > 2n).$$

We test the validity of the statement. For $n=3$ and $n=5$ the statement is true. In fact, it is not hard to prove that if $n>2$, the statement is valid. However, when $n=1$ we get $2^1 > 2$ which is obviously false. This one instance is sufficient to establish that the statement is false. We say that $n=1$ yields a *counterexample* to the assertion. ∎

The proof we used to establish the validity of the statement of Example 1.5.1 is called a *direct proof*. Often, if we experience difficulty in proving that $p \to q$, we can try to prove the logically equivalent statement $\neg q \to \neg p$, the so-called *contrapositive of* $p \to q$; or, we assume that both p and $\neg q$ are true and try to arrive at a contradiction, thus proving that the truth of p and the falsehood of q cannot co-exist, i.e., the truth of p implies the truth of q. This latter method of proof is called *a proof by contradiction*.

1.5.3 *Example*

Prove that if n^2 is an odd integer, then n is odd.

Proof. The contrapositive of "n^2 odd $\to n$ odd" is "n even $\to n^2$ even". Therefore, let n be even. Then $n = 2k$ for some k in $\mathbf{Z}$. Thus

$$n^2 = (2k)^2 = 4k^2 = 2(2k^2)$$

which is obviously even. ∎

Finally we give an example of a proof by contradiction.

1.5.4 *Example*

Prove that $\sqrt{2}$ is irrational. (An *irrational number* is a real number which cannot be expressed as a quotient of two whole numbers. See Appendix A.)

Proof. Suppose the statement is false. Then

$$(*) \sqrt{2} = p/q$$

where p and q are positive integers which we may assume have no common factors. Squaring both sides we get $2 = p^2/q^2$ and so $2q^2 = p^2$. It follows that p^2 must be even. Hence, p must be even and so $p = 2k$ for some integer k. Therefore, $2q^2 = p^2 = 4k^2$ and so $q^2 = 2k^2$ proving that q is even. But we assumed that p and q have no common factors and yet we have just proved that they do. This contradiction indicates that something is wrong. Our argument is solid, so it must be the assumption that $\sqrt{2}$ is rational that is false. Hence $\sqrt{2}$ is irrational. ■

1.6 Exercises

1) Let p: It is raining; q: It is cloudy. Translate each of the following into symbolic form

    (i)   If it rains, then it is cloudy.
    (ii)   If it is cloudy, then it rains.
    (iii) It is cloudy if and only if it is raining.
    (iv) If it does not rain, then it is not cloudy.
    (v)   It is cloudy only if it is raining.
    (vi) It is not cloudy only if it is not raining.

2) Construct the truth tables for

    (i) $p \to (q \vee r)$.
    (ii) $(p \vee r) \wedge (p \to q)$.
    (iii) $[p \to (q \to r)] \to [(p \to q) \to (p \to r)]$.
    (iv) $(p \vee q) \leftrightarrow (\neg r \wedge \neg s)$.
    (v) $(p \wedge q) \to \neg[\neg p \wedge (r \vee s)]$.
    (vi) $\neg[(\neg p \wedge \neg q) \wedge (p \vee r)]$.
    (vii) $p \to [(r \vee q) \leftrightarrow \neg(r \wedge s)]$.

Which of these statements are tautologies?

3) Find a simpler statement having the same truth table as the one found in 2(vi).

4) Show that the truth table in 2(vii) can be constructed much more easily by identifying the cases in which the statement is false.

5) A compound statement in p and q must have one of 16 possible truth tables. Find all these tables and for each table find a statement whose truth table is the given one.

6) Give an example of an implication which is valid, but whose converse is false.

7) Prove that the statements in Examples 1.3.3 are all tautologies.

8) The contrapositive of the statement $(\forall x \in U)(P(x) \to Q(x))$ is defined to be $(\forall x \in U)(\sim Q(x) \to \sim P(x))$. Prove that a statement and its contrapositive are logically equivalent.

9) Let the universal set be the set of all nonzero real numbers. Determine the truth value of each of the following statements:

    (i) $(\exists x \forall y)(xy = 1)$;
    (ii) $(\forall x \exists y)(xy = 1)$;
    (iii) $(\forall x \forall y)([xy > 1] \to [(x > 1) \vee (y > 1)])$;
    (iv) $(\exists x \forall y)([xy > 0] \to [x + y > 0])$;
    (v) $(\forall x \forall y)([xy > 0] \to [x + y > 0])$.

10) Prove that if p is a positive integer which is not divisible by any positive integer less than or equal to $\sqrt{p}$, then p has no positive factors other than p and 1, that is, p is prime.

11) Determine the truth value of each of the following:

    (i)   If the moon is made of blue cheese, then $2 + 3 = 6$;

(ii) If it snows at the North Pole, then dogs are bipeds;

(iii) If there are 12 months in the year, then $4 + 5 = 9$.

12) Negate each of the following:

(i) $(\forall x \in \mathbf{R})([x(x+1) < 0] \to [x < -1])$;

(ii) $(\exists x, y \in \mathbf{Z})([(x^2 > y^2) \wedge (xy > 0)] \leftrightarrow [(x > y) \vee (-x > -y)])$;

(iii) $(\forall x \exists y \forall z, x, y, z \in \mathbf{Z})([xyz \le 0] \to [[(x \ge 0) \wedge (y \ge 0)] \vee [z \le 0]])$.

Chapter 2

Set Theory

2.1 Definitions

The concept of a set and the language of set theory are of fundamental importance in mathematics. In this chapter we shall familiarize ourselves with the ideas and notation of set theory. We shall use these freely in subsequent chapters.

2.1.1 *Definition*

A *set* is a designated collection of objects. The designation is effected either by listing the members of the collection or by stipulating some property possessed by each member of the collection and not possessed by any object outside the collection.

2.1.2 *Example*

Consider the collection consisting of the numbers $1, 2$, and 3. We denote this by $\{1, 2, 3\}$ and think of the curly brackets as a "bag" containing the members of the set, which we call *elements*. Thus, the set above consists of three elements, namely, $1, 2$, and 3.

To indicate that a particular element belongs to a given set, we use the symbol "$\in$". In this example, $1 \in \{1,2,3\}$ while $4 \notin \{1,2,3\}$. The reader will infer that "$\notin$" means "is not a member of". Note that the order in which the elements of a set are listed is immaterial. ∎

Often we have occasion to refer to a specific set several times. In such a situation, to avoid writing out a long-winded expression every time we refer to it, it is convenient to give the set a name, usually an upper case letter. In the example above, we could let A stand for the set in question. We convey this by writing $A = \{1, 2, 3\}$. Also abstract, unspecified sets are usually denoted by upper case letters. There is one exception to this, namely, *the empty set* which contains nothing and which is denoted by ϕ. This set can be thought of as an empty bag (think of it as $\{\}$). We denote the number of elements in a set A by $|A|$ so that, for example, $|\phi| = 0$.

In the example above, we have designated the set by listing all the elements belonging to it. Sometimes the set in question is either very big or even infinite. In the former case it would be time-consuming and tedious to list all the elements while in the latter it would be impossible to do so. We therefore resort to using a property which singles out the elements of the set.

2.1.3 *Example*

Suppose we happen to be interested in the set of all Canadians. The property "Canadian" is used together with the so-called "set builder notation" as follows:

$$\{x \mid x \text{ is Canadian}\}.$$

The vertical line $"\mid"$ is to be read "such that". Therefore, this expression is read as: "The set of all x such that x is Canadian."

At this point let us establish notation for some of the well-known sets:

2.1.4 *Notation*

(i) $\mathbf{N} = \mathbf{Z}^+ = \{n \mid n \text{ is a positive whole number}\} = \{1, 2, 3, \ldots\}$.   The notation $\{1, 2, 3, \ldots\}$ is often used when the elements of the set in question can be listed in a sequence having an obvious pattern. We list enough elements of the sequence to establish the pattern followed by "$\ldots$" which means "and so on". This set is called *the set of natural numbers* or *the set of positive integers*.

(ii) $\mathbf{Z} = \{z \mid z \text{ is a whole number}\}$ consists of $0, \pm1, \pm2, \pm3, ., \pm n, \dots$. and is called *the set of integers*. The German word for "number" is "Zahl", hence the zee.

(iii) $\mathbf{Q} = \{x \mid x = p/q \text{ where } p \text{ and } q \text{ are integers with } q \neq 0\}$. This is the set of *rational numbers*. The "$\mathbf{Q}$" stands for "quotients".

(iv) $\mathbf{R} = \{x \mid x \text{ is a real number}\}$ consists of all the numbers, rational and irrational which appear on the number line.

(v) $\mathbf{C} = \{z \mid z = x + iy \text{ where } i = \sqrt{-1} \text{ and } x \text{ and } y \text{ are real numbers}\}$ is the set of *complex numbers* (see Appendix A).

2.2 Relations Between Sets

2.2.1 *Definitions*

(i) When talking about sets, we always have in mind a big *universal set* from which all the elements in the sets occurring in the discourse are taken. For example, if we are interested in certain sets of whole numbers, we might take $\mathbf{Z}$ or $\mathbf{R}$ as the universal set.

(ii) Let A and B be sets whose elements are in a universal set U. Then, if every element of A is an element of B, we say that A *is contained in* B or A *is a subset of* B and write $A \subseteq B$. If $A \subseteq B$ and $A \neq B$, we say that A *is properly contained in* B or A *is a proper subset of* B and write $A \subset B$. (Note the similarity to $<$).

Remarks

(i) The symbol "$\subseteq$" is a "rounded" version of "$\leq$" which is the symbol for "less than or equal." Observe also that, whatever the set A, $A \subseteq A$ and that to prove $A \subset B$ we must prove that every element of A is an element of B *and* there is an element of B which is *not* in A.

(ii) If a set A is not a subset of a set B, then we write: $A \not\subseteq B$. To prove that $A \not\subseteq B$, we must show that there exists an x in A which is not in B. The reader may wish to verify this by negating the statement

$$(\forall x \text{ in some universal set})(x \in A \rightarrow x \in B).$$

(iii) The empty set ϕ mentioned above, is a subset of every set. Indeed, if it were not the case that, for example, ϕ is a subset of A, then by (ii), there exists $x \in \phi$ with $x \notin A$. However, since ϕ contains no elements, the existential part of the statement above is false. Therefore, it is not the case that ϕ is not a subset of A and so ϕ *is* a subset of A.

(iv) In the case $x \in A$ iff $x \in B$, then clearly A and B contain the same elements and we write $A = B$. (We have used "*iff*" as shorthand for "*if, and only, if*" and shall do so throughout the text)

The reader can easily verify that "$\subseteq$"possesses the following properties:

2.2.2 *Theorem*

(i) For all sets A, $A \subseteq A$ (Reflexivity);

(ii) For all sets A, B, C, $A \subseteq B$ and $B \subseteq C$ imply $A \subseteq C$ (Transitivity);

(iii) For all sets A and B, if $A \subseteq B$ and $B \subseteq A$, then $A = B$ (Antisymmetry)

Proof. Exercise. ∎

Later on, we shall be studying relations abstractly. Reflexivity, transitivity and antisymmetry are among the various properties a relation can possess.

2.3 Operations Defined on Sets — Or New Sets from Old

Given a collection of sets, we can combine them in much the same way we combined statements to form compound statements.

2.3.1 *Definitions*

(i) Given two sets A and B, their *union*, denoted by $A \cup B$ is defined by:

$$A \cup B = \{x \mid \text{either } x \in A \text{ or } x \in B \text{ or } x \text{ is in both}\}.$$

(ii) The *intersection* of two sets A and B, consists of those elements common to both and is denoted by $A \cap B$. Formally,

$$A \cap B = \{x \mid x \in A \text{ and } x \in B\}.$$

(iii) If A is a set contained in some universal set U, the *complement* of A (relative to U), denoted by A' (when U is apparent), is defined by:

$$A' = \{x \mid x \in U \text{ and } x \notin A\}.$$

(v) The *difference* of two sets A and B, denoted by $A \setminus B$, is the set consisting of all the elements of A which are *not* in B. Formally,

$$A \setminus B = \{x \in U \mid x \in A \text{ and } x \notin B\}.$$

Union, intersection and difference are examples of *binary operations* ("bi" means two sets are involved), while complementation is an example of a *unary operation* (only one set is involved). Later on, abstract binary operations will be the focus of our investigations.

To help us visualize the relationships between sets and the operations defined above, we draw representations of sets. These pictorial representations are called *Venn diagrams*. We let a rectangle represent the universal set and the subsets are represented by circles as shown below. The darkest shaded area represents $A \cap B$ and the shaded area $A \cup B$.

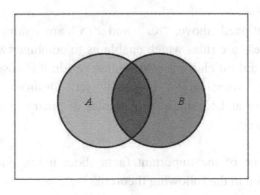

Fig. 2.1. $A \cup B$ and $A \cap B$.

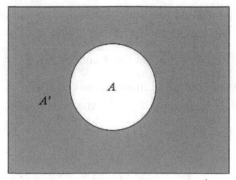

Fig. 2.2 The shaded area represents A'.

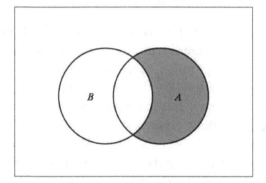

Fig. 2.3 The shaded area represents $A \setminus B$.

Note. As mentioned above, "$\cup$" and "$\cap$" are examples of *binary operations*. These are rules which enable us to combine two elements of a set to obtain a third element of the same set. In this case the elements are subsets of some given universal set. Ordinary addition and multiplication (+ and $\times$) are also examples of binary operations on the set of real numbers.

We collect some of the important facts about union, intersection and complementation in the following theorem.

2.3.2 *Theorem*

Let A, B and C be sets contained in some universal set. Then the following hold

   (i) $A\cup(B\cup C)=(A\cup B)\cup C$; $A\cap(B\cap C)=(A\cap B)\cap C$;
(Associativity of $\cup$ and $\cap$);
   (ii) $A\cap(B\cup C)=(A\cap B)\cup(A\cap C)$ ($\cap$ is distributive over $\cup$);
$A\cup(B\cap C)=(A\cup B)\cap(A\cup C)$ ($\cup$ is distributive over $\cap$);
   (iii) $(A')'=A$;.
   (iv) $(A\cup B)'=A'\cap B'$; $(A\cap B)'=A'\cup B'$ (DeMorgan's Laws)
   (v) $A\cup B=B\cup A$; $A\cap B=B\cap A$ (Commutativity of $\cup$ and $\cap$).

Proof. We prove only one of these statements, leaving the rest to the reader.
   (ii) Let $x\in A\cap(B\cup C)$; then $x\in A$ and $x\in B\cup C$. Hence,
$(x\in A)$ and $(x\in B$ or $x\in C)$. Therefore
$(x\in A)$ and $(x\in B$ or $x\in C)\}$ iff $(x\in A$ and $x\in B)$ or $(x\in A$ and $x\in C)$

$$\text{iff } (x\in A\cap B) \text{ or } (x\in A\cap C)$$
$$\text{iff } x\in (A\cap B)\cup(A\cap C).$$

At this point, we have proved that
$$A\cap(B\cup C)\subseteq(A\cap B)\cup(A\cap C).$$

Conversely, let $x\in (A\cap B)\cup(A\cap C)$; then, reading the above line of logical equivalences backwards, we conclude that $x\in A\cap(B\cup C)$ and so

$$(A\cap B)\cup(A\cap C)\subseteq A\cap(B\cup C).$$

It now follows that the two sets are equal.■

A note on the proof. In the proof above we showed that, given an arbitrary element from the first set, it is a member of the second and conversely. This is standard procedure when trying to prove two sets equal: we prove each of the two sets in question is contained in the other – a process which involves two distinct steps.

We observe that, in view of 2.3.2(i), we do not need to use brackets when forming the union or intersection of any number of sets. By contrast, the operation "\" is not associative and so brackets have to be used. For example, the expression $A \setminus B \setminus C \setminus D$ is ambiguous. It could mean one of the following:

$$(A \setminus B) \setminus (C \setminus D), \{(A \setminus B) \setminus C\} \setminus D, A \setminus \{(B \setminus C) \setminus D\}, \{A \setminus (B \setminus C)\} \setminus D,$$

$$A \setminus \{B \setminus (C \setminus D\}.$$

The reader is invited to provide examples to show that these five sets are indeed different from one another.

2.3.3 *Notation*

It is essential to have notation to express the union or intersection of any finite or infinite number of sets. For example, if we have sets $A_1, A_2, A_3, \ldots\ldots, A_n$ and we want to form the union of these sets, it would be tedious and long-winded to write $A_1 \cup A_2 \cup A_3 \cup \ldots\ldots \cup A_n$. We write instead $\bigcup_{i=1}^{n} A_i$. In a similar fashion, for the intersection we write $\bigcap_{i=1}^{n} A_i$. For an infinite collection of sets, we choose an *indexing set*, say I, which is nothing but a set of labels which we use to distinguish one set from another and write $\bigcup_{i \in I} A_i$ and $\bigcap_{i \in I} A_i$.

2.4 Exercises

1) Draw Venn diagrams to illustrate the statements of Theorem 2.3.2(ii), (iii), (iv).

2) Let A, B and C be sets contained in some universal set U. Prove the following and illustrate with Venn diagrams:

   (i) $(A')' = A$
   (ii) $A \setminus B = A \cap B'$;

(iii) $(A \setminus B) \setminus C = A \setminus (B \cup C)$;

(iv) $A \setminus (B \setminus C) = (A \setminus B) \cup (A \setminus C)$;

(v) $A \cup (B \setminus C) = (A \cup B) \setminus (C \setminus A)$;

(vi) $A \cap (B \setminus C) = (A \cap B) \setminus (A \cap C)$.

3) Prove that the following are equivalent:

(i) $A \subseteq B$;

(ii) $A \cap B = A$;

(iii) $A \cup B = B$.

4) Given a set A, *the power set of* A, denoted $P(A)$, is the set of all subsets of A. For example, if

$A = \{1, 2, 3\}$, then $P(A) = \{\phi, \{1\}, \{2\}, \{3\}, \{1, 2\}, \{1, 3\}, \{2, 3\}, \{1, 2, 3\}\}$.

(Observe that $\phi \in P(A)$ but that $\phi \subseteq A$. Similarly, $1 \in A$ but $\{1\} \subseteq A$ and $\{1\} \in P(A)$.) List the elements of $P(B)$ where $B = \{1, 2, 3, 4\}$. If C is a set of m elements, how many elements are there in $P(C)$? (The usual notation for the number of elements in a set C is $|C|$.)

5) Let U be some universal set and define an operation Δ (called *the symmetric difference*) on $P(U)$ by: $A \Delta B = (A \setminus B) \cup (B \setminus A)$ for all A, $B \in P(U)$.

Prove the following and illustrate (i) below with a Venn diagram:

(i) For all $A, B, C \in P(U)$, $A \Delta (B \Delta C) = (A \Delta B) \Delta C$;

(ii) For all $A \in P(U)$, $A \Delta A = \phi$;

(iii) For all $A \in P(U)$, $A \Delta \phi = A$;

(iv) For all $A, B \in P(U)$, $A \Delta B = B \Delta A$.

If $U = \{1, 2, 3\}$, compute $A \Delta B$ for all $A, B \in P(U)$.

6 Let S be the set of all strings of 0's and 1's of length four. For example, $0110 \in S$. Let A be a set of four elements. What can you say about the sizes of S and $P(A)$? Generalize this to the set of all strings of 0's and 1's of length n and the set of all subsets of a set of n elements. Can you prove the generalization?

7) If A and B are finite sets, prove that $|A \cup B| = |A| + |B| - |A \cap B|$. If A, B, C and D are sets, find formulas similar to the one above for $|A \cup B \cup C|$ and $|A \cup B \cup C \cup D|$. What would be the appropriate generalization if there are n sets? (Recall that $|A|$ denotes the number of elements in A.)

8) Let U be a fixed universal set and let $p, q, r, \ldots$ be properties possessed by all, some or none of the elements of U. If for some $x \in U$, property p holds, we write $p(x)$; if it does not we write $\neg p(x)$. Letting $P = P = \{x \in U \mid p(x)\}, Q = \{x \in U \mid q(x)\}$, etc., prove the following:

(i) $P \cup Q = \{x \in U \mid p(x) \vee q(x)\}$;
(ii) $P \cap Q = \{x \in U \mid p(x) \wedge q(x)\}$;
(iii) $P' = \{x \in U \mid \neg p(x)\}$.

9) Let I and J be disjoint indexing sets, that is, $I \cap J = \phi$, and let A_i and B_j be collections of sets indexed by I and J, respectively. Prove:

(i) If $A_i \subseteq X$ for all $i \in I$, then $\displaystyle\bigcup_{i \in I} A_i \subseteq X$.

(ii) If $A_i \supseteq Y$ for all $i \in I$, then $\displaystyle\bigcap_{i \in I} A_i \supseteq Y$. $\displaystyle\bigcap_{i \in I} A_i \supseteq Y$).

(iii) $\displaystyle(\bigcup_{i \in I} A_i) \cap (\bigcup_{j \in J} B_j) = \bigcup_{i \in I, j \in J} (A_i \cap B_j)$.

(iv) $\displaystyle(\bigcap_{i \in I} A_i) \cup (\bigcap_{j \in J} B_j) = \bigcap_{i \in I, j \in J} (A_i \cup B_j)$.

(v) $\displaystyle(\bigcup_{i \in I} A_i)' = \bigcap_{i \in I} A_i'$.

(vi) $\displaystyle(\bigcap_{i \in I} A_i)' = \bigcup_{i \in I} A_i'$.

10) If A and B are subsets of a universal set U, prove that the following are equivalent:

(i) $A \subseteq B$
(ii) $A \cap B' = \phi$
(iii) $A' \cup B = U$.

11) Prove or disprove, $P(A) \cup P(B) = P(A \cup B)$.

12) Prove $A \subseteq B$ iff $A' \supseteq B'$.

13) If $A \cap B = A \cap B'$, prove that $A = \phi$.

14) If $A \cup B = A \cup B'$, prove that $B \subseteq A$.

15) Prove that $A \cap B' = \phi$ iff $A \subseteq B$.

16) Under what circumstances will $P(A) \cap P(B) = P(A \cap B)$? Prove your assertion.

17) Prove that

$$[A \cup (B \cap C)] \cap [B \cup (A \cap C)] = (A \cap B) \cup (A \cap C) \cup (B \cap C)$$
$$= [B \cap (A \cup C)] \cup [C \cap (A \cup B)].$$

18) Prove that

$$\left| \bigcup_{i=1}^{n} A_i \right| = \sum_{i=1}^{n} |A_i| - \sum_{1 \le i < j \le n} |A_i \cap A_j| + \sum_{1 \le i < j < k \le n} |A_i \cap A_j \cap A_k|$$
$$- \sum_{1 \le i < j < k \le n} |A_i \cap A_j \cap A_k \cap A_l| + \dots (\text{See Problem 7}).$$

19) If $|A_1| = 9, |A_2| = 6, |A_3| = 11, |A_4| = 12, |A_1 \cap A_2| = 4, |A_1 \cap A_3| = 7,$ $|A_2 \cap A_3| = 4, |A_1 \cap A_2 \cap A_3| = 2, |A_4| = 12, |(A_1 \cup A_2 \cup A_3) \cap A_4| = 6$, how many elements are there in $|A_1 \cup A_2 \cup A_3 \cup A_4|$?

20) Let $A_1, A_2, \dots, A_n$ be subsets of a universal set U. Write $A_i^1 = A_i$ and $A_i^0 = A_i'$. Show that each A_i is uniquely expressible in the form $\bigcup_{i=1}^{2^{n-1}} A_1^{\delta_i^1} \cap A_2^{\delta_i^2} \cap A_3^{\delta_i^3} \cap \dots \cap A_n^{\delta_i^n}$ where δ_i^j is 0 or 1.

21) Let $S = \{A \mid A \text{ is a set and } A \notin A\}$. Is $S \in S$ or is $S \notin S$? This is at the root of Russell's Paradox.

22) Prove or disprove each of the following:

    (i) If $A \cap B = A \cap C$, then $B = C$;
    (ii) If $A \cap B = A \cap C$ and $A \cup B = A \cup C$, then $B = C$;
    (iii) If $A \triangle B = A \triangle C$, then $B = C$.

23) Prove that $(A \cap B) \cup (B \cap C)' = A \cup B' \cup C'$.

24) Prove or disprove (by giving a counterexample) each of the following:

    (i) $A \setminus B = A \setminus C$ implies $B = C$;
    (ii) $A \cap C = B \cap C$ and $A \setminus C = B \setminus C$ imply $A = B$;
    (iii) $A \cup C = B \cup C$ and $A \setminus C = B \setminus C$ imply $A = B$.

Chapter 3

Cartesian Products, Relations, Maps and Binary Operations

3.1 Introduction

Relations and maps occur in almost every branch of mathematics. We shall first study relations in general and then, as special cases of relations, equivalence relations and maps (sometimes also called mappings or functions). Finally we shall introduce binary operations as special kinds of maps. The framework we use is the Cartesian or cross product of sets.

3.2 Cartesian Products

3.2.1 *Definition*

Let A and B be sets. The *Cartesian or cross product* of A and B, denoted $A \times B$ is the set of all ordered pairs (a,b) where a is an element of A and b is an element of B. In symbols

$$A \times B = \{(a,b) \mid (a \in A) \wedge (b \in B)\}.$$

Remark. When we say that a pair is *ordered*, we mean that the order in which the members of the pair are written matters. For example, the ordered pair $(1,2)$ is different from the ordered pair $(2,1)$. If $A \neq B$, it then follows that $A \times B \neq B \times A$.

3.2.2 *Examples*

(i) Let $A = \{a,b,c\}$ and $B = \{1,2\}$. Then

$$A \times B = \{(a,1),(b,1),(c,1),(a,2),(b,2),(c,2)\}.$$

We can illustrate this set pictorially as follows:

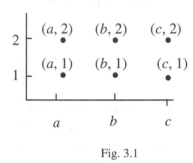

Fig. 3.1

(ii) The set $\mathbf{R} \times \mathbf{R}$, the cross product of the set of real numbers by itself, is the Cartesian plane of co-ordinate geometry. We sometimes call the elements of this set 2-dimensional vectors. ∎

We are now ready to define the abstract notion of a relation.

3.2.3 *Definition*

A *relation from* A to B is a subset of $A \times B$. If $A = B$, then, rather than saying "a relation from A to A ", we say a *"relation on A "*.

3.2.4 *Example*

(i) Let $A = B = \mathbf{R}$ and let L be the relation defined on $\mathbf{R}$ by

$$L = \{(x,y) \in \mathbf{R} \times \mathbf{R} \,|\, x - y \le 0\}.$$

Clearly, this is none other than the less-than-or-equal-to relation between pairs of real numbers.

(ii) Let $A = \{a,b,c,d\}$, $B = \{1, 2, 3\}$ and $S = \{(a, 2), (b, 1), (c, 2)\}$. This relation can be represented pictorially as follows:

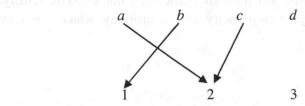

Fig. 3.2. The graph of a relation

Such a pictorial representation is called a *graph of the relation.*

(ii) This same relation can be represented by a matrix (called *the matrix of the relation*).

$$\begin{bmatrix} 0100 \\ 1010 \\ 0000 \end{bmatrix}$$

In this matrix, the first column is labeled by a, the second by b, etc., while the rows are labeled by $1, 2, 3$. Observe that there is a 1 at the intersection of column b and row 1 to indicate that b is related to 1. There is a 0 at the intersection of column c with row 3 to indicate that c is not related to 3. ∎

3.2.5 *Notation*

If S is a relation from A to B, it is customary to write aSb to mean $(a,b) \in S$.

3.2.6 *Example*

Let $T \subseteq \mathbf{R} \times \mathbf{R}$ be the equality relation. Then $T = \{(a,a) | a \in \mathbf{R}\}$ In this case, aTa is written as $a = a$. ∎

There are four properties which we are particularly interested in, namely, reflexivity, symmetry, antisymmetry and transitivity which we now define.

3.2.7 *Definition*

A relation R on a set S is:

  (i) *Reflexive* if aRa for all $a \in S$;
  (ii) *Symmetric* if aRb implies bRa for all $a,b \in S$;
  (iii) *Transitive* if aRb and bRc imply aRc for all a, b, c in S;
  (iv) *Antisymmetric* if aRb and bRa imply $a = b$ for all $a,b,c \in S$.

3.2.8 *Example*

  (i) Let $A = \{a,b,c,d,e\}$ and consider the relation represented by the graph below.

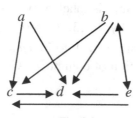

Fig. 3.3

The arrows can be thought of as one-way roads (except for the road from b to e) joining the towns a,b,c,d and e. We see that whenever there is a circuitous route from one town to another, there is also a direct one. This relation is therefore transitive. For example, we can get from a to d via c, but we can also go directly from a to d. It is not reflexive, symmetric or antisymmetric.

(ii) Consider the relation $\leq$ defined on $\mathbf{R}$. It has the following properties:

(i) For all $a \in \mathbf{R}, a \leq a$ (Reflexivity);

(ii) For all $a, b, c \in \mathbf{R}$, $a \leq b$ and $b \leq c$ imply $a \leq c$ (Transitivity);

(iii) For all $a, b \in \mathbf{R}$, $a \leq b$ and $b \leq a$ imply $a = b$ (Antisymmetry). ∎

Note that the relation $\subseteq$ on the set of subsets of a given set has these very same properties.

3.2.9 *Definition*

A relation P is a *partial order* if it is reflexive, transitive and antisymmetric. Often, the set together with the partial order is referred to as a *poset*.

3.2.10 *Example*

Let $D = \{1, 2, 3, 6\}$ and let L be the relation defined by: aLb if there exists an integer k such that $b = ka$ (see Definition 3.2.9 below). For example, $2L6$. The relation L is a partial order and its graph is:

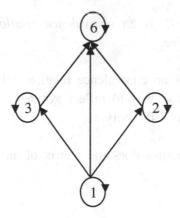

Fig. 3.4

Note that the loops around the numbers indicate that the relation is reflexive. ■

Question. What properties would the matrix of a partial order have?

The next type of relation will play an important role in later chapters. We start with a definition and then give an example.

3.2.11 *Definition*

Let a and b be integers, $a \neq 0$. We say that a *divides* b and write $a|b$ if there exists an integer k such that $b = ka$. If a does not divide b we write $\neg(a|b)$. (Thus $15|45$ but $\neg(7|18)$.)

3.2.12 *Example*

Let $E_4 = \{(a,b) \in \mathbf{Z} \times \mathbf{Z} \mid 4 \mid (a-b)\}$. We leave it to the reader to verify that E_4 has the following properties:
  (i) For all $a \in \mathbf{Z}$, aE_4a (Reflexivity);
  (ii) For all $a,b \in \mathbf{Z}$, if aE_4b, then bE_4a (Symmetry);
  (iii) For all $a,b,c \in \mathbf{Z}$, aE_4b and bE_4c imply aE_4c (Transitivity). ■

3.2.13 *Definition*

An abstract relation E is an *equivalence relation* if it is reflexive, symmetric and transitive.

Note. In the graph of an equivalence relation, all "roads" joining the various "towns" are two-way to reflect symmetry and at each "town" there is a loop to reflect reflexivity.

Question. What properties does the matrix of an equivalence relation have?

3.2.14 *Definition and notation*

Let E be an equivalence relation on a set A and let $a \in A$. Then $[a]_E = \{x \in A \mid xEa\}$ is called the *equivalence class of a mod E*.

3.2.15 *Example*

In Example 3.2.11 let us find the various equivalence classes. We omit the subscript and write $[a]$ rather than $[a]_{E_4}$. This is usually done when the equivalence relation in question is apparent.

$$[0] = \{0, \pm 4, \pm 8, \pm 12, \dots, \pm 4n, \dots\} = \{4k \mid k \in \mathbf{Z}\}; [1] = \{4k+1 \mid k \in \mathbf{Z}\};$$
$$[2] = \{4k+2 \mid k \in \mathbf{Z}\}; [3] = \{4k+3 \mid k \in \mathbf{Z}\}.$$

The reader is invited to show that these do indeed constitute the set of all the equivalence classes mod E_4.

We observe the following:

(i) $\mathbf{Z} = [0] \cup [1] \cup [2] \cup [3]$;
(ii) If $[a] \neq [b]$, then $[a] \cap [b] = \phi$,

that is, the equivalence classes cover the set and no two distinct equivalence classes have any elements in common. ∎

3.2.16 *Definition*

A *partition* of a set A is a nonempty collection of nonempty subsets $A_i, i \in I$, called the *cells* of the partition, such that

(i) $A = \bigcup_{i \in I} A_i$;
(ii) $A_i \cap A_j = \phi$ for all $i, j \in I, i \neq j$.

In terms of this definition, we may say that sets [0], [1], [2] and [3] in the above example form a partition of $\mathbf{Z}$.

There is a close connection between partitions and equivalence relations, which we now establish.

3.2.17 *Theorem*

If E is an equivalence relation on a set A, then the equivalence classes of E form a partition of A.

Conversely, given a partition

$$\Pi = \{\, A_i, i \in I \,\} \text{ of } A,$$

we can define an equivalence relation E_Π as follows: $aE_\Pi b$ if a and b belong to the same cell of the partition Π. Moreover, the equivalence classes of E_Π are precisely the A_i.

Proof. We prove only the first part, leaving the second as an exercise for the reader. Denote the equivalence class of $a \bmod E$ by $[a]$.

We first show that $A = \bigcup_{a \in A} [a]$. Let $b \in A$ and consider

$$[b] = \{x \in A \mid xEb\}.$$

Since bEb by reflexivity, it follows that $b \in [b]$ and so $b \in \bigcup_{a \in A} [a]$, thus proving our first assertion.

We must next prove that if

$$[b] \neq [c] \text{ then } [b] \cap [c] = \phi,$$

or, contrapositively,

$$[b] \cap [c] \neq \phi \text{ implies } [b] = [c].$$

So let us assume that $[b] \cap [c] \neq \phi$ and pick $d \in [b] \cap [c]$. We must show that $[b] = [c]$.

If $x \in [b]$, by definition, xEb. Now

$$d \in [b] \cap [c] \text{ implies } d \in [b]$$

and so *dEb*. By symmetry, *bEd* and by transitivity,

$$(xEb) \text{ and } (bEd) \text{ imply } xEd.$$

Also, $d \in [c]$, and so *dEc*. By transitivity again, we have *xEc* since *xEd* and *dEc* and so $x \in [c]$, proving that $[b] \subseteq [c]$.

We can prove the reverse inclusion in a similar manner. ∎

3.3 Maps

The reader has already encountered the notion of a map under the name of function in high school and in courses in calculus. Generally, a function was represented by a formula in which one substituted one or more real numbers for the variables to obtain the value of this function at the substituted values. Our definition below is much more general than this although the idea is basically the same.

3.3.1 *Definition*

A *map* Φ *from* A *to* B, denoted $\Phi : A \to B$, is a relation from A to B with the following properties:

(i) Given any $a \in A$, there exists $b \in B$ such that $(a,b) \in \Phi$.
(ii) $(a,b) \in \Phi \}$ and $(a,c) \in \Phi$ imply $b = c$.

That is, each element of A is related to *exactly one element of* B. Note that we ARE NOT saying that each element of B is related to exactly one element of A.

The set A is called the *domain* of the map and the set B the *codomain.*

This is the formal definition of a map, but in practice, we rarely think of it this way. Since to each element of A there is a unique element of B associated with it, instead of writing $(a,b) \in \Phi$, we write $\Phi(a) = b$ and think of Φ as somehow "sending" or "transforming" a to b. $\Phi(a)$ is called the *image of a under* Φ.

An example should make things clear.

3.3.2 *Example*

Let $A = \{1,2,3,4\}$, $B = \{a,b,c,d,e\}$ and $\Phi = \{(1,b), (2,b), (3,a), (4,d)\}$.
We would write, $\Phi(1) = b, \Phi(4) = d$,etc. We can also draw a diagram to
help us visualize the map.

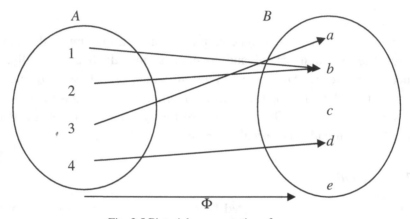

Fig. 3.5 Pictorial representation of a map

Recall that relations can also be represented by a matrix. Here is the
matrix of this relation which happens to be a map.

$$\begin{bmatrix} 0010 \\ 1100 \\ 0000 \\ 0001 \\ 0000 \end{bmatrix}$$

Observe that in every column there is exactly one 1 since to each
element of A there corresponds exactly one element of B. ∎

3.3.3 *Definition*

Let $f : A \rightarrow B$ be a map and $X \subseteq A, Y \subseteq B$.

   (i) *The image of X under f* is the subset of B, denoted $f(X)$, and defined by:

$$f(X) = \{b \in B \mid \text{ there exists } x \in X \text{ such that } f(x) = b\}.$$

Informally, $f(X)$ is the set of elements in B which are "hit" (via f) by elements of X. $f(A)$ is called *the image of f* and is denoted by Im f.

   (ii) *The inverse image of Y under f* is the subset $f^{-1}(Y)$ of A defined by:

$$f^{-1}(Y) = \{a \in A \mid f(a) \in Y\}.$$

Informally, $f^{-1}(Y)$ consists of all those elements in A which are "sent" to elements of Y by f.

We draw diagrams to illustrate (i) and (ii) above.

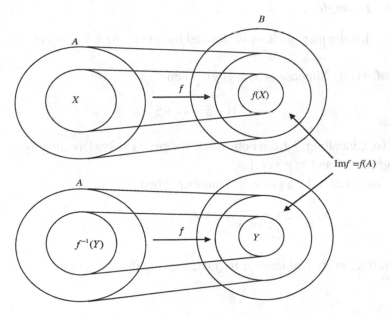

Fig. 3.6 Image and inverse image under the map f

3.3.4 *Example*

Refer to Example 3.3.3. $\Phi(\{1,2,3\}) = \{a,b\}$ while $\Phi^{-1}(\{b\}) = \{1,2\}$.

3.3.5 *Definition*

(i) A map $f : A \to B$ is said to be *one-to-one* or $1{:}1$ if different elements of A map to different elements of B. More precisely, f is one-to-one if for all x and y in A, $x \neq y$ implies $f(x) \neq f(y)$. Generally, to prove that a map is $1{:}1$, we prove the contrapositive, namely, $f(x) = f(y)$ implies $x = y$.

(ii) A map $f : A \to B$ is *onto* if $f(A) = B$.

(iii) A *bijective* map is both $1{:}1$ and onto.

Note. The word "*injective*" is often used instead of "$1{:}1$"while "*surjective*" is used as a variant of "onto".

3.3.6 *Example*

Prove that the map $g : \mathbf{R} \to \mathbf{R}$ defined by $g(x) = 3x + 5$ is bijective.

Proof. $(1{:}1)$: Suppose $g(x) = g(y)$. Then

$$3x + 5 = 3y + 5$$

and so, cancelling 5 from both sides, we get $3x = 3y$. Dividing by 3, we get $x = y$ and so g is $1{:}1$.

Onto: Let y be a given real number. Then

$$g\left(\frac{y-5}{3}\right) = 3\left(\frac{y-5}{3}\right) + 5 = y - 5 + 5 = y.$$

Therefore, every real number is "hit", proving that g is onto.

The reader may wonder how we obtained the expression $(y-5)\big/3$. We let $y = 3x + 5$ and solve for x in terms of y. ∎

If f is a map from A to B and g a map from B to C, we can define a map from A to C as follows: First apply f to an element of A to obtain an element of B; next, apply g to this element and obtain an element of C. Thus, given an element of A, we obtain an element of C passing via B. More formally:

3.3.7 *Definition*

If $f : A \rightarrow B$ and $g : B \rightarrow C$ are maps then the *composite of* f *and* g, denoted by $g \circ f$, is defined by

$$(g \circ f)(a) = g(f(a)) \quad \text{for all } a \in A.$$

More precisely, we first compute $f(a) = b \in B$, say, and then we compute $g(b)$.

The following diagram is often helpful in picturing what is going on.

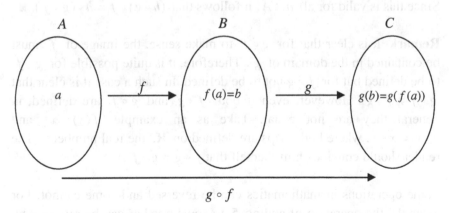

Fig. 3.7 Composition of maps

3.3.8 *Definition*

We say that two maps f and g from A to B are *equal* and write $f = g$ if for all $a \in A$, $f(a) = g(a)$.

We see from the definition that to prove two maps are equal, we must prove that *they agree everywhere in the domain*; to prove them unequal, therefore, it is sufficient to show that they disagree at some point in the domain.

3.3.9 *Theorem*

Composition of maps is associative. More precisely, if f is a map from A to B, g a map from B to C and h a map from C to D, then $(h \circ g) \circ f = h \circ (g \circ f)$.

Proof. By definition,

$$[(h \circ g) \circ f](a) = (h \circ g)(f(a) = h(g[f(a)]) = h[(g \circ f)(a)]$$
$$= [h \circ (g \circ f)](a).$$

Since this is valid for all $a \in A$, it follows that $(h \circ g) \circ f = h \circ (g \circ f)$. ∎

Remark. It is clear that for $g \circ f$ to make sense, the image of f must be contained in the domain of g. Therefore, it is quite possible for $g \circ f$ to be defined but for $f \circ g$ not to be defined. In such a case it is clear that $g \circ f \neq f \circ g$. However, even if both $f \circ g$ and $g \circ f$ are defined, in general they are not equal. Take as an example $f(x) = x^2$ and $g(x) = x - 1$, where both maps are defined on **R**, the real numbers. (The reader should convince him-/herself that $f \circ g \neq g \circ f$.)

Some operations in mathematics can be reversed and some cannot. For example, the operation of adding 5 to some number can be reversed by subtracting 5, thus recovering the original number. On the other hand, multiplication by 0 is irreversible since any number times 0 is 0.

Analogously, if for a given map f we are told that the image under f of a certain element is, say, b, we cannot, in general, recover the pre-image. Referring to Example 3.3.2, if we know that the image of a certain element under Φ is b, we cannot recover the pre-image with certainty since it could be either 1 or 2.

We shall now investigate, in the case of maps, ideas analogous to those discussed above.

3.3.10 *Definition*

The *identity map on the set* A is the map $I_A : A \to A$ defined by $I_A(a) = a$ for all $a \in A$. When the set in question is obvious, we omit the subscript and just write I.

The identity map must surely be as reversible as a map can get since it does nothing. It is the "I-dle" map! It is analogous to multiplication of real numbers by 1. Indeed, if $f : A \to B$, it is easily checked that

$$f \circ I_A = f = I_B \circ f \ .$$

3.3.11 *Definition*

Let $f : A \to B$. A map $g : B \to A$ is said to be *a left inverse of f* if $g \circ f = I_A$. In this case, f is *a right inverse of g*. Maps which possess left inverses are said to be *left invertible* and those with right inverses, *right invertible*.

Not all maps are left (right) invertible. The following theorem tells us precisely how to identify which are left invertible and which are right invertible.

3.3.12 *Theorem*

(i) $f : A \to B$ is left invertible if, and only if, f is 1:1;

(ii) $f : A \to B$ is right invertible if, and only if, f is onto.

Proof.

(i) Only if) Suppose f is left invertible with g as a left inverse and let $f(a) = f(b)$. Then, applying g to both sides, we have $g(f(a)) = g(f(b))$. Hence, by definition, $(g \circ f)(a) = (g \circ f)(b)$. But $g \circ f = I_A$ and so $I_A(a) = I_A(b)$. Therefore, $a = b$, proving that f is 1:1.

If) Conversely, suppose f is 1:1. We define a map $g : B \to A$ as follows: Let $b \in B$. If $b \notin f(A)$, define $g(b)$ to be an arbitrary element of A. If $b \in f(A)$, then, because f is 1:1, there is a unique element $a \in A$ such that $f(a) = b$. Define $g(b) = a$. (Observe that we have some latitude in defining g if $f(A) \neq B$.) It is now clear that $g \circ f = I_A$. The diagram below should be of help in following this proof.

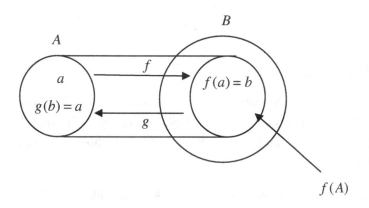

Fig. 3.8

(ii) Only if) If f is right invertible, there is a map $h : B \to A$ such that $f \circ h = I_B$. Let $b \in B$. Then

$$(f \circ h)(b) = I_B(b) = b,$$

and so $f[h(b)] = b$. Setting $h(b) = a$, we have $f(a) = b$, proving that every element of B is "hit". Hence f is onto.

If) Conversely, assume that f is onto. Define a map $h : B \rightarrow A$ as follows : Given that f is onto, $f^{-1}(\{b\}) \neq \phi$. Thus (see diagram below), given $b \in B$ pick any $a \in f^{-1}(\{b\})$ and set $h(b) = a$. It is then clear that $f \circ h = I_B$. The diagram below illustrates the proof.

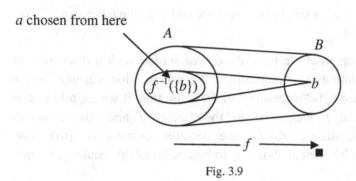

Fig. 3.9

3.3.13 *Corollary*

A map has both a left and a right inverse if, and only if, it is bijective. Moreover, in such a case, the left and right inverses are identical.

Proof. The first statement follows immediately from Theorem 3.3.12. As for the second, let $f : A \rightarrow B$ have a left inverse $g : B \rightarrow A$ and a right inverse $h : B \rightarrow A$. Then

$$h = I_A \circ h = (g \circ f) \circ h = g \circ (f \circ h) = g \circ I_B = g$$

using the associativity of the composition of maps (Theorem 3.3.9). Hence $h = g$. ∎

Note. What we have proved above is that there exists $g : B \to A$ such that $g \circ f = I_A$ and $f \circ g = I_B$ if, and only if f is bijective. Moreover, by the proof above, such a g, if it exists, must be unique.

3.3.14 *Definition*

Let $f : A \to B$. If there exists $g : B \to A$ such that $f \circ g = I_B$ and $g \circ f = I_A$, then f is said to be *invertible* and g is the *inverse* of f. We denote g by f^{-1}.

Let us recap what we have done. We started with a discussion of reversible operations in mathematics. We can see that a map f which has a left inverse g is "reversible" in the sense that, if we are told that b is in the image of f, we can "undo" the effect of f and retrieve the pre-image. Indeed, since $g \circ f = I_A$, the required element is $g(b)$. For, suppose $f(x) = b$. Then applying g to both sides of this equality, we get

$$g[f(x)] = (g \circ f)(x) = I_A(x) = x = g(b).$$

If f has an inverse and we are given b, we do not even have to be told that b is in the image of f to find the element which maps to it. This element is, of course, $f^{-1}(b)$ (not to be confused with $f^{-1}(\{b\})$). Observe that we can write $f^{-1}(b)$ if, and only if, f is bijective whereas $f^{-1}(\{b\})$ makes sense no matter what f is.

3.4 Binary Operations

Our next and final topic in this chapter is *binary operations*. The notion of a binary operation is an abstraction of ordinary addition and multiplication between real numbers.

Let us examine what is involved in these operations to distill the essence of the idea. For addition, we have a "rule" which allows us to associate with a given ordered pair (x, y) of real numbers a third real number denoted $x + y$. We thus have a map $+: \mathbf{R} \times \mathbf{R} \to \mathbf{R}$ where instead of writing

$+(x, y)$ for the image of (x, y) under $+$, we write $x + y$. A similar analysis shows that ordinary multiplication of real numbers may also be viewed as a map $\times: \mathbf{R} \times \mathbf{R} \to \mathbf{R}$ where again, instead of writing $\times(x, y)$ we write $x \times y$ or xy.

Bearing this discussion in mind, we arrive at the following definition.

3.4.1 *Definition*

A *binary operation* on a set S is a map $\#$ from $S \times S$ to S.

Instead of writing $\#(x, y)$ for the image of (x, y) under $\#$, we write $x \# y$ or just xy. When the image of (x, y) under an operation is written xy, we say that the operation is denoted by *juxtaposition* or that it is denoted *multiplicatively*. We also say that xy is the *product* of x and y, notwithstanding the fact xy is not, in general, a product in the usual sense.

This definition of a binary operation is rather formal and in practice we think of an operation on a set S as a rule which allows us to "combine" two elements of S to obtain a unique third element of S.

3.4.2 *Example*

Let S be a set and define an operation on S by $ab = a$ for all $a, b \in S$. ∎

3.4.3 *Example*

Let S denote the set of all maps from the set $\{1, 2, 3, 4\}$ to itself. Composition of maps is obviously an operation on S which we shall denote by juxtaposition rather than with the usual symbol $\circ$.

We examine this example more closely, in part to introduce notation which we shall be using regularly.

Consider the map $f : \{1,2,3,4\} \to \{1,2,3,4\}$ defined as follows: $f(1) = 3$, $f(2) = 4$, $f(3) = 4$, $f(4) = 1$. We denote this map by the following self-evident notation:

$$\begin{bmatrix} 1\ 2\ 3\ 4 \\ 3\ 4\ 4\ 1 \end{bmatrix}$$

or more briefly as

$$[3\ 4\ 4\ 1].$$

Let us compute the product (i.e., the composition) of the two maps [2 3 1 3] and [3 4 4 1] in that order; that is, apply [2313] first and then [3441]:

$$[3\ 4\ 4\ 1][2\ 3\ 1\ 3] = [4\ 4\ 3\ 4].$$

The reader may find the order in which the maps are written somewhat confusing. Bear in mind, however, that maps are written to the left of the element they are acting on.

Computing the product in the reverse order, we get

$$[2\ 3\ 1\ 3][3\ 4\ 4\ 1] = [1\ 3\ 3\ 2]$$

which shows that this operation is not *commutative*; that is, we cannot change the order in which the elements are "multiplied" and expect to get the same answer (unlike ordinary multiplication). On the other hand, by Theorem 3.3.9, this operation *is* associative. The reader should verify this by computing

$$([3\ 4\ 4\ 1][2\ 3\ 1\ 3])[4\ 1\ 2\ 3] \text{ and } [3\ 4\ 4\ 1]([2\ 3\ 1\ 3][4\ 1\ 2\ 3]).$$

We note in passing that there are many operations in mathematics which are not associative, including subtraction.

Among the elements of S is the identity map $I = [1234]$. Recall that $If = fI = f$ for all $f \in S$. We call I an *identity for the given operation*.

There are subsets of maps of some interest. We examine two of them.

(i) The set of all constant maps.
(ii) The set of all bijections.

(i) This is quite a small set and consists of [1111], [2222], [3 3 3 3], [4 4 4 4]. It is easily verified that

$$[i\ i\ i\ i][j\ j\ j\ j] = [i\ i\ i\ i].$$

We note that $C = \{[1\ 1\ 1\ 1], [2\ 2\ 2\ 2], [3\ 3\ 3\ 3], [4\ 4\ 4\ 4]\}$ is *"closed" under the given operation*; that is, given two elements of C, their product is in C. As an example of a subset of S which is not closed under the operation, take $T = \{[2\ 3\ 1\ 1], [3\ 1\ 2\ 2], [4\ 3\ 1\ 4]\}$. The elements [2311] and [3 1 2 2] are in T but their product [1 2 3 3] is not.

(ii) A bijection on a set is usually called a *permutation of the set*. In this case, the set of permutations of $\{1, 2, 3, 4\}$ is denoted by S_4. There are $4! = 24$ elements in S_4. (Why?) Observe that $I \in S_4$ and, for each $f \in S_4$, $f^{-1} \in S_4$. Also, S_4 is closed under the operation of composition of maps since the composition of two bijections is a bijection. (Proof?) ∎

There is very little that can be said about general abstract binary operations. Later on we shall be studying sets endowed with one or more binary operations having certain properties. Among these are properties possessed by some of the well-known operations such as ordinary addition and multiplication. In anticipation of our future work, we give a list of the relevant properties.

3.4.4 *Definition*

Let S be a set and $*$ a binary operation on S

(i) $*$ is *associative* if for all a, b, c in S $(a*b)*c = a*(b*c)$.

(ii) $*$ is *commutative* if for all a, b in S $a*b = b*a$.

(iii) An element $l \in S$ is a $*$- *left identity* (or a *left identity for* $*$) if for all $s \in S$, $l*s = s$.

(iv) An element $r \in S$ is a $*$- *right identity* (or a *right identity for* $*$) if for all $s \in S$, $s*r = s$.

(v) An element $e \in S$ is a (two-sided) *identity* for $*$ if it is both a left and a right identity.

(vi) An element $z \in S$ is a *left zero* for $*$ if for all $s \in S$, $z * s = z$

(vii) An element $z' \in S$ is a *right zero* for $*$ if for all $s \in S$, $s * z' = z'$.

(viii) An element is a *zero* if it is both a left and a right zero.

3.4.5 *Examples*

(i) On the set S define an operation, denoted by juxtaposition, by $ab = b$ for all $a,b \in S$. Every element of S is both a left identity and a right zero for this operation. The operation is associative but not commutative if S has at least two elements.

(ii) On the set $\mathbf{Z}$ of integers define an operation $*$ by $a * b = (2 + a)(2 + b) - 2$. This operation is associative, commutative, -1 is the identity and -2 the zero. ∎

3.4.6 *Definition*

A set S together with one or more binary operations is called an *algebraic system*. If $*, \circ, \bullet, ..$ are binary operations defined on S, we shall write $(S, *, \circ, \bullet, ...)$ to indicate that on the set S are defined the binary operations $*, \circ, \bullet, ...$ When the operation is denoted by juxtaposition, we usually write $(S, .)$. For example, we would write $(\mathbf{Z}, +, .)$ to denote the integers under the usual operations of addition and multiplication.

Depending on what properties the operations possess, these systems are given special names.

3.4.7 *Definition*

A *semigroup* is an algebraic system $(S, *)$ in which the operation $*$ is associative.

We shall now prove a number of theorems of a general nature on binary operations. These will be quite useful in later chapters when we undertake the study of specific algebraic systems.

3.4.8 *Theorem*

If $(S, .)$ is an algebraic system having a left identity e and a right identity f, then $e = f$ and so e is the unique identity.

Proof. Since e is a left identity, $ef = f$. On the other hand, since f is a right identity, $ef = e$. Therefore $e = f$ and so e is an identity. That it is unique follows from the fact that if g is another identity, it is, in particular a right identity and so, by the proof above, $e = g$. ∎

3.4.9 *Definition*

Let $(S, .)$ be an algebraic system with an identity e (e is *the* identity by Theorem 3.4.8).

(i) An element x' is a left inverse of x if $x'x = e$.
(ii) An element x'' is a right inverse if $xx'' = e$.
(iii) An element x' is an inverse of x if $x'x = xx' = e$.

3.4.10 *Example*

Let $(S, \circ)$ be the set of all maps from $\mathbf{Z}^+$ to $\mathbf{Z}^+$ under composition of maps. Since composition of maps is associative, $(S, \circ)$ is a semigroup with identity.

By Theorem 3.3.12, an element $f \in S$ has a left inverse but not a right inverse if and only if f is injective but not surjective; it has a right inverse which is not a left inverse if and only if it is surjective but not injective. An element of S has an inverse if and only if it is bijective.

Each element of S which is injective but not surjective has many left inverses. For example, consider the map $f \in S$ defined by $f(n) = 2n$ for all $n \in \mathbf{Z}^+$. This is clearly injective and one left inverse is given by g where

$$g(n) = \begin{cases} n/2 & \text{if } n \text{ is even} \\ 1 & \text{if } n \text{ is odd.} \end{cases}$$

A moment's thought will convince the reader that any map h defined by

$$h(n) = \begin{cases} n/2 & \text{if } n \text{ is even} \\ \text{anything} & \text{if } n \text{ is odd} \end{cases}$$

is a left inverse of f. There are therefore infinitely many left inverses of f. Similar remarks apply to right inverses. Note, however, the following theorem. ∎

3.4.11 *Theorem*

Let $(S, .)$ be a semigroup with identity e. If an element a of S has a left inverse a' and a right inverse a'', then $a' = a''$ and so a' is the unique (2-sided) inverse of a.

Proof. In the line of equalities below, we use associativity once:

$$a'' = ea'' = (a'a)a'' = a'(aa'') = a'e = a'.$$

Therefore $a'' = a'$. If b is another inverse of a, then in particular, b is a right inverse of a and so by the argument given above, $a' = b$. ∎

3.4.12 *Notation*

If an element a has an inverse, its (unique) inverse is denoted by a^{-1} (in multiplicative notation).

We end this chapter with a theorem which we shall find useful later on.

3.4.13 *Theorem*

Let $(S, .)$ be a semigroup. Suppose:

(i) S has a left identity e;
(ii) Corresponding to each $s \in S$ there exists $s' \in S$ such that $s's = e$;
i.e., each element has a left inverse relative to e. Then e is a (2-sided) identity and each element of S has a (2-sided) inverse.

Proof. Let s be an arbitrary element of S and s' an element as in (ii). Let s'' be such that $s''s' = e$. Such an element exists by (ii). Then, using associativity, we have

$$(*) \quad s = es = (s''s')s = s''(s's) = s''e \text{ and so } s = s''e.$$

But $ee = e$ since e is a left identity and therefore, multiplying both sides of the equality $s = s''e$ by e on the right, we get

$$se = (s''e)e = s''(ee) = s''e = s,$$

that is, $se = s$, showing that e is also a right identity and so a (2-sided) identity. (Note that s is an arbitrary element of S.) From (*) we have

$$s = s''e = s'', \text{ and so } s's = ss' = e.$$

Therefore, s has a (2-sided) inverse and the theorem is proved.∎

3.5 Exercises

1) Give an example of each of the following.

(i) A relation which is reflexive and transitive but not symmetric;
(ii) A relation which is reflexive and symmetric but not transitive;
(iii) A relation which is transitive and symmetric but not reflexive;
(iv) A relation which is antisymmetric and transitive but not reflexive.

2) How many relations can be defined on a set of n elements? Prove your answer.

3) Which of reflexive, transitive and symmetric hold for the following relations defined on the set C of living Canadians:

    (i) is married to;
    (ii) is a blood relative of;
    (iii) lives within a mile of;
    (iv) born in the same town as;
    (v) is an acquaintance of an acquaintance of an acquaintance of an....

4) If $|A| = p, |B| = q$, how many elements does $A \times B$ have?

5) Define a relation $|$ on the set of natural numbers $\mathbf{N}(= \mathbf{Z}^+)$ by $a|b$ if $b = ka$ for some $k \in \mathbf{N}$. (This relation is, of course, "is a factor of" or "divides".) Prove that $|$ is a partial order relation on $\mathbf{N}$.

6) Let $F = \{(a,b)|a,b \in \mathbf{Z}$ and $b \neq 0\}$. On F define a relation $\sim$ by: $(a,b) \sim (c,d)$ if $ad = bc$. Prove that $\sim$ is an equivalence relation. Describe the equivalence class of $(1,2)$.

7) Let T denote the set of all relations on a set A. If α and β are elements of T, define the relation $\alpha \circ \beta$ by: $a(\alpha \circ \beta)b$ if there exists $c \in A$ such that $a\alpha c$ and $c\beta b$.

    (i) Prove that if $\alpha, \beta, \gamma \in T$, then $(\alpha \circ \beta) \circ \gamma = \alpha \circ (\beta \circ \gamma)$;
    (ii) If $\alpha, \beta \in T$ and $M(\alpha), M(\beta)$ are the matrices of α and β respectively, then prove that $M(\alpha \circ \beta) = M(\beta)M(\alpha)$ if we interpret $1 + 1 = 1 = 1 + 0 = 0 + 1, 0 + 0 = 0, 1.0 = 0 = 0.1 = 0.0$ and $1.1 = 1$ when computing the matrix product $M(\beta)M(\alpha)$.

8) Let $|A| = p, |B| = q$. If $p \leq q$, how many 1:1 maps are there from A to B? Note that if $p > q$, there are none.

9) Let $S(p,q)$ denote the number of onto maps from a set of p elements to a set of q elements. (Observe that $S(p,q) = 0$ if $p < q$). Prove:

   (i) $S(p,p) = p! = p(p-1)(p-2)...3.2.1.$
   (ii) $S(p,1) = 1.$
   (iii) if $p > q$, then $S(p,q) = q[S(p-1,q) + S(p-1,q-1)].$
   (iv) Using (i) — (iii), compute $S(5,3)$.

10) Let $f : A \to B$ and define a relation, denoted $\ker f$, by: $a_1 (\ker f) a_2$ a_2 if $f(a_1) = f(a_2)$ ("ker" is short for "kernel").

   (i) Prove $\ker f$ is an equivalence relation.
   (ii) What is another way of writing $[a]_{\ker f}$?
   (iii) Denote the set of equivalence classes mod $\ker f$ by $A/_{\ker f}$ and

define a map $\phi : A/_{\ker f} \to B$ by $\phi([a]_{\ker f}) = f(a)$. Prove that ϕ is 1:1.

11) Let m be a fixed positive integer. Define a relation "$\equiv$ mod m" on $\mathbf{Z}$, the set of integers, by: $a \equiv b \bmod m$ if $b - a = km$ for some integer k. (The expression "$\equiv$ mod m" is read "congruence modulo m".) Prove that this is an equivalence relation and determine the equivalence classes.

12) If $|A)| = p, |B| = q$, how many maps are there from A to B? (The set of maps from A to B is denoted by B^A. This hints at the answer.)

13) With the notation of 12) above, let $h \in A^A$. for some set A. Prove:

   (i) h is 1:1 iff, for all $f, g \in A^A$, $h \circ f = h \circ g$ implies $f = g$.
   (ii) h is onto iff, for all $f, g \in A^A$, $f \circ h = g \circ h$ implies $f = g$.

14) Let f be the map from $\mathbf{Z}^+$ to $\mathbf{Z}^+$ defined by $F(m) = m + 2$. Find at least three left inverses of f.

15) Let $f : \mathbf{Z} \rightarrow \{0,1,2,3,4\}$ be defined as follows:
$$f(m) = \begin{cases} 0 \text{ if } m = 5k \text{ for some } k \in \mathbf{Z}; \\ 1 \text{ if } m = 5k + 1 \text{ for some } k \in \mathbf{Z}; \\ 2 \text{ if } m = 5k + 2 \text{ for some } k \in \mathbf{Z}; \\ 3 \text{ if } m = 5k + 3 \text{ for some } k \in \mathbf{Z}; \\ 4 \text{ if } m = 5k + 4 \text{ for some } k \in \mathbf{Z}. \end{cases}$$

Define at least three right inverses of f.

16) Let $f : A \rightarrow B$. If X and Y are subsets of A and V and W subsets of B, prove the following:

(i) $f(X \cup Y) = f(X) \cup f(Y)$;
(ii) $f(X \cap Y) \subseteq f(X) \cap f(Y)$ and show by means of an example that, in general, $f(X \cap Y) \neq f(X) \cap f(Y)$;
(iii) $f^{-1}(V \cup W) = f^{-1}(V) \cup f^{-1}(W)$;
(iv) $f^{-1}(V \cap W) = f^{-1}(V) \cap f^{-1}(W)$;
(v) $f^{-1}[f(X)] \supseteq X$;
(vi) $f[f^{-1}(V)] \subseteq V$;

17) Let $f : A \rightarrow B$.

(i) Prove that f is 1:1 iff for all $X \subseteq A$, $f^{-1}[f(X)] = X$.
(ii) Prove that f is onto iff for all $V \subseteq B$, $f[f^{-1}(V)] = V$.

18) If $f : A \rightarrow B$ is bijective, prove that $(f^{-1})^{-1} = f$.

19) With the notation of Problem 12), prove that if $f \in B^A, g \in C^B$ with both f and g injective, then $g \circ f$ is injective. Is the converse true? Explain. Same questions if "injective" is replaced by "surjective".

20) [Hard!] Prove that the number of partitions of a set of n elements is $\sum_{k=1}^{n} \frac{1}{k!} S(n,k)$. Here $S(n,k)$ is the number of onto maps from a set of n elements to a set of k elements. (See Problem 9.) Using this result and Problem 9, find the number of partitions of a set of 3, 4 and 5 elements. For the sets of three and four elements, list the partitions and check that the formula is correct.

21) Define an operation $\circ$ on the set $\mathbf{Q}$ of rationals by $a \circ b = a + b - ab$.

    (i) Prove $\circ$ is associative and commutative.

    (ii) Does $(\mathbf{Q}, \circ)$ have an identity? If so, which elements have inverses?

22) Let S be any nonempty set and define on $S \times S$ an operation, denoted by juxtaposition, by $(a,b)(c,d) = (a,d)$. Prove $(S \times S, .)$ is a semigroup in which, for all $x, y \in S \times S, xyx = x$.

23) If T is a semigroup with the property that for all $x, y \in T$, $xyx = x$, prove that each element of T is an *idempotent* (an element x is an idempotent if $xx = x$).

24) Let X be a nonempty set and $\{X_i | i \in I\}$ a partition of X. A subset T of X is a *cross-section* of the partition if for each $i \in I$ we have $|X_i \cap T| = 1$.

    (i) How many cross-sections are there of the partition $1,2,3|4,5|6,7,8,9$ of the set $\{1,2,3,4,5,6,7,8,9\}$?

    (ii) Let S be the semigroup of all maps from a set X to itself under composition of maps as operation. Prove that an element $f \in S$ is an idempotent (see Problem 23) if and only if:

    (a) Im f is a cross-section of the partition determined by the equivalence classes of the relation $\ker f$;

    (b) For all classes K of $\ker f$, $f(K) \subseteq K$. (See Problem 10.)

(c) If in (ii) above $X = \{1,2,3\}$, how many idempotents does S have?

25) Let $(S,\bullet)$ and $(T,*)$ be semigroups. On $S \times T$ define an operation, denoted by juxtaposition, by

$$(s,t)(s',t') = (s \bullet s', t * t').$$

Prove $S \times T$ is a semigroup under this operation. If S and T each have an identity, prove that $S \times T$ also has an identity and find a necessary and sufficient condition for (s,t) to possess an inverse.

26) Let $\Re(S)$ denote the set of all relations on a given set S. Recall that if $\lambda, \mu \in \Re(S)$, we define $\lambda \circ \mu$ by $a(\lambda \circ \mu)b$ if there exists c such that $a\lambda c$ and $c\mu b$.

   (i) Prove that $(\Re(S), \circ)$ is a semigroup.

   (ii) Prove that λ is idempotent if λ is reflexive and transitive. Conversely, prove that if λ is idempotent then λ is transitive (but not necessarily reflexive). Give an example of an idempotent relation which is not reflexive.

   (iii) Prove that the equality relation is the identity for the semigroup and that the empty relation is the zero.

Chapter 4

The Integers

4.1 Introduction

The set of *integers* consists of all the whole numbers, positive, negative and zero. As indicated earlier, this set will be denoted by $\mathbf{Z}$. The set of *positive integers* or *natural numbers* is the set $\mathbf{N} = \mathbf{Z}^+ = \{1, 2, 3, ...\}$ and $\mathbf{N}^0 = \{0, 1, 2, 3, ...\}$ is the set of *nonnegative integers*. Both $\mathbf{N}$ and $\mathbf{N}^0$ are clearly subsets of $\mathbf{Z}$.

In this chapter we prove some basic properties of the integers. We assume as an axiom that every nonempty set of the positive integers contains a least integer and show how this gives rise to the principle of mathematical induction, a powerful tool we shall often use in this and subsequent chapters.

After developing properties of divisibility, we prove the Fundamental Theorem of Arithmetic which states that every positive integer is uniquely expressible as a product of primes and deduce that there are infinitely many primes. We end the chapter with a discussion of congruences and show how they can be viewed as equations in certain algebraic systems. It is hoped that a familiarization with these concrete algebraic systems will help the reader make the transition to the more abstract structures to be studied in subsequent chapters.

4.2 Elementary Properties

There are two familiar operations defined on $\mathbf{Z}$, namely, addition and multiplication. We shall assume that the reader is familiar with the following properties which we adopt as axioms:

4.2.1 *Axioms for the integers*

(i) For all $a,b \in \mathbf{Z}$, if $a,b \in \mathbf{Z}$ then $a+b \in \mathbf{Z}$. (Closure).

(ii) For all $.a,b,c \in \mathbf{Z}$, $(a+b)+c = a+(b+c)$. (Associativity of addition).

(iii) There exists an integer, namely 0, such that for all $a \in \mathbf{Z}$, $a+0 = 0+a = a$. (Existence of additive identity)

(iv) Given $a \in \mathbf{Z}$, there exists $b \in \mathbf{Z}$ such that $a+b = b+a = 0$. (Existence of additive inverses).

(v) For all $a,b \in \mathbf{Z}$, $a+b = b+a$ (Commutativity of addition).

The next four properties dealing with multiplication match properties (i), (ii), (iii) and (v) above and so we number them (I) (II), etc.

(I) $a,b \in \mathbf{Z}$ implies $ab \in \mathbf{Z}$. (Closure).

(II) For all $a,b,c \in \mathbf{Z}, (ab)c = a(bc)$. (Associativity of multiplication).

(III) There exists an integer, namely 1, such that $a.1 = 1.a = a$. (Existence of multiplicative identity).

(V) For all $a,b \in \mathbf{Z}$, $ab = ba$. (Commutativity of multiplication).

The final property links the two operations of addition and multiplication.

(vi) For all $a,b,c \in \mathbf{Z}$, $a(b+c) = ab+ac$. (We say that multiplication is distributive over addition).

The property for multiplication analogous to property (iv) for addition would be: Given $a \in \mathbf{Z}$, there exists $b \in \mathbf{Z}$ such that $ab = 1$.

This property evidently does not hold in $\mathbf{Z}$, but a somewhat weaker one does, namely:

(vii) If $a \in \mathbf{Z}$ and $a \neq 0$, then for all $b,c \in \mathbf{Z}$, $ab = ac$ implies $b = c$. (Cancellation law for multiplication).

Remark. We also have a cancellation law for addition. It is in fact a consequence of Property (iv). Indeed, if $a+b = a+c$, then adding the additive inverse of a to both sides of this equation and using associativity yields $0+b = 0+c$ and so $b = c$.

4.2.2 *Notation*

(i) Given $a \in \mathbf{Z}$, the unique integer b such that $a+b=0$ is denoted by $-a$;

(ii) We shorten $a+(-b)$ to $a-b$ and call the operation "—" *subtraction*. Observe that subtraction is not associative. For example, $(2-3)-4 \neq 2-(3-4)$.

In addition to the above, we shall assume that the reader is familiar with the rules governing multiplication of signed quantities and the manipulation of inequalities.

A further intuitively obvious property possessed by the integers but probably not encountered in explicit form by the reader is the so-called *Well Ordering Principle* (WOP) for the integers which states that *every nonempty subset of the positive integers contains a least member* (which is easily seen to be unique). In addition to all the properties listed above, we shall assume as an axiom that the *positive integers are well ordered by the usual ordering,* which is another way of saying that the Well ordering Principle holds for the integers.

Remark. The real numbers are not well ordered by the usual ordering. For example, consider the set $\mathbf{R}^+$ of all positive real numbers. It is clear that this set contains no least member. It is true that 0 is less than every positive real number, but $0 \notin \mathbf{R}^+$, as is required by well ordering.

4.2.3 *Definition*

A nonempty subset S of the integers is *bounded below* if there exists an integer m such that for all $s \in S$, $m \leq s$. Each such m is a *lower bound* for S.

In the following lemma we extend the Well Ordering Principle to subsets of the integers which are bounded below. (A *lemma* is a preliminary result which is needed in a subsequent theorem but which is proved independently so as not to disrupt the flow of the proof of the main result).

4.2.4 *Lemma*

Let S be a nonempty subset of the integers which is bounded below. Then S contains a least member.

Proof. Let m be a lower bound for S and consider the set
$$T = \{s - m + 1 \mid s \in S\}.$$
If $t \in T$, then $t = s - m + 1$ for some $s \in S$. But $m \leq s$ and so $m - 1 < s$. Therefore $t = s - m + 1 > 0$, proving that T is a subset of the positive integers. By the WOP, T contains a least element l, say. Since $l \in T$, there exists $s_1 \in S$ such that $l = s_1 - m + 1$.

We show that $s_1 = l + m - 1$ is the least member of S. Indeed, if $s \in S$, then $s - m + 1 \in T$ and so
$$l = s_1 - m + 1 \leq s - m + 1.$$
Thus $s_1 \leq s$, proving that s_1 is the least element of S. ■

An immediate consequence of the lemma is the following theorem.

4.2.5 *Theorem*

Let S be a nonempty subset of the integers which is bounded below. Let l be the least integer in S and suppose S has the following property:

(*) for all $m \in \mathbf{Z}$ such that $m \geq l$, $m \in S$ implies $m + 1 \in S$.

Then S contains all integers greater than or equal to l.

Proof. Observe first that we can claim the existence of l on account of Lemma 4.2.4. Suppose the theorem is false Then there exists an integer $m > l$ such that $m \notin S$ and so
$$T = \{z \in \mathbf{Z} \mid z \geq l, z \notin S\}$$
is a nonempty subset of $\mathbf{Z}$ which is bounded below by l. Thus, by Lemma 4.2.4, T contains a least element t_0, say. Now $t_0 > l$ since $l \in S$ and so $t_0 - 1 \geq l$. Also, $t_0 - 1 \notin T$ since $t_0 - 1 < t_0$ and t_0 is least in T. Hence, $t_0 - 1 \in S$. By (*), $(t_0 - 1) + 1 \in S$ and so $t_0 \in S$, a contradiction. Therefore, S contains all integers greater than or equal to l. ■

This theorem in the form given in the corollary below is called the *First Principle of Mathematical Induction.*

4.2.6 *Corollary (First Principle of Mathematical Induction)*

Let $P(z)$ be a statement about an integer z. ($P(z)$ may be true for some integers and false for others.) Assume:

(i) $P(l)$ is true for some integer l.

(ii) For all integers $z \geq l$, the truth of $P(z)$ implies the truth of $P(z + 1)$.

Then $P(z)$ is true for all integers $z, z \geq l$.

Proof. Let $T = \{z \in \mathbf{Z} \mid z \geq l$ and $P(z)$ is true$\}$. Then l is a lower bound for T and by (i), $l \in T$. Moreover, by (ii), $z \in T$ implies $z+1 \in T$. By Theorem 4.2.5, $T = \{z \in \mathbf{Z} \mid z \geq l\}$ and the theorem is proved. ∎

Note. Think of the integers as an infinite ladder, extending infinitely in both directions, down and up. Then the corollary says that if:

(i) You can reach the lth rung.

(ii) If, whenever you can reach the mth rung for some m, $m \geq l$, you can reach the $(m+1)^{\text{st}}$ rung, then you can climb this infinite ladder from the lth rung up.

In the following, we give examples of how the First Principle of Mathematical Induction is applied.

4.2.7 *Example*

Prove that if x is a real number greater than -1, then $(1+x)^n \geq 1+nx$ for all $n \in \mathbf{N}$.

Proof. Let $P(x)$ denote the statement $(1+x)^n \geq 1+nx$. We must now prove (i) and (ii) of Corollary 4.2.6.

(i) $P(1)$ is the statement $(1+x)^1 \geq 1+1.x$ which is obviously true. (This step is sometimes referred to as the *basis for the induction*).

(ii) Now assume that $P(m)$ is true for some $m \geq 1$. (This is called the *induction hypothesis*). Thus we are assuming that

$$(1+x)^m \geq 1 + mx$$

is true for this particular value m. Since $x > -1$, it follows that $1+x$ is positive. Therefore, multiplying both sides of this inequality by $1+x$, we get

$$(1+x)^{m+1} \geq (1+x)(1+mx).$$

Multiplying the right-hand side out, we find that

$$(1+x)^{m+1} \geq 1 + x + mx + mx^2 = 1 + (m+1)x + mx^2 \geq 1 + (m+1)x,$$

the last inequality being a consequence of the fact that $mx^2 \geq 0$. Hence,

$$(1+x)^{m+1} \geq 1 + (m+1)x$$

is true; that is, $P(m+1)$ is true. Therefore, by the First Principle of Mathematical Induction, $P(n)$ is true for all integers greater than or equal to 1 and so, $P(n)$ is true for all **N**. ∎

We have written out the proof in excruciating detail. Once the reader has gotten the hang of it, such statements as "let $P(n)$ denote etc." will be unnecessary. Also it is not necessary to assert that the statement is true. Merely writing it down will carry that implication.

In the next example, we shall write out the proof with fewer details.

4.2.8 *Example*

Prove that $n! \geq 2^{n+1}$ for all integers $n \geq 5$.

Proof. Recall that $n! = n(n-1)(n-2).....3.2.1$. Observe first that this statement is false for $n = 1, 2, 3, 4$.

(i) Set $n = 5$. We get $5! \geq 2^6$, that is, $120 \geq 64$. Therefore, the statement is valid for $n = 5$.

(ii) Assume that for some integer $m \geq 5$, $m! \geq 2^{m+1}$. Multiplying both sides of this inequality by $m+1$ we get

$$(m+1)! \geq (m+1)2^{m+1}.$$

But $m+1 \geq 6$ and so

$$(m+1)! \geq (m+1)2^{m+1} \geq 6.2^{m+1} \geq 2^{(m+1)+1},$$

proving that

$$(m+1)! \geq 2^{(m+1)+1}.$$

Therefore, by the First Principle Mathematical Induction, the inequality is valid for all integers $n \geq 5$. ∎

We refer to a proof using the First Principle of Mathematical Induction as a *proof by induction*.

Another consequence of well ordering is the *Division Algorithm*.

4.2.9 *Theorem (The Division Algorithm)*

Let a and b be integers with $b \neq 0$. Then there exist unique integers q and r with $0 \leq r < |b|$ such that

$$a = bq + r.$$

We call q the *quotient* and r the *remainder*. ($|b|$ is the *absolute value* of b, that is, $|b| = b$ if $b \geq 0$ and $|b| = -b$ if $b < 0$).

Proof. *Existence.* Let $S = \{a - bt \mid t \in \mathbf{Z}\}$ and set $T = S \cap \mathbf{N}^0$. T is bounded below by 0 and $T \neq \phi$. By well ordering, T contains a least integer, r, say.

Since $r \in T$, there exists $q \in \mathbf{Z}$ such that $r = a - qb$ or $a = bq + r$. Suppose, if possible that $r \geq |b|$. Then $r = |b| + p$ for some $p \geq 0$ But then

$$p = a - b(q \pm 1) \in S \cap \mathbf{N}^0.$$

By the minimality of r in $S \cap \mathbf{N}^0$, it follows that $p \geq r$. But since, $b \neq 0$ we deduce $p < r$, a contradiction. Therefore, $0 \leq r < |b|$.

Uniqueness. Suppose $a = bq_1 + r_1 = bq_2 + r_2$ with $0 \leq r_1, r_2 < |b|$. Then

$$b(q_1 - q_2) = r_2 - r_1.$$

But $|r_2 - r_1| < |b|$ while, if $q_1 - q_2 \neq 0$, we have $|r_2 - r_1| = |b(q_1 - q_2)| \geq |b|$, a contradiction. Hence $q_1 - q_2 = 0$, proving that $q_1 = q_2$.

It then follows from $r_2 - r_1 = b(q_1 - q_2)$ that $r_1 = r_2$, thus establishing uniqueness. ∎

We shall be using the Division Algorithm in the very next theorem and later, after we have discussed greatest common divisors.

4.2.10   *Definition*

A nonempty set S of integers is said to be *closed under subtraction* if, for all $a,b \in \mathbf{Z}$, $a,b \in S$ implies $a - b \in S$.

Obviously $\mathbf{Z}$ itself is closed under subtraction as is the set of all multiples of a fixed integer; for example, the set of all multiples of 3. The interesting fact is that these are the only kinds of subsets of $\mathbf{Z}$ which are closed under subtraction.

4.2.11   *Theorem*

If S is a nonempty subset of $\mathbf{Z}$ closed under subtraction, then there exists a unique nonnegative integer n such that $S = \mathbf{Z}n = \{zn \mid z \in Z\}$.

Proof. Choose any $x \in S$. Since S is closed under subtraction, $0 = x - x \in S$. If $S = \{0\}$, take $n = 0$ and we have $\{0\} = \mathbf{Z}0$.

Assume $S \neq \{0\}$ and let x be a nonzero element of S. Then since $0 \in S$, we have $-x = 0 - x \in S$ and so S contains positive integers. Let $T = S \cap \mathbf{Z}^+ \neq \phi$.. By well ordering of the natural numbers, T contains a least member n, say.

Now $n \in S$ implies $0 - n = -n \in S$. Therefore, $n - (-n) = 2n \in S$. Similarly, $0 - 2n = -2n \in S$ and so $n - (-2n) = 3n \in S$. Continuing in the same vein, we prove that $\mathbf{Z}n \subseteq S$. (The reader should try a proof by induction to establish this.)

For the reverse containment, let $x \in S$. By the Division Algorithm, $x = qn + r$ where q and r are integers and $0 \leq r < n$. By the first part, $qn \in S$ and so, since S is closed under subtraction, $x - qn = r \in S$. But n was chosen least positive in S and so, since $r < n$, r cannot be positive. Hence $r = 0$, proving that $x = qn$. Therefore, $x \in \mathbf{Z}n$ and so $S \subseteq \mathbf{Z}n$. This, combined with the first part of the proof, yields the result. The uniqueness of n left as an exercise. ∎

4.3 Divisibility

4.3.1 *Definition and notation*

Let a and b be integers, $a \neq 0$. We write $a|b$ (and read "a divides b" or "a is a factor of b") if $b = qa$ for some $q \in \mathbf{Z}$.

The following are some easily proved properties of division. The reader is urged to prove them. They will be used without fanfare from now on.

4.3.2 *Theorem*

For all a, b and c in $\mathbf{Z} \setminus \{0\}$ the following hold:

(i) $a | b$ and $b | a$ imply $a = \pm b$.
(ii) $a|b$ and $b|c$ imply $a|c$;
(iii) 0 is the only integer divisible by every nonzero integer;
(iv) $a|b$ implies $\pm a|\pm b$;
(v) $(a|b$ and $a|c$ imply $a|(\pm b \pm c)$
(vi) $a|b$ and $a|(\pm b \pm c)$ imply $a|c$.

Proof. Excrcise. ∎

4.3.3 *Definition*

Let a and b be integers, not both zero. A *positive* integer d is a *greatest common divisor* (gcd) *of a and b* if:

(i) $d|a$ and $d|b$.
(ii) For all $c \in \mathbf{Z}$, $c|a$ and $c|b$ imply $c|d$.

This definition makes considerable demands on d and it is conceivable that our requirements are too stringent with the result that d may not exist for certain pairs of integers. The next theorem proves that gcd's do exist and that, moreover, the positive integer d of Definition 4.3.2 is unique. We shall therefore be able to speak of *the* gcd of a and b rather than *a* gcd of a and b.

4.3.4   *Theorem*

Let a and b be integers, not both zero. Then there exists a unique positive integer d satisfying (i) and (ii) of Definition 4.3.2. Moreover, there exist integers p and q such that $d = pa + qb$.

Proof. Let
$$S = \{xa + yb \,|\, x, y \in Z\}.$$
Since $a = 1.a + 0.b$ and $b = 0.a + 1.b$, we have $a, b \in S$ and so $S \neq \{0\}$. Also, it is easy to check that S is closed under subtraction. It then follows from Theorem 4.2.10 that $S = \mathbf{Z}d$ for some positive integer d.

Since $a, b \in S \,(= \mathbf{Z}d)$, it follows that $a = ud$ and $b = vd$ for some $u, v \in \mathbf{Z}$. Therefore, $d \,|\, a$ and $d \,|\, b$. Moreover, since $d \in S$, there exist integers p and q such that
$$d = pa + qb.$$
Suppose that $c \,|\, a$ and $c \,|\, b$. Then $a = kc$ and $b = lc$ for some integers k and l. Substituting in the equation $d = pa + qb$ we get
$$d = kcp + lcq = c(kp + lq),$$
and so $c \,|\, d$. Therefore, d is a gcd of a and b.

For uniqueness, assume d_1, d_2 are gcd's of a and b. Then, by virtue of d_1 being a gcd of a and b and d_2 dividing both a and b, we deduce $d_2 \,|\, d_1$. Similarly, by interchanging the roles of d_1 and d_2, we find $d_1 \,|\, d_2$. By Theorem 4.3.1 (i), $d_1 = \pm d_2$. Since both are positive, it follows that $d_1 = d_2$. ∎

4.3.5   *Notation*

We denote the greatest common divisor of a and b by $\gcd(a,b)$.

This theorem is an *existence theorem* which does not answer the question: How do we compute gcd's? There is an algorithm which was known to Euclid (and so called the Euclidean Algorithm) which yields the gcd of two integers very efficiently. This algorithm has the added bonus that, from the computations involved, we can find integers p and q such that $d = pa + qb$. Rather than explain the algorithm in full generality, we apply it to a particular case in the following.

4.3.6 *Example*

Find the gcd(180, 462).

Use the Division Algorithm repeatedly, always dividing the smaller of the two underlined numbers into the bigger as follows:

$$462 = 2 \times \underline{180} + \underline{102}$$
$$180 = 1 \times \underline{102} + \underline{78}$$
$$102 = 1 \times \underline{78} + \underline{24}$$
$$78 = 3 \times \underline{24} + 6$$
$$24 = 4 \times \underline{6} + \underline{0}.$$

The *last nonzero remainder*, namely 6, is the gcd of 180 and 462.

Why is 6 = gcd(180,462)? Recall that we must first show that 6 divides both 180 and 462. We work from the bottom line of the above computation up to the top line.

Clearly 6 divides 24. From the next line and Theorem 4.3.1(v), 6 divides 78. Going up one more line and applying Theorem 4.3.1(v) it follows that 6 divides 102, etc., yielding finally that 6 divides both 180 and 462.

We must now prove the second property of a gcd, namely, we must show that if c divides both 180 and 462, then c divides 6. This time we work down the computation, starting with the first line.

So, suppose c divides both 180 and 462. Then from the first line and Theorem 4.3.1(vi), c divides 102. From the second line and Theorem 4.3.1(vi), c divides 78, etc., yielding at the last step that c divides 6.

(The reader should fashion a general proof that the procedure described above applied to two integers a and b yields their gcd. The argument is identical to that given. The only difficulty which may be encountered is the notation).

To find what linear combination of 180 and 462 will yield their gcd, we proceed as follows: The penultimate line of the computation gives

$$(*)\ 6 = 78 - 3 \times 24 .$$

From the next line up we get $24 = 102 - 78$. Substituting back in (*) we have

$$(**)\ 6 = 78 - 3 \times [102 - 78]$$
$$= 4 \times 78 - 3 \times 102 .$$

From the next line up we obtain

$$78 = 180 - 102 .$$

Substituting in (**) we find that

$$6 = 4 \times [180 - 102] - 3 \times 102$$

and so

$$(***)\ 6 = 4 \times 180 - 7 \times 102 .$$

Finally, from $462 = 2 \times 180 + 102$ we get

$$102 = 462 - 2 \times 180$$

which, when substituted in (***), gives

$$6 = 4 \times 180 - 7 \times [462 - 2 \times 180].$$

Hence,

$$6 = 18 \times 180 + (-7) \times 462 . \blacksquare$$

Remark. We define $\gcd(a_1, a_2, ..., a_n)$ to be that positive number d such that

(i) $d \mid a_i, i = 1, 2, ..., n$.

(ii) If $c \mid a_i, i = 1, 2, ..., n$, then $c \mid d$. (See Problem 38.)

The next concept is closely related to the gcd.

4.3.7 *Definition*

Let a and b be nonzero integers. Then *a least common multiple of a and b* is a *positive* integer m satisfying:

  (i) $a|m$ and $b|m$;
  (ii) for all $c \in \mathbf{Z}$, $a|c$ and $b|c$ imply $m|c$.

As in the case of gcd's, we show that least common multiples exist and that they are unique.

4.3.8 *Theorem*

Let a and b be nonzero integers and let $d = \gcd(a,b)$. Then $\frac{|ab|}{d}$ is the unique least common multiple of a and b, which we shall henceforth denote by $\operatorname{lcm}(a,b)$.

Proof. *Existence.* It is clearly sufficient to prove the theorem in the case a and b are both positive. Thus let $m = \frac{ab}{d}$. Then

$$m = \left[\frac{a}{d}\right]b = a\left[\frac{b}{d}\right].$$

(Note that $\frac{a}{d}$ and $\frac{b}{d}$ are integers since d divides both a and b).

Therefore $a|m$ and $b|m$, satisfying (i) of Definition 4.3.6.

  Suppose that $a|c$ and $b|c$. Then:

$$(*) \quad c = ka = lb$$

where k and l are integers. There exist integers p and q such that

$$d = pa + qb,$$

from which, dividing by d, we get

$$1 = p\frac{a}{d} + q\frac{b}{d}.$$

Multiply through by c to obtain

$$c = pc(\tfrac{a}{d}) + qc(\tfrac{b}{d}).$$

Substituting for c from (*) we find

$$c = pl(\tfrac{ab}{d}) + qk(\tfrac{ab}{d}) = plm + qkm = m(pl + qk),$$

thus proving that $m|c$. Therefore, $\dfrac{ab}{d} = m$ is *an* lcm of a and b.

Uniqueness. Exercise. ∎

We shall now work our way towards a proof of the *Fundamental Theorem of Arithmetic*. To do so we require an additional tool and a few concepts.

The tool we need is called the *Second (or Strong) Principle of Mathematical Induction*. It is closely related to the first with a subtle difference. In the first we are only allowed to assume the truth of a proposition for some integer m, and, based on this, we try to deduce the truth of the proposition for $m+1$

In the Second Principle we may assume the truth of the proposition for all integers greater than or equal to some lower bound and less than some m. Based on this stronger hypothesis, we then attempt to prove that the statement is true for m.

It appears on the surface that the First Principle is stronger than the Second since the induction hypothesis of the former is weaker. (The weaker the hypothesis, the stronger the theorem! Why?) It turns out that in fact the First Principle, the Second Principle and the Well Ordering Principle are equivalent.

4.3.9 *Theorem*

(Compare with Theorem 4.2.5.) Let S be a nonempty subset of the integers which is bounded below and let l be the least integer in S. (See Lemma 4.2.4.) Suppose S has the following property:

(*) for all $m \in \mathbf{Z}, \{z \in \mathbf{Z} \mid l \le z < m\} \subseteq S$ implies $m \in S$.

Then S contains all integers greater than or equal to l.

(The metaphor of the ladder reads: If the lth rung is attainable and if, whenever we can reach the lth rung and all rungs above the lth one and strictly below the mth one, we can reach the mth one, then we can reach any rung from the lth one up.)

Proof. Assume the theorem is false. Then

$$T = \{z \in \mathbf{Z} \mid z \ge l\} \cap S' \ne \phi.$$

Clearly T is bounded below by l and so by Lemma 4.2.4, T contains a least member t_0, say. But t_0 is minimal in T and so $\{z \in \mathbf{Z} | l \le z < t_0\}$ $\subseteq S$. By (*), $t_0 \in S$, a contradiction since $t_0 \notin S$. ■

Compare the following with Corollary 4.2.6.

4.3.10 Corollary (Second Principle of Mathematical Induction)

Let $P(z)$ be a statement involving the integer z. Assume:
(i) $P(l)$ is true for some $l \in \mathbf{Z}$.
(ii) For all integers $m > l$, the truth of $P(l), P(l+1), ..., P(m-1)$ implies the truth of $P(m)$.
Then the statement $P(z)$ is true for all integers $z \ge l$.

Proof. Exercise. ■

We shall often come across situations where the First Principle will not work while the Second will. Indeed, this is the case in the proof of the Fundamental Theorem of Arithmetic, which we shall tackle shortly.

4.4   The Fundamental Theorem of Arithmetic

We need a couple of definitions and one more tool to tackle the proof of The Fundamental Theorem of Arithmetic

4.4.1 Definition

(i) A *positive* integer $p \neq 1$ is a *prime* if the only factors of p are ± 1, $\pm p$. If an integer is not prime, it is said to be *composite*. A partial list of primes is $\{2,3,5,7,11,13,17,19,...\}$.

(ii) Two integers a and b are said to be *relatively prime* if $\gcd(a,b) = 1$.

4.4.2 *Theorem*

Let a, b and c be integers, $a \neq 0$. If

$$a \mid bc \text{ and } a \text{ and } b \text{ are relatively prime, then } a \mid c.$$

Proof. Since $\gcd(a,b) = 1$, there exist integers p and q such that $pa + qb = 1$. Multiply through by c to get

$$(*) \quad c = pac + qbc.$$

But $bc = ta$ since a divides bc. Substituting in (*), we find

$$c = pac + qta = a(pc + qt).$$

Therefore, a divides c. ∎

4.4.3 *Corollary*

If p is a prime and $p \mid bc$, then $p \mid b$ or $p \mid c$.

Proof. If $p \mid b$ we are done. Assume, therefore, that p does not divide b. Then p and b are relatively prime (why?) and so by Theorem 4.4.2, $p \mid c$. ∎

4.4.4 *Corollary*

If p is prime and $p \mid a_1 a_2 a_3 \ldots a_n$, then $p \mid a_i$ for some i, $1 \leq i \leq n$.

Proof. Exercise (use the previous corollary and induction). ∎

We are now ready to prove the Fundamental Theorem of Arithmetic (FTA).

4.4.5 *Theorem (Fundamental Theorem of Arithmetic)*

Every integer greater than 1 is either a prime or a product of primes. This factorization is unique up to the order in which the factors occur.

Proof. *Existence*. We proceed using the Second Principle of Mathematical Induction.

If the integer is 2, the theorem is clearly valid. Assume inductively that if n is an integer and $2 \le m < n$, then either m is a prime or m is a product of primes.

If n is prime, we are done. Assume therefore that n is composite, say $n = rs$ where both r and s are integers greater than 1 and consequently both less than n. By the induction hypothesis,

$$r = p_1 p_2 p_3 \ldots p_k \text{ and } s = q_1 q_2 q_3 \ldots q_l$$

where the p_i and q_j are primes. Hence,

$$n = rs = p_1 p_2 p_3 \ldots p_k q_1 q_2 q_3 \ldots q_l$$

is a factorization of n as a product of primes. By the Second Principle of Mathematical Induction, the existence half of the theorem is proved.

Uniqueness. If the integer is 2, the uniqueness is clear. Assume inductively that if n is an integer and $2 \le m < n$, then m is *uniquely* expressible as a product of primes.

If n is prime, there is nothing to prove. Assume therefore that n is composite and suppose

$$n = p_1 p_2 p_3 \ldots p_k = q_1 q_2 q_3 \ldots q_l$$

where the p_i and q_j are primes. (Note that both k and l are greater than 1 since n is composite).

Since $p_1 | q_1 q_2 q_3 \ldots q_l$, by Corollary 4.4.4, p_1 divides one of the q's. By renumbering if necessary, we may assume $p_1 | q_1$. But since q_1 is prime, it follows that $p_1 = q_1$. Hence

$$p_1 p_2 p_3 \ldots p_k = p_1 q_2 q_3 \ldots q_l.$$

Cancelling p_1 from both sides, we get

$$p_2 p_3 \ldots p_k = q_2 q_3 \ldots q_l.$$

This is an integer greater than 1 and less than n (why?) and so, by the induction hypothesis, $k - 1 = l - 1$ (whence $k = l$) and, after renumbering, if necessary, $p_j = q_j, j = 2, 3, \ldots, k$. The uniqueness now follows by induction. ∎

Remark. We would not have been able to apply the First Principle in the proof of the Fundamental Theorem. The reader should pinpoint where the First Principle would fail.

The following theorem was known to, and proved by, Euclid.

4.4.6 *Theorem*

There are infinitely many primes.

Proof. Suppose the contrary and let $p_1, p_2, p_3, ..., p_n$ be a complete list of all the primes. Consider the integer

$$m = p_1 p_2 p_3 ... p_n + 1.$$

By the Fundamental Theorem, m is divisible by some prime (since m can be expressed as a product of primes). Since none of the listed primes divides m, it follows that there is a prime not listed, a contradiction. Therefore, there are indeed infinitely many primes. ∎

4.5 The Algebraic System $(\mathbf{Z}_n, +, \cdot)$ and Congruences

In this section we shall give a fairly extensive treatment of congruences. Recall (Problem 11, Chapter 3) that the relation "congruence mod n", denoted "$\equiv$ mod n" is the relation defined on $\mathbf{Z}$ as follows:

$a \equiv b \bmod n$ if $a - b = kn$ for some $k \in \mathbf{Z}$.

The positive integer n is called the *modulus*.

This relation is an equivalence relation (see proof below) and the equivalence class to which the integer a belongs is denoted by $[a]_n$. We write just "$\equiv$" rather than "$\equiv \bmod n$" and $[a]$ or $\bar{a}$ rather than $[a]_n$ when there is no ambiguity.

We begin with a few elementary properties of congruences. *In all that follows, the congruences considered will be* mod n *for some fixed positive integer* n *and we shall use the abbreviated notation alluded to above.*

4.5.1 *Theorem*

Congruence mod n is an equivalence relation on $\mathbf{Z}$.

Proof. Reflexivity: $a \equiv a \bmod n$ since n divides $a - a = 0$.

Symmetry: If $a \equiv b \bmod n$, then $a - b = kn$ for some $k \in \mathbf{Z}$. Therefore $b - a = (-k)n$ and so $b \equiv a \bmod n$.

Transitivity: Suppose $a \equiv b \bmod n$ and $b \equiv c \bmod n$. Then

$$a - b = kn \text{ and } b - c = ln.$$

Adding these two equations, we find

$$a - c = (k + l)n$$

and so $a \equiv c \bmod n$. ∎

4.5.2 Theorem

If $a,b,c,d \in \mathbf{Z}$ with $a \equiv b$ and $c \equiv d$, then $a + c \equiv b + d$ and $ac \equiv bd$.

Proof. The hypotheses imply that $a - b = sn$ and $c - d = tn$ where s, $t \in \mathbf{Z}$. Adding the left- and right-hand sides of the two equations we get

$$(a - b) + (c - d) = (s + t)n$$

Therefore,

$$(a + c) - (b + d) = (s + t)n,$$

and so

$$a + c \equiv b + d.$$

Next,

$$ac - bd = ac - bc + bc - bd = (a - b)c + b(c - d) = (sc + bt)n.$$

Therefore,

$$ac \equiv bd. ∎$$

Since congruence $\bmod n$ is an equivalence relation, it partitions $\mathbf{Z}$ into mutually disjoint equivalence classes. The next theorem supplies us with two criteria, each of which enables us to determine when two equivalence classes are equal.

4.5.3 Theorem

Let a and b be any two integers. Then:

(i) $[a] = [b]$ if, and only if, $n \mid (a - b)$.

(ii) $[a] = [b]$ if, and only if, a and b yield the same remainder on division by n.

Proof. We prove only (ii). Suppose $[a] = [b]$. By the Division Algorithm,

$$a = qn + r_1 \text{ and } b = sn + r_2$$

where $0 \le r_1 < n$ and $0 \le r_2 < n$. Subtracting the two equations, we get

$$a - b = (q - s)n + r_1 - r_2.$$

But $a \equiv b$ since $[a] = [b]$, and so, $a - b = kn$ for some $k \in \mathbf{Z}$. Therefore,

$$kn = (q - s)n + r_1 - r_2.$$

This implies that $n|(r_1 - r_2)$. Since $0 \le r_1 < n$ and $0 \le r_2 < n$, it follows that $0 \le |r_1 - r_2| < n$. This last inequality together with $n|(r_1 - r_2)$ implies that $r_1 = r_2$ (why?).

Conversely, suppose a and b have the same remainder on division by n. Then, by the Division Algorithm, $a = qn + r$ and $b = sn + r$. Hence, $a - b = (q - s)n$ and so $a \equiv b$ proving that $[a] = [b]$ by (i). ∎

4.5.4 *Corollary*

There are exactly n equivalence classes mod n, namely, $[0], [1], [2], \ldots,$ $[n-1]$.

Proof. There are exactly n possible remainders on division by n, namely, $0, 1, 2, \ldots, n - 1$. The classes determined by these remainders are distinct since n does not divide $i - j$ if $0 \le i < j < n$ and by the preceding theorem, there are no more classes. ∎

4.5.5 *Notation*

$\mathbf{Z}_n = \{[0], [1], [2], \ldots, [n-1]\} = \{\overline{0}, \overline{1}, \overline{2}, \ldots, \overline{n-1}\}.$

Observe: $[a] = \overline{a} = \{a + kn | k \in \mathbf{Z}\}.$

We turn $\mathbf{Z}_n$ into an algebraic system by defining two operations $\oplus$ and $\bullet$ on the set $\{[0],[1],[2],...,[n-1]\}$. We call $\oplus$ "addition" and $\bullet$ "multiplication". We shall first describe the operations informally in words.

Think of each of the classes as a bag of elements. To add bag A to bag B, pick an arbitrary element a out of A and an arbitrary element b out of B. Since both a and b are integers, we can add them ordinarily to obtain the integer $a+b$. Find the bag to which $a+b$ belongs, say C, and set $A \oplus B = C$.

To multiply bag A by bag B, proceed as above and find the bag D to which ab belongs. Set $A \bullet B = D$. Symbolically, we have the following definition.

4.5.6 *Definition*

$$[a] \oplus [b] = [a+b] \text{ and } [a] \bullet [b] = [ab].$$

(From now on we shall denote $\bullet$ by juxtaposition).

4.5.7 *Example*

In $\mathbf{Z}_{12}$ let us add $[8]$ and $[5]$. According to our definition,

$$[8] \oplus [5] = [8+5] = [13] = [1].$$

Multiplying these two bags together, we get

$$[8].[5] = [8 \times 5] = [40] = [4].$$

In the verbal description of these operations given above, we have a choice of elements from each bag: we dip into bag A, pick *any* element, dip into bag B, pick *any* element, add (or multiply) these two elements, etc.

What if we picked different representatives from $[8]$ amd $[5]$? Recall that

$$[8] = \{...,-28,-16,-4,8,20,32,....\} \text{ and } [5] = \{....,-19,-7,5,17,29,...\}.$$

Suppose we pick $-16 \in [8]$, $-7 \in [5]$ and carry out the instructions for addition. We get the bag containing -23. But $-23 \equiv 1$ and so $[-23] = [1]$, which is the same answer we got before using 8 and 5 as our choices from the respective bags.

If we carry out the instructions for multiplication using -16 and -7 rather than 8 and 5, the answer is the bag containing $-16 \times -7 = 112$. Since $112 \equiv 4$ we again obtain the same answer as before. ∎

If $\oplus$ and $\bullet$ worthy of being called "operations", we should end up with the same result whatever choice of elements we make from the respective classes. From our computations above, it appears that this is so. We now prove in full generality that $\oplus$ and $\bullet$ are *well defined*, that is, we end up with the same answer no matter which elements we choose from our respective equivalence classes to effect the computation.

4.5.8 *Theorem*

$\oplus$ and $\bullet$ are well defined on $\mathbf{Z}_n$.

Proof. We must show that if $a' \in [a]$ and $b' \in [b]$, then $[a'+b'] = [a+b]$ and $[a'b'] = [ab]$.

Since $a' \in [a]$) and. $b' \in [b]$ we have $a \equiv a'$ and $b \equiv b'$. Therefore, by Theorem 4.5.2,

$$a'+b' \equiv a+b \text{ and } a'b' \equiv ab.$$

By Theorem 4.5.3, $[a'+b'] = [a+b]$ and $[a'b'] = [ab]$. ∎

To simplify notation, we set $[a] = \overline{a}$ and write $+$ instead of $\oplus$ and juxtaposition or . instead of $\bullet$. With this new notation, the definitions of addition and multiplication read

$$\overline{a} + \overline{b} = \overline{a+b} \text{ and } \overline{a}.\overline{b} = \overline{ab}.$$

Note that the two plus signs occurring in the first equation, although denoted by the same symbol, mean quite different things. The first equation should be read: "The class of a plus the class of b is the class of $a+b$; the second should likewise be read: "The class of a times the class of b is the class of ab."

Below is a list some of the properties of the system $(\mathbf{Z}_n, +, .)$. We shall prove only one, leaving the rest to the reader.

4.5.9 *Theorem*

The following hold in $(\mathbf{Z}_n, +, .)$:
  (i) If $\overline{a}, \overline{b} \in \mathbf{Z}_n$, then $\overline{a} + \overline{b} \in \mathbf{Z}_n$ (Closure under +).
  (ii) For all $\overline{a}, \overline{b}, \overline{c} \in \mathbf{Z}_n$, $(\overline{a} + \overline{b}) + \overline{c} = \overline{a} + (\overline{b} + \overline{c})$ (Associativity)
  (iii) There exists a +-identity, namely $\overline{0}$, called the *zero* of $\mathbf{Z}_n$.
  (iv) Each element $\overline{a} \in \mathbf{Z}_n$ has a +-inverse, namely, $(\overline{-a})$.
  (v) For all $\overline{a}, \overline{b} \in \mathbf{Z}_n$, $\overline{a} + \overline{b} = \overline{b} + \overline{a}$ (Commutativity of +).
  (I) If $\overline{a}, \overline{b} \in \mathbf{Z}_n$, then $\overline{a}.\overline{b} \in \mathbf{Z}_n$ (Closure under .).
  (II) For all $\overline{a}, \overline{b}, \overline{c} \in \mathbf{Z}_n$, $(\overline{a}.\overline{b}).\overline{c} = \overline{a}.(\overline{b}.\overline{c})$ (Associativity of .).
  (III) There exists a .-identity, namely, $\overline{1}$.
  (IV) For all $\overline{a}, \overline{b} \in \mathbf{Z}_n$, $\overline{a}.\overline{b} = \overline{b}.\overline{a}$ (Commutativity of .).
  (vi) For all $\overline{a}, \overline{b}, \overline{c} \in \mathbf{Z}_n$, $\overline{a}.(\overline{b} + \overline{c}) = \overline{a}.\overline{b} + \overline{a}.\overline{c})$
(. is distributive over +).

Proof. We prove (vi) only. If $\overline{a}, \overline{b}, \overline{c} \in \mathbf{Z}_n$, then

$$\overline{a}(\overline{b} + \overline{c}) = \overline{a}\overline{(b+c)} = \overline{a(b+c)} = \overline{ab+ac} = \overline{a}.\overline{b} + \overline{a}.\overline{c}.$$

The first equality above follows from the definition as does the second. The third equality follows from the fact that ordinary multiplication of integers is distributive over ordinary addition, and the last is a consequence of definitions. ■

The reader should compare these properties with those of the integers listed at the beginning of this chapter. (S)he will see that they are identical except for Property (vii) for the integers which does not hold in general in $(\mathbf{Z}_n, +, .)$. This means that any theorem or computation which is valid for the integers will also hold for $(\mathbf{Z}_n, +, .)$, provided we use only the common properties listed above.

Our next theorem tells us under what circumstances Property (vii) for $\mathbf{Z}$ is valid in $\mathbf{Z}_n$. First we state a simple result whose proof is left to the reader.

4.5.10 *Theorem*

If $\overline{a}, \overline{b} \in \mathbf{Z}_n$ and $\overline{a} = \overline{b}$, then $\gcd(a, n) = \gcd(b, n)$.

Proof. Exercise. [Hint: Let $b = a + kn$ for some integer k and show that $\gcd(a,n) \mid \gcd(b,n)$ and $\gcd(b,n) \mid \gcd(a,n)$]. ∎

4.5.11 *Notation*

For $\bar{a} \in \mathbf{Z}_n$ let $\lambda(\bar{a}) = \gcd(a,n)$. By the previous result, the value of λ is independent of the choice of representative of the class.

4.5.12 *Definition*

An element a of a commutative algebraic system $(S,.)$ is *cancellable* if, for all elements x and y in S, $ax = ay$ implies $x = y$.

Note. When we say an element in $\mathbf{Z}_n$ is cancellable, we tacitly mean that it is multiplicatively cancellable since every element of $\mathbf{Z}_n$ is additively cancellable; that is, from $\bar{a} + \bar{x} = \bar{a} + \bar{y}$ we can deduce $\bar{x} = \bar{y}$.

4.5.13 *Theorem*

The following are equivalent in $\mathbf{Z}_n$:
   (i)   An element $\bar{a}$ of $\mathbf{Z}_n$ has a multiplicative inverse;
   (ii) $\lambda(\bar{a}) = 1$;
   (iii) $\bar{a}$ is cancellable

Proof. (i) implies (ii). Suppose $\bar{a}$ has a multiplicative inverse, $\bar{b}$, say. Then $\bar{a}.\bar{b} = \bar{1}$ and so
$$ab = 1 + kn$$
for some $k \in \mathbf{Z}$. Thus $ab - kn = 1$. From this it follows that if $s \mid a$ and $s \mid n$, then $s \mid 1$. Therefore, $s = \pm 1$ and so $\lambda(\bar{a}) = 1$.

   (ii) implies (i). Assume $\lambda(\bar{a}) = 1$. Then $\gcd(a,n) = 1$ and so there exist integers p and q such that $pa + qn = 1$. Hence
$$\bar{1} = \overline{pa + qn}.$$
By definition,
$$\overline{pa + qn} = \bar{p}.\bar{a} + \bar{q}.\bar{n} .$$
But $\bar{n} = \bar{0}$ and so $\bar{1} = \bar{p}.\bar{a}$, proving that $\bar{p}$ is the multiplicative inverse of $\bar{a}$.

(i) implies (iii) Assume $\bar{a}$ has an inverse $\bar{b}$ and suppose $\bar{a}.\bar{x} = \bar{a}.\bar{y}$. Then $\bar{b}(\bar{a}.\bar{x}) = \bar{b}(\bar{a}.\bar{y})$ and so $(\bar{b}\bar{a}).\bar{x} = (\bar{b}.\bar{a}).\bar{y}$. Therefore,

$$\bar{x} = \bar{1}.\bar{x} = \bar{1}.\bar{y} = \bar{y},$$

and so $\bar{a}$ is cancellable.

(iii) implies (ii). We prove the contrapositive. Assume therefore that $\lambda(\bar{a}) = s \neq 1$. Then there are integers k and l with

$$a = ks \text{ and } n = ls \text{ where } 0 < l < n \text{ since } s > 1.$$

Clearly $\bar{a}.\bar{l} = \bar{0}$ since $n \mid al$ and so $\bar{a}.\bar{l} = \bar{a}.\bar{0}$. But then we have

$$\bar{a}.\bar{l} = \bar{a}.\bar{0} \text{ and } \bar{l} \neq \bar{0}$$

which shows that $\bar{a}$ is not cancellable. ∎

One of the central tasks of abstract algebra is the classification of algebraic systems and their elements. We show how the elements of $\mathbf{Z}_n$ fall neatly into two classes.

4.5.14 *Definition*

(i) An element $\bar{a} \in \mathbf{Z}_n$ is a *zero divisor* if there exists a *nonzero* element $\bar{b} \in \mathbf{Z}_n$ such that $\bar{a}.\bar{b} = \bar{0}$. (For example, $\bar{0}$ is a zero divisor).

(ii) An element $\bar{a} \in \mathbf{Z}_n$ is a *unit* if it has a multiplicative inverse.

4.5.15 *Notation*

We shall denote the set of units in $\mathbf{Z}_n$ by $U(\mathbf{Z}_n)$ and the set of zero divisors by $J(\mathbf{Z}_n)$.

4.5.16 *Theorem*

$\mathbf{Z}_n = U(\mathbf{Z}_n) \cup J(\mathbf{Z}_n)$ and $U(\mathbf{Z}_n) \cap J(\mathbf{Z}_n) = \phi$.

Proof. Suppose $\bar{a}$ is not a zero divisor and let $\bar{a}.\bar{x} = \bar{a}.\bar{y}$. Then $\bar{a}(\bar{x} - \bar{y}) = \bar{0}$. Since $\bar{a}$ is not a zero divisor, it follows that $\bar{x} - \bar{y} = \bar{0}$ and so $\bar{x} = \bar{y}$. This means that $\bar{a}$ is cancellable. By Theorem 4.5.13, $\bar{a}$ is a unit and so

$$\mathbf{Z}_n = U(\mathbf{Z}_n) \cup J(\mathbf{Z}_n).$$

Suppose that $\bar{b} \in U(\mathbf{Z}_n)$. Then, $\bar{b}.\bar{c} = \bar{0}$ implies $\bar{b}^{-1}(\bar{b}.\bar{c}) = \bar{b}^{-1}\bar{0} = \bar{0}$, that is, $\bar{c} = \bar{0}$. Therefore, $\bar{b}$ is not a zero divisor. Hence,

$$U(\mathbf{Z}_n) \cap J(\mathbf{Z}_n) = \phi. \blacksquare$$

The following corollary is an immediate consequence of the previous three theorems. The reader is invited to prove it.

4.5.17 *Corollary*

The following are equivalent:
   (i) $\bar{a}$ is cancellable;
   (ii) $\lambda(\bar{a}) = 1$;
   (iii) $\bar{a}$ has a multiplicative inverse;
   (iv) $\bar{a} \notin J(\mathbf{Z}_n)$

Proof. Exercise.■

4.5.18 *Corollary*

Every nonzero element of $\mathbf{Z}_n$ is a unit if, and only if, n is prime.

Proof. Only if] Assume n is not prime. Then $n = rs$ where $0 < r < n$ and $0 < s < n$ and so $\bar{r} \neq \bar{0}, \bar{s} \neq \bar{0}$. However, $\bar{r}.\bar{s} = \bar{n} = \bar{0}$, showing that both $\bar{r}$ and $\bar{s}$ are zero divisors. By the previous result, $\bar{r}$ and $\bar{s}$ are nonzero nonunits.

If] Suppose n is prime. Then $\lambda(\bar{a}) = 1$ for each $a, 0 < a < n$. Hence, by the previous result, $\bar{a}$ is a unit.■

From the previous results we can see that when n is prime, $(\mathbf{Z}_n, +, .)$ is very much like the rationals in that in either system we can add, subtract, multiply and divide. (Subtraction is, of course, addition of the additive inverse and division is multiplication by the multiplicative inverse.) Therefore, in any computation in $\mathbf{Z}_n$ we can operate as though we are in $\mathbf{Q}$ when n is prime.

The situation is somewhat different when n is not prime, because then there are nonzero zero divisors which do not satisfy the cancellation law for multiplication. This is in sharp contrast to the situation in **Q** or even **Z** where there are no nonzero zero divisors. Therefore, care must be taken in this case.

Below are examples which illustrate the foregoing theory.

4.5.19 *Example*

(i) $\mathbf{Z}_{12} = \{\overline{0}, \overline{1}, \overline{2}, ..., \overline{11}\}$, $U = \{\overline{1}, \overline{5}, \overline{7}, \overline{11}\}$ $J = \{\overline{0}, \overline{2}, \overline{3}, \overline{4}, \overline{6}, \overline{9}, \overline{10}\}$. Also $\overline{5}^{-1} = \overline{5}$, $\overline{7}^{-1} = \overline{7}$; $\overline{2}.\overline{6} = \overline{0} = \overline{3}.\overline{4}$;

(ii) $\mathbf{Z}_{16} = \{\overline{0}, \overline{1}, ..., \overline{15}\}$, $U = \{\overline{a} \mid 0 \le a < 16, a \text{ odd}\}$ and $J = \{\overline{a} \mid 0 \le a < 16, a \text{ even}\}$. ∎

4.6 Congruences in **Z** and Equations in $\mathbf{Z}_n$

Equations of the form $ax + ny = b$ where $a, n,$ and b are given integers and integral solutions x and y are sought, are called linear *Diophantine equations* after Diophantus of Alexandria (ca 250 CE) who first studied them. It is an easy matter to see that solving such an equation is equivalent to solving the linear congruence $ax \equiv b \bmod n$ and this in turn is equivalent to solving the linear equation $\overline{a}.\overline{x} = \overline{b}$ in $\mathbf{Z}_n$. In this section we develop an algorithm for the solution of equations of this form.

4.6.1 *Example*

We can solve some linear equations in $\mathbf{Z}_{12}$ but not others. Take for example $\overline{5}\overline{x} + \overline{8} = \overline{3}$. We proceed just as though we were computing in **Z**. We get

$$\overline{5}\overline{x} = \overline{3} - \overline{8} = \overline{3} + \overline{4} = \overline{7}.$$

"Dividing" by $\overline{5}$, (which we can do since $\overline{5}$ has a multiplicative inverse), we obtain $\overline{x} = \overline{5}.\overline{7} = \overline{35} = \overline{11}$.

On the other hand, consider $\overline{8}\overline{x} = \overline{3}$. Here $\overline{8}$ has no multiplicative inverse and so we must be careful. Suppose in fact that $\overline{a}$ is a solution. Then $\overline{8}\overline{a} = \overline{3}$. Multiplying both sides by $\overline{3}$ we get $\overline{24}\,\overline{a} = \overline{9}$ and so $\overline{0} = \overline{9}$, a contradiction. Therefore no solution exists. ∎

Lest the reader believe that the reason no solution exists is that $\overline{8}$ is a zero divisor, consider the equation $\overline{8}\overline{x} = \overline{4}$. Here there are four solutions, namely, $\overline{2}, \overline{5}, \overline{8}, \overline{11}$ so the situation is more complicated than we might be led to believe. We shall give the full story below.

Note. Solving the equation
$$\overline{5}\overline{x} + \overline{8} = \overline{3} \text{ in } \mathbf{Z}_{12}$$
is equivalent to solving
$$5x + 8 \equiv 3 \bmod 12,$$
which in turn is equivalent to solving the linear Diophantine equation
$$5x + 12y = -5.$$
We have chosen to emphasize equations in $\mathbf{Z}_n$ rather than congruences because, conceptually, equations are easier to grasp.

Before giving a systematic treatment for the solution of linear equations in $\mathbf{Z}_n$, we give two examples, one of solving a pair of simultaneous linear equations and one of solving a quadratic equation to illustrate that when the modulus is prime, we operate as though we are computing in the rationals.

4.6.2 *Example*

Consider the following simultaneous equations in $\mathbf{Z}_{11}$:
$$\overline{2}\overline{x} + \overline{5}\overline{y} = \overline{7}$$
$$\overline{4}\overline{x} + \overline{3}\overline{y} = \overline{8}.$$
Multiplying the first equation by $\overline{2}$ and subtracting the second from the result yields
$$(*)\,\overline{7}\overline{y} = \overline{6}.$$
The multiplicative inverse of $\overline{7}$ is $\overline{8}$. Multiplying both sides of (*) by $\overline{8}$, we get $\overline{y} = \overline{4}$. On substituting this value for $\overline{y}$ in the first equation, we find that
$$\overline{2}\overline{x} + \overline{20} = \overline{7}.$$

Therefore,

$$\overline{2}\overline{x} = \overline{-13} = \overline{9}.$$

"Dividing" by $\overline{2}$, that is, multiplying by the multiplicative inverse of $\overline{2}$, namely $\overline{6}$, we get $\overline{x} = \overline{10}$. ∎

4.6.3 *Example*

We solve the following quadratic equation in $\mathbf{Z}_7$:

$$\overline{2}\overline{x}^2 + \overline{3}\overline{x} + \overline{5} = \overline{0}.$$

As above, since 7 is prime, we can calculate as though we are in the rationals. Using the usual quadratic formula, we get

$$\overline{x} = \frac{-\overline{3} \pm \sqrt{\overline{9} - \overline{4}.\overline{2}.\overline{5}}}{\overline{4}}.$$

Division by $\overline{4}$ is multiplication by $\overline{4}^{-1} = \overline{2}$ and so

$$\overline{x} = (-\overline{3} \pm \sqrt{\overline{2} - \overline{5}}).\overline{2} = \overline{1} \pm \overline{2}\sqrt{\overline{4}} = \overline{1} \pm \overline{4}.$$

Hence $\overline{x} = \overline{5}$ or $\overline{x} = \overline{4}$.

The reader should substitute these two values back in the equation and verify that they are indeed solutions. ∎

If n is prime, it can be easily shown that the solution of the equation

$$\overline{a}.\overline{x}^2 + \overline{b}\overline{x} + \overline{c} = \overline{0}, \overline{a} \neq \overline{0} \text{ in } \mathbf{Z}_n$$

is given by the formula

$$\overline{x} = (\overline{2a})^{-1}(-\overline{b} \pm \sqrt{\overline{b}^2 - \overline{4}.\overline{a}.\overline{c}})$$

provided that $(\overline{2a})^{-1}$ exists in $\mathbf{Z}_n$. If $n = 2$, then $\overline{2} = \overline{0}$ and so the inverse does not exist. For any other prime, the formula works. There is, of course, no solution in $\mathbf{Z}_n$ if $\sqrt{\overline{b}^2 - \overline{4}.\overline{a}.\overline{c}}$ does not exist in $\mathbf{Z}_n$ The reader should verify that if the constant term in the above equation is replaced by $\overline{4}$, the resulting quadratic has no solution in $\mathbf{Z}_7$.

We now focus our attention on the solution(s) of equations of the form $\overline{a}\overline{x} = \overline{b}$ where $\overline{a}$ and $\overline{b}$ are given and we have to solve for $\overline{x}$.

4.6.4 *Theorem*

Fix $\bar{a} \in \mathbf{Z}_n$ and let $d = \lambda(\bar{a})$. Set $a = a_0 d$, $n = n_0 d$. Then the equation $\bar{a}.\bar{x} = \bar{b}$ has a solution if, and only if, $d \mid b$. Moreover, if $d \mid b$ and $\bar{c}$ is any solution, then the full solution set S is

$$\{\overline{c + kn_0} \mid k = 0, 1, 2, ..., d-1\}.$$

Proof. Only if] Assume $\bar{a}.\bar{x} = \bar{b}$ has a solution Then there exists $\bar{c} \in \mathbf{Z}_n$ such that $\bar{a}.\bar{c} = \bar{b}$. Hence $\overline{ac} = \bar{b}$ and so, $ac = b + kn$ for some integer k. Since $d \mid a$ and $d \mid n$, it follows that $d \mid b$.

If] Suppose conversely that $d \mid b$. Then $b = sd$ for some integer s. But $d = pa + qn$ for some integers p and q and so
$$sd = spa + sqn.$$
Therefore,
$$\bar{b} = \overline{spa + sqn} = \overline{\bar{a}.sp} + \overline{sqn} = \overline{\bar{a}.sp} + \bar{0} = \overline{\bar{a}.sp}.$$
Hence, $\overline{sp}$ is a solution.
Since
$$\bar{a}.\overline{(c + kn_0)} = \bar{a}.\bar{c} + \overline{a_0 kn_0 d} = \bar{a}.\bar{c} + \bar{0} = \bar{b}$$
$\overline{c + kn_0}$ is a solution of $\bar{a}.\bar{x} = \bar{b}$ for all integers k. In particular, every element of
$$\{\overline{c + kn_0} \mid k = 0, 1, 2, ..., d-1\}$$
is a solution.

We show next that the elements of $\{\overline{c + kn_0} \mid k = 0, 1, 2, ..., d-1\}$ are distinct. Suppose that

$$0 \le k, l < d \text{ and } c + kn_0 = c + l.n_0 + sn.$$
Then
$$kn_0 = l.n_0 + sdn_0.$$
Cancelling n_0 and re-arranging, we get $k - l = sd$. Therefore, $d \mid (k-l)$. But $|k - l| < d$ and so $k = l$. Hence S contains at least d elements.

Finally, suppose $\bar{g}$ is a solution. Then
$$\overline{ag} = \bar{b} = \bar{a}.\bar{c} \text{ and so } a(g - c) = sn.$$

Therefore
$$a_0 d(g - c) = s n_0 d \ .$$
Cancelling d, we get
$$a_0(g - c) = s n_0 .$$
But $\gcd(a_0, n_0) = 1$ (why?) and so $g - c = t n_0$.

By the Division Algorithm, $t = ud + r$ where $0 \le r < d$.
Hence, $g - c = (ud + r) n_0$ and so
$$g = c + u d n_0 + r n_0 = c + r n_0 + un.$$
Therefore, $\overline{g} = \overline{c + r n_0}$ since $\overline{un} = \overline{0}$ and the theorem is proved. ■

This theorem tells us under what circumstances a general linear equation over $\mathbf{Z}_n$ has at least one solution. It also tells us how many solutions there are and what they are provided we can find one. All that remains to do, therefore, is to show how to find that one solution.

4.6.5 *Theorem*

Let $\overline{a}, \overline{b}$ be given elements of $\mathbf{Z}_n$ and set $a = a_0 d, b = b_0 d, n = n_0 d$ where $d = \lambda(\overline{a})$. If we denote the elements of $\mathbf{Z}_{n_0}$ by using double bars over the representatives rather than a single bar, then $\overline{c} \in \mathbf{Z}_n$ is a solution of $\overline{a}.\overline{x} = \overline{b}$ if, and only if, $\overline{\overline{c}} \in \mathbf{Z}_{n_0}$ is a solution of $\overline{\overline{a_0}}.\overline{\overline{y}} = \overline{\overline{b_0}}$.

Proof. If $\overline{c}$ is a solution of $\overline{a}.\overline{x} = \overline{b}$, then $\overline{a}.\overline{c} = \overline{b}$, and so
$$a_0 dc - b_0 d = k n_0 d \text{ for some integer } k.$$
Therefore, cancelling d, we get $a_0 c - b_0 = k n_0$, proving that $\overline{\overline{c}}$ is a solution of $\overline{\overline{a_0}}.\overline{\overline{y}} = \overline{\overline{b_0}}$. The converse is easily proved by retracing the steps above. ■

Observe that, $\lambda(\overline{\overline{a_0}}) = 1$ since $\gcd(a_0, n_0) = 1$. Therefore since $1 | b_0$, it follows by Theorem 4.6.4 that $\overline{\overline{a_0}}.\overline{\overline{y}} = \overline{\overline{b_0}}$ has exactly one solution, namely, $\overline{\overline{a_0}}^{-1} \overline{\overline{b_0}}$.

Algorithm for solving $\overline{a}.\overline{x} = \overline{b}$.

(i) Make sure that $\overline{a}.\overline{x} = \overline{b}$ has a solution by checking that $d = \lambda(\overline{a})$ ($= \gcd(a,n)$) divides b.

(ii) Write $a = a_0 d, b = b_0 d, n = n_0 d$ as above.

(iii) Find $\overline{a_0}^{-1}$ in $\mathbf{Z}_{n_0}$ proceeding as in Theorem 4.5.13.

(iv) The solution to $\overline{a_0}.\overline{y} = \overline{b_0}$ is then $\overline{a_0}^{-1}\overline{b_0} = \overline{c}$, say. (There is only one solution. Why?)

(v) The full solution set is $\{\overline{c + kn_0} \,|\, 0 \le k < d\} \subseteq \mathbf{Z}_n$.

We illustrate this algorithm with the following example.

4.6.6 *Example*

Solve $\overline{84}.\overline{x} = \overline{54}$ in $\mathbf{Z}_{198}$.

(i) In this case $d = \lambda(\overline{a}) = 6 = \gcd(84,198)$. Since 6 divides 54, the equation has a solution.

(ii) $84 = 14 \times 6, 54 = 9 \times 6$, and $198 = 33 \times 6$ that is,

$a_0 = 14, b_0 = 9, n_0 = 33$.

(iii) We find $\overline{14}^{-1}$ in $\mathbf{Z}_{33}$ by applying the Euclidean Algorithm:

$$33 = 2.14 + 5$$
$$14 = 2.5 + 4$$
$$5 = 4 + 1.$$

Working from the bottom up, we obtain
$$1 = 5 - 4 = 5 - (14 - 2 \times 5) = 3 \times 5 - 14 = 3 \times (33 - 2 \times 14) - 14$$
$$= 3 \times 33 + (-7) \times 14.$$

Hence $\overline{14}^{-1} = \overline{-7} = \overline{26}$.

(iv) The solution to $\overline{14}.x = \overline{9}$ in $\mathbf{Z}_{33}$ is $\overline{26}.\overline{9} = \overline{3}$.

(v) The solution set of $\overline{84}.\overline{x} = \overline{54}$ is $\{\overline{3}, \overline{36}, \overline{69}, \overline{102}, \overline{135}, \overline{168}\}$. Note that the representatives of the classes increase by 33 starting with 3. Also observe that if we added 33 to the last entry, we would get 201 which is in the same class as 3 and so we wouldn't get anything new. This is to be expected since we know that there are exactly 6 solutions.

The equation $\overline{84}.\overline{x} = \overline{54}$ in $\mathbf{Z}_{198}$ can also be written as

$$84x \equiv 54 \bmod 198.$$

If written in this form, the solution set is infinite, namely, the union of the six classes given above. Thus the solution set is

$$\bigcup_{0 \le l \le 5} \{3 + 33l + 198k \mid k \in \mathbf{Z}\} = \{3 + 33k \mid k \in \mathbf{Z}\}. \blacksquare$$

4.7 Exercises

In Exercises 1—5, use induction to prove that each identity is valid for all positive integers n..

1) $1 + 3 + 5 + ... + (2n - 1) = n^2$.

2) $1^2 + 2^2 + 3^2 + ... + n^2 = \dfrac{n(n+1)(2n+1)}{6}$.

3) $1^3 + 2^3 + 3^3 + ... + n^3 = \dfrac{n^2(n+1)^2}{4}$.

4) $1.2 + 2.3 + 3.4 + ... + n(n+1) = \dfrac{n(n+1)(n+2)}{3}$.

5) $\dfrac{1}{2^2 - 1} + \dfrac{1}{3^2 - 1} + \dfrac{1}{4^2 - 1} + ... + \dfrac{1}{(n+1)^2 - 1} = \dfrac{3}{4} - \dfrac{1}{2(n+1)} - \dfrac{1}{2(n+2)}$.

6) Use induction to prove that if p is prime and $p \mid a_1 a_2 a_3 ... a_n$, then $p \mid a_i$ for some i, $1 \le i \le n$.

7) If a_i, $i = 1, 2, ..., 2^n$ are positive real numbers, use induction to prove

$$(a_1 a_2 a_3 ... a_{2^n})^{\frac{1}{2^n}} \le \dfrac{a_1 + a_2 + a_3 + ... + a_{2^n}}{2^n}, \text{ for all natural numbers } n.$$

8) Prove that, for all natural numbers n, $6.7^n - 2.3^n$ is divisible by 4.

9) Show that postage of 4 cents or more can be achieved by using 2-cent and 5-cent stamps only.

10) Use induction to show that n straight lines in the plane divide the plane into $\dfrac{(n^2 + n + 2)}{2}$ regions. Assume that no two lines are parallel and that no three lines have a common point.

11) In Theorem 4.2.11, prove that $S \supseteq \{kn | k \in Z\}$ by induction.

12) Prove Theorem 4.2.10.

13) Prove that the Euclidean Algorithm works in general.

14) In each part, find the gcd of the pair of given integers and express the gcd as a linear combination of the integers

   (i) $12,138$;
   (ii) $210,130$;
   (iii) $34,126$;
   (iv) $17,164$;
   (v) $1211, -6203$.

15) For a given pair of integers a and b, let $a \vee b$ denote the bigger of a and b and $a \wedge b$ the smaller of a and b. (For example $-5 \vee -3 = -3$; $2 \wedge -5 = -5$.)
Prove that
$$a \wedge (b \vee c) = (a \wedge b) \vee (a \wedge c),$$
that is, $\wedge$ is distributive over $\vee$.
   Is $\vee$ distributive over $\wedge$? Prove or give a counterexample.

16) Let
$$a = p_1^{m_1} p_2^{m_2} p_3^{m_3} ... p_k^{m_k} \text{ and } b = p_1^{n_1} p_2^{n_2} p_3^{n_3} ... p_k^{n_k}$$

where the p_i are distinct primes and $m_i \geq 0, n_i \geq 0$. Set $r_i = m_i \wedge n_i$ and $s_i = m_i \vee n_i$. (See previous question). Prove that

$$\gcd(a,b) = p_1^{r_1} p_2^{r_2} p_3^{r_3} \ldots p_k^{r_k} \text{ and } \text{lcm}((a,b)) = p_1^{s_1} p_2^{s_2} p_3^{s_3} \ldots p_k^{s_k}.$$

Hence prove that

$$\gcd(a, \text{lcm}(b,c)) = \text{lcm}(\gcd[a,b], \gcd[a,c]).$$

17) Prove the uniqueness of lcm's. (Theorem 4.3.7.)

18) Prove Corollary 4.3.9.

19) Prove the properties listed in Theorem 4.5.9.

20) Prove Theorem 4.5.10.

21) Prove that the set U of units in $\mathbf{Z}_n$ is closed under multiplication. Same question for the set J of zero divisors. Are these sets closed under addition? Prove or give counterexamples.

22) Compute the multiplicative inverses of the following elements

(i) $\bar{3} \in \mathbf{Z}_{25}$
(ii) $\overline{12} \in \mathbf{Z}_{31}$
(iii) $\bar{5} \in \mathbf{Z}_{112}$.

23) Prove that if p is prime, $(\bar{a} + \bar{b})^{p^K} = \bar{a}^{p^K} + \bar{b}^{p^K}$ in $\mathbf{Z}_p$ for all positive integers k.

24) Prove that for all positive integers n, $\overline{10}^n = \bar{1}$ in $\mathbf{Z}_9$.

25) Using the previous question, prove that a positive integer is divisible by 9 if, and only if, the sum of its digits is divisible by 9.

26) If the decimal expansion of an integer m is $a_n a_{n-1} a_{n-2} \ldots a_1 a_0$, prove that m is divisible by 4 if, and only if, 4 divides $a_1 a_0$. Prove also that m is divisible by 8 if, and only if, 8 divides $a_2 a_1 a_0$. (Example: 8 divides 315628957264 since 8 divides 264.)

27) On $\mathbf{Z}_m \times \mathbf{Z}_n$ define two operations + and . by

$$(\overline{a},\overline{\overline{b}}) + (\overline{c},\overline{\overline{d}}) = (\overline{a+c},\overline{\overline{b+d}}) ; \quad (\overline{a},\overline{\overline{b}}).(\overline{c},\overline{\overline{d}}) = (\overline{ac},\overline{\overline{bd}})$$

Define a map $\alpha : \mathbf{Z}_{mn} \to \mathbf{Z}_m \times \mathbf{Z}_n$ by

$$\alpha(\tilde{a}) = (\overline{a},\overline{\overline{a}}) \text{ for all } \tilde{a} \in \mathbf{Z}_{mn}.$$

Prove the following:

(i) α is well defined;

(ii) $\alpha(\tilde{a}+\tilde{b}) = \alpha(\tilde{a})+\alpha(\tilde{b})$ and $\alpha(\tilde{a}.\tilde{b}) = \alpha(\tilde{a}).\alpha(\tilde{b})$;

(iii) If m and n are relatively prime, α is a bijection; (Hint: To prove α is onto you must show that, given integers a and b, there is an integer c such that $c \equiv a \bmod m$ and $c \equiv b \bmod n$.)

(iv) Describe the units of $\mathbf{Z}_m \times \mathbf{Z}_n$.

28) (The Euler φ-function.) For a given positive integer m, define $\varphi(m)$ to be the number of positive integers less than m and relatively prime to m. (For example, $\varphi(12) = 4$).) If m and n are relatively prime, prove that $\varphi(mn) = \varphi(m)\varphi(n)$. (Hint: Note that $\varphi(m)$ is the number of units in $\mathbf{Z}_m$. Apply the preceding question.)

29) Prove by induction on k that if $p_i, i=1,2,...,k$ are primes, then

$$\varphi(p_1^{m_1} p_2^{m_2} ... p_k^{m_k}) = \varphi(p_1^{m_1})\varphi(p_2^{m_2})....\varphi(p_k^{m_k}).$$

30) Prove that if p is prime and n a positive integer, then

$$\varphi(p^n) = p^n - p^{n-1}.$$

31) Combining 29) and 30), prove that $\varphi(n) = n \prod_{p\in P}(1-\frac{1}{p})$ where P is the set of distinct primes dividing n. (The symbol $\prod_{p\in P}$ means "take the product of the expression in brackets over all primes dividing n." For example, $\varphi(12) = 12(1-\frac{1}{2})(1-\frac{1}{3}) = 12 \times \frac{1}{2} \times \frac{2}{3} = 4$. Here $P = \{2,3\}$).)

32) Let x, y, and z be relatively prime positive integers such that $x^2 + y^2 = z^2$. Prove that one of x and y is odd and the other even. Also prove that one of x and y is divisible by 3. (Hint: If $x^2 + y^2 = z^2$, then $\bar{x}^2 + \bar{y}^2 = \bar{z}^2$ in $\mathbf{Z}_3$.)

33) Solve the following linear equations:

   (i) $\overline{3}.\bar{x} = \overline{2}$ in $\mathbf{Z}_7$;
   (ii) $\overline{15}.\bar{x} = \overline{10}$ in $\mathbf{Z}_{25}$;
   (iii) $\overline{980}.\bar{x} = \overline{1500}$ in $\mathbf{Z}_{1600}$;
   (iv) $\overline{5}.\bar{x} = \overline{9}$ in $\mathbf{Z}_{35}$.

Convert each of the above equations into congruences and give their solutions.

34) Prove that the familiar quadratic formula for the solution of a quadratic equation is valid in $\mathbf{Z}_p$ where p is an odd prime.

35) (Fermat's Little Theorem.) If p is prime and $\gcd(a, p) = 1$, then $\bar{a}^{p-1} = \overline{1}$ in $\mathbf{Z}_p$; equivalently, $a^{p-1} \equiv 1 \bmod p$.

36) (Wilson's Theorem.) p is prime if, and only if, $(p-1)! \equiv -1 \bmod p$.

37) Prove that in $\mathbf{Z}_p$, p prime, every element is a square if, and only if, $p = 2$.

38) Let a_i, $i = 1, 2, ..., n$ be integers, not all zero.

   (i) Formulate a definition for the gcd of a_i, $i = 1, 2, ..., n$ and prove that it exists, is unique and that

$$\gcd(a_1, a_2, ..., a_n) = \sum_{i=1}^{n} x_i a_i$$

where the x's are integers. How would the lcm of $a_i, i = 1, 2, ..., n$ be defined? Find a formula for it.

(ii) Prove that $\gcd(a_1, a_2, ..., a_n) = \gcd(a_1, \gcd(a_2, a_3, ..., a_n))$.

(iii) Compute $\gcd(30, 45, 105, 210)$ and express it in the form $30a + 45b + 105c + 210d$, a, b, c, d integers.

Chapter 5

Groups

5.1 Introduction

As early as 2000 BCE the Babylonians knew how to solve quadratic equations and by the 16th century, formulae for the roots of cubics and quartics were known. In spite of the efforts of the finest mathematicians of the day, however, the solution of the general quintic proved to be elusive until, in the 1820's, the famous Norwegian mathematician, Niels Henrik Abel, using sets of permutations of the roots of equations (we would call them groups of permutations nowadays), proved the insolvability of the quintic. More precisely, he proved that there does not exist a formula involving the usual arithmetic operations (addition, subtraction, multiplication, division and the extraction of roots) for the roots of a general quintic.

Although the abstract notion of a group was not formulated until the 1880's Abel is given credit for showing the important role these sets of permutations play and we honor him by calling a commutative group abelian.

In the 1830's, Evariste Galois, an iconoclastic but brilliant French mathematician, developed what we now call Galois Theory in which groups play an important role. He showed why the general quintic cannot be solved in terms of the usual arithmetic operations listed above by associating a group with each equation and showing that in order for a polynomial equation to have a solution of the prescribed form it is necessary and sufficient for the associated group to have a certain property which we now call solvability. With the same tools which form the essence of Galois Theory, he was also able to solve several problems from antiquity dealing with geometrical constructions using only straightedge and compass.

Since group theory emerged as an independent discipline, it has been found indispensable within mathematics and in applications in fields other than mathematics such as physics and computer science. The basic reason that it is such a powerful tool rests on the fact that it is eminently useful in describing symmetry in nature and mathematics.

5.2 Definitions and Elementary Properties

5.2.1 *Definition*

A *group* is a set G together with a binary operation, usually denoted by juxtaposition, satisfying the following axioms:

(i) $a,b \in G$ implies $ab \in G$ (Closure);

(ii) For all $a,b,c \in G$, $(ab)c = a(bc)$. (Associativity);

(iii) There exists an element $e \in G$ such that $ae = ea = a$ for all $a \in G$. (Existence of an identity);

(iv) Given $a \in G$, there exists $a' \in G$ such that $aa' = a'a = e$. (Existence of inverses).

Notes.

(i) Although closure is implicit in the definition of an operation, we have chosen to list it for emphasis.

(ii) We have already proved in Theorem 3.4.7 that there is only one identity. We may therefore speak of *the* identity.

(iii) Theorem 3.4.10 establishes that there is only one inverse to a given element. We denote the inverse of an element a by a^{-1} (see (vii) below).

(iv) Since associativity holds, we omit brackets. Thus, rather than writing $[(ab)c][d(ef)]$, we write $abcdef$.

(v) If we add the axiom: For all $a,b \in G$, $ab = ba$, the group is said to be *abelian* or *commutative*.

(vi) To verify that an associative system is a group, according to Theorem 3.4.11, it is sufficient to show the existence of a left identity, and, for each element, the existence of a left inverse relative to this left identity.

(vii) When an abstract group is abelian we often denote the operation by "+". When we do so we say that the *group is denoted additively* (as opposed to *multiplicatively* when the operation is denoted by juxtaposition). In the case of additive notation, the identity is denoted by 0 and the inverse of an element a is denoted by $-a$.

5.2.2 Definition

(i) If a group G is denoted multiplicatively, $g \in G$ and n is a positive integer, then define g^n to mean $gg...g$ (n times). If m is a negative integer, set $g^m = g^{-1}g^{-1}...g^{-1}$ ($|m|$ times) and $g^0 = e$ where e is the identity of G.

(ii) If G is denoted additively and n is a positive integer, set $ng = g + g +...+ g$ (n times). If m is negative, set $mg = (-g) + (-g) + ... + (-g)$ ($|m|$ times) and $0g = 0$. (In this last equation, the zero on the left-hand side is the zero integer, while that on the right is the identity of the group.)

We gather together in the next theorem some of the rules of computation in a group. We give the rules in multiplicative notation, leaving it up to the reader to formulate them in additive notation.

5.2.3 Theorem

Let $(G,.)$ be a group. Then

(i) For all $a, x \in G$, if either $ax = e$ or $xa = e$, then $x = a^{-1}$);

(ii) For all $a, b \in G$, $(ab)^{-1} = a^{-1}b^{-1}$;

(iii) For all $a, b, c \in G$, if either $ab = ac$ or, $ba = ca$, then $b = c$. (Cancellation);

(iv) For all $m, n \in \mathbf{Z}$ and for all $g \in G$, $g^m g^n = g^{m+n}$, $(g^m)^n = g^{mn}$ and $(g^{-1})^{-1} = g$.

Proof.

(i) Suppose $ax = e$. Then $a^{-1}ax = a^{-1}e$ and so $x = ex = a^{-1}$. Similarly if $xa = e$.

(ii) $(ab)(b^{-1}a^{-1}) = a(bb^{-1})a^{-1} = aea^{-1} = aa^{-1} = e$. By (i), $(ab)^{-1} = b^{-1}a^{-1}$.

(iii) Suppose $ab=ac$. Then $a^{-1}ab=a^{-1}ac$ and so $eb=ec$. Thus $b=c$. Similarly if $ba=ca$.

We leave (iv) as exercises for the reader. ∎

5.2.4 *Examples*

(i) The integers **Z** under ordinary addition is an abelian group.

(ii) The real numbers **R** under multiplication just fail to be a group because 0 has no inverse.

(iii) The nonzero real numbers $\mathbf{R}^*$ under ordinary multiplication form an abelian group.

(iv) The set of $m\times n$ real matrices, denoted by $\mathbf{R}^{m\times n}$, is an abelian group under addition of matrices.

(v) The set of invertible $n\times n$ real matrices, denoted by $GL(n,\mathbf{R})$ under ordinary matrix multiplication is a (nonabelian) group. ($GL(n,\mathbf{R})$ is read: "The general linear group of dimension n over **R**".)

(vi) $(\mathbf{Z}_n,+)$ is a finite abelian group of *order n*. (The *order of a group G*, denoted $|G|$, is the number of elements in G. Examples (i) through (v) are groups of *infinite order*.)

(vii) If A is any nonempty set and G is the power set of A, define on G an operation Δ by $X\Delta Y=(X\cap Y')\cup(X'\cap Y)$ (see Problem 5 of Chapter 2). Then (G,Δ) is an abelian group. Each element of this group is its own inverse and the identity is the empty set.

(viii) If X is any nonempty set and S_X denotes the set of all bijections on X under composition of mappings, then S_X is a group called the *symmetric group on X*. If $X=\{1,2,...,n\}$, the group is denoted by S_n and is called the *symmetric group on n symbols*. The order of S_n is $n!$. For example, the order of S_4 is $4!=24$.

(ix) $(\mathbf{Z}_n,.)$ is not a group. Indeed, $\overline{0}$ has no inverse and even if $\overline{0}$ is excluded in the case n is composite, there are zero divisors and no zero divisor can have a multiplicative inverse (why?). If p is prime, however, $\mathbf{Z}_p^*=\{\overline{1},\overline{2},....,\overline{p-1}\}$ is an abelian group of order $p-1$ under multiplication.

(x) Let

$$D_4 = \left\{ \begin{bmatrix} 1 & 0 \\ 0 & 1 \end{bmatrix}, \begin{bmatrix} 0 & -1 \\ 1 & 0 \end{bmatrix}, \begin{bmatrix} -1 & 0 \\ 0 & -1 \end{bmatrix}, \begin{bmatrix} 0 & 1 \\ -1 & 0 \end{bmatrix}, \begin{bmatrix} 1 & 0 \\ 0 & -1 \end{bmatrix}, \begin{bmatrix} -1 & 0 \\ 0 & 1 \end{bmatrix}, \begin{bmatrix} 0 & 1 \\ 1 & 0 \end{bmatrix}, \begin{bmatrix} 0 & -1 \\ -1 & 0 \end{bmatrix} \right\}.$$

Under ordinary matrix multiplication D_4, is a group called the *dihedral group of order* 8. The significance of the subscript 4 will become apparent later when we shall have more to say about it. We shall see that the elements of D_4 represent certain transformations of the plane which leave any square with center at the origin and sides parallel to the coordinate axes fixed. The reader should convince him-/herself that this is indeed a group.

(xi) The set of all $n \times n$ nonsingular matrices such that $A^{-1} = A^T$ over the reals is a group under multiplication called the *orthogonal group of dimension n over the reals* and denoted by $O(n, \mathbf{R})$.(See Appendix B.) These represent *isometries* of $\mathbf{R}^n$ fixing the origin, that is, maps of $\mathbf{R}^n$ to itself which preserve distance. Note that $O(n, \mathbf{R})$ is contained in $GL(n, \mathbf{R})$ (see Example (v) above). ∎

To exhibit the algebraic structure of a group we draw the so-called Cayley Table of the group. We give a couple of examples.

5.2.5 *Examples* (i)

Table 5.1   Cayley Table of $(\mathbf{Z}_6, +)$

+	0	1	2	3	4	5
0	0	1	2	3	4	5
1	1	2	3	4	5	0
2	2	3	4	5	0	1
3	3	4	5	0	11	2
4	4	5	0	1	2	3
5	5	0	1	2	3	4

Observe that we have omitted the bars over the numbers. The elements listed in the first row and first column are the elements to be added. The sum of the element in the ith row with the element in the jth column appears at the intersection of the ith row and the jth column.

(ii) The Cayley Table of S_3. Recall that this is the set of all permutations of 1, 2, 3 under composition of maps.

Table 5.2 Cayley Table of S_3

∘	[123]	[132]	[321]	[213]	[231]	[312]
[123]	[123]	[132]	[321]	[213]	[231]	[312]
[132]	[132]	[123]	[231]	[312]	[321]	[213]
[321]	[321]	[312]	[123]	[231]	[213]	[132]
[213]	[213]	[231]	[312]	[123]	[132]	[321]
[231]	[231]	[213]	[132]	[321]	[312]	[123]
[312]	[312]	[321]	[213]	[132]	[123]	[231]

5.2.6 *Examples*

We introduce a whole collection of examples which we obtain from geometrical considerations.

(i) We start with one of the simplest geometrical objects, namely a line segment

A   B

What are the *rigid motions* which can be performed on the segment *AB* so that at the end of the application of each motion the segment occupies the same position it was in before the motion? Note that we make a distinction between position and orientation. If we pick the segment up, rotate it through 180 degrees about its midpoint and place it down so that the end marked *B* lands where *A* was and vice-versa, then we say that the segment is in the same position (but with different orientation). With this in mind, we see easily that there are two so-called *symmetries* of a line segment (i.e., rigid motions which leave the position of the geometrical object unchanged), namely, the identity *I* which leaves the segment as it is and a rotation of 180 degrees about its midpoint. Call this symmetry *R*. If we apply *R* twice, the segment ends up in its original position with the initial orientation, i.e., the effect is the same as applying *I*. We denote this symbolically by $R^2 = I$. These two symmetries constitute the *group of symmetries* of a line segment. The Cayley Table of this group is displayed below.

Table 5.3

	I	*R*
I	*I*	*R*
R	*R*	*I*

(ii) Consider next the group of symmetries of a rectangle which is not a square.

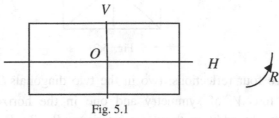

Fig. 5.1

In this case we have four symmetries. First, I, which leaves the rectangle where it is. Next, two 180 rotations, one about the vertical axis of symmetry, labeled V and the other about the horizontal axis of symmetry, labeled H. In addition there is a rotation R of 180 degrees about the center O. This can be clockwise or counter clockwise- the effect is the same. We compute explicitly the product of the two symmetries R and H in the order H followed by R. The resulting symmetry is denoted by RH. The symmetry V produces the same result and we write $RH = V$.

The reader should check that the following is the Cayley Table of the symmetries of a rectangle which is not a square.

Table 5.4

	I	*V*	*H*	*R*
I	*I*	*V*	*H*	*R*
V	*V*	*I*	*R*	*H*
H	*H*	*R*	*I*	*V*
R	*R*	*H*	*V*	*I*

This group (for it is indeed a group) is called the Klein 4-group, denoted K_4. We shall encounter several instances of this group later on.

(iii) We now compute the Cayley Table of the symmetries of a square. There are a total of eight symmetries, four of which are marked as shown in the diagram below.

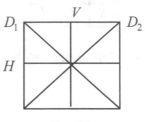

Fig. 5.2.

There are four reflections: two in the two diagonals D_1, D_2; one in the vertical line V of symmetry and one in the horizontal line H of symmetry. In addition there are rotations R_1, R_2, R_3 of 90, 180, 270 degrees, respectively, in the counter-clockwise direction about the center of the square. The identity can be viewed as a rotation of 0 (or 360) degrees.

Table 5.5 Cayley table of the symmetries of a square

	I	R_1	R_2	R_3	H	V	D_1	D_2
I	I	R_1	R_2	R_3	H	V	D_1	D_2
R_1	R_1	R_2	R_3	I	D_1	D_2	V	H
R_2	R_2	R_3	I	R_1	V	H	D_2	D_1
R_3	R_3	I	R_1	R_2	D_2	D_1	H	V
H	H	D_2	V	D_1	I	R_2	R_3	R_1
V	V	D_1	H	D_2	R_2	I	R_1	R_3
D_1	D_1	H	D_2	V	R_1	R_3	I	R_2
D_2	D_2	V	D_1	H	R_3	R_1	R_2	I

(iv) Symmetries of a circle. Rotations about the centre and reflections in any diameter are clearly symmetries. If counterclockwise rotations are taken as positive and clockwise rotations as negative, then, taking into account sign, a rotation by an angle of α followed by a rotation of β results in a rotation of $\alpha + \beta$. Using obvious notation, we have

$$R_\beta R_\alpha = R_{\alpha+\beta}.$$

Since every angle can be represented by a real number γ where $0 \le \gamma < 2\pi$, we compute the sum $\alpha + \beta$ modulo 2π.

Let M_α denote a reflection in a diameter making an angle α with the positive x-axis measured in the counterclockwise direction. This can be thought of as a reflection in a mirror placed vertically with the bottom edge aligned with the diameter. The set

$$G = \bigcup_{0 \le \alpha < 2\pi} R_\alpha \cup \bigcup_{0 \le \alpha < \pi} M_\alpha$$

is an infinite group under composition of maps. The rules for computation are:

(a) $M_\alpha^2 = I$.

(b) $R_\alpha R_\beta = R_{\beta+\alpha} = R_{\alpha+\beta} = R_\beta R_\alpha$.

(c) $M_\alpha M_\beta = R_{2(\alpha-\beta)}$.

(d) $M_\alpha R_\beta = M_{\alpha-\beta/2}$.

(e) $R_\alpha M_\beta = M_{\beta+\alpha/2}$.

The first rule is clear since reflecting the circle twice in a given diameter brings the circle back to its original position and so has the same effect as the identity, I.

The second is also quite obvious: a rotation of α followed by a rotation of β is a rotation of $\alpha + \beta$. Naturally, we have to take into account the direction of the rotations.

The third is the composition of two reflections in mirrors appropriately placed as described above and can be derived by means of a simple geometric argument which the reader should attempt. The fifth can be established as follows: Multiply both sides of the third equation on the right by M_β. Then bearing in mind that $M_\beta^2 = I$, we get

$$M_\alpha = R_{2(\alpha-\beta)}M_\beta.$$

Now set $\gamma = 2(\alpha - \beta)$. Solving for α we have $\alpha = \beta + \dfrac{\gamma}{2}$ from which we get

$$R_\gamma M_\beta = M_{\beta+\frac{\gamma}{2}}.$$

The fourth rule is similarly proved. ∎

5.3   Alternative Axioms for Groups

In this section we introduce several defining properties of groups which are sometimes easier to verify than those of Definition 5.2.1.

In a group we can solve equations of the form $ax = b$ and $ya = b$ where a and b are given and x and y are sought. For example, using the table above, we find that the solutions to

$$Hx = R_1 \text{ and } yD_1 = V$$

are $x = D_2$ and $y = R_1$.

The following theorem states that if we can always solve such equations in a semigroup, then the semigroup is in fact a group.

5.3.1   *Theorem*

Let $(G,.)$ be a semigroup. Then G is a group if, and only if, given a and b in G, the equations $ax = b$ and $ya = b$ can be solved for x and y in G.

Proof. Suppose G is a group. Then multiplying the first equation on the left by a^{-1} and the second by a^{-1} on the right, we get, respectively,

$$a^{-1}ax = a^{-1}b \text{ and } yaa^{-1} = ba^{-1}$$

(We omit brackets because, by hypothesis, the operation is associative.) The first equation yields $x = a^{-1}b$ and the second $y = ba^{-1}$. These elements do indeed satisfy the equations above.

Conversely, suppose that equations of the form

$$ax = b \text{ and } ya = b$$

have solutions in G for all $a, b \in G$.

If e is a solution of $ya = a$, then $ea = a$. We show that e is a left identity because at this point we know only that e acts as a left identity for the particular element a.

Let b be any element of G and let c be a solution of $ax = b$. Then $ac = b$ and, multiplying this equation by e on the left and remembering associativity, we get $eac = eb$. It then follows from $ea = a$ that $ac = eb$. But $ac = b$ and so $eb = b$, proving that e is a left identity. Furthermore, given $a \in G$, by hypothesis, $ya = e$ has a solution $a' \in G$. Therefore, $a'a = e$, and so each element of G has a left inverse relative to the left identity e. By Theorem 3.4.11 G is a group. ∎

Remarks.

(i) Although we do not assume the uniqueness of the solutions to $ax = b$ and $ya = b$, this does indeed follow. (Why?)

(ii) It follows from the previous theorem and the remark above that in the Cayley Table of a group, every element of the group occurs once and once only in every row and column of the *body* of the table, i.e., the rows and columns of the table, excluding the first row and first column.

5.3.2 *Notation*

If $(S,.)$ is an algebraic system and a is an element of S, then the set $\{sa \mid s \in S\}$ is denoted by Sa. Similarly, $aS = \{as \mid s \in S\}$.

5.3.3 *Corollary*

Let $(S,.)$ be a semigroup. Then S is a group if, and only if, for all $a \in S$, $aS = S = Sa$.

Proof. It is always the case that $aS \subseteq S$ and $Sa \subseteq S$ by closure.

Suppose now that S is a group and let $b \in S$. Then, by Theorem 5.3.1 there are elements x and y of S such that $ax = b$ and $ya = b$. Hence $b \in aS$ and $b \in Sa$, and so

$$aS = S = Sa.$$

Conversely, assume the condition holds. Then, given a and b in S, it follows from $aS = S$ that the equation $ax = b$ has a solution in S. Similarly, from $Sa = S$, we conclude that the equation $ya = b$ has a solution in S. By Theorem 5.3.1, S is a group. ∎

We saw in Theorem 5.2.3(iii) that the cancellation laws hold in a group. On the other hand, a semigroup satisfying the cancellation laws is not, necessarily, a group. For example, consider the natural numbers under either addition. or multiplication. If, however, the cancellation laws hold in a *finite* semigroup, then the semigroup is a group.

5.3.4 *Theorem*

If $(S,.)$ is a finite semigroup in which the cancellation laws are valid, then S is a group.

Proof. By Corollary 5.3.2, it is sufficient to prove that for all a in S,

$$aS = Sa = S.$$

Therefore, suppose S has n elements, say $S = \{a_1, a_2, a_3, ..., a_n\}$ and fix some $a \in S$. Then $aS = \{aa_1, aa_2, aa_3, ..., aa_n\}$ and if $aa_i = aa_j$, it follows by cancelling a on the left that $a_i = a_j$. Hence if $i \neq j$, $aa_i \neq aa_j$ and so aS contains n elements. But $aS \subseteq S$ and S contains n elements. Therefore, $aS = S$.

Similarly, we can show that $Sa = S$ using the hypothesis that we can cancel on the right. Therefore by Corollary 5.3.3, S is a group. ∎

5.4 Subgroups

The set $\mathbf{Q}^*$ of nonzero rational numbers under ordinary multiplication is a group contained in the group $\mathbf{R}^*$ of nonzero real numbers under multiplication. Whenever a subset of a group is itself a group under the operation inherited from the "parent" group, then the subset is called a subgroup. More formally:

5.4.1 *Definition*

Let $(G,.)$ be a group. A nonempty subset H is a *subgroup* of G if H is a group under the operation defined on G. If $H \neq G$, then H is a *proper subgroup* of G We write $H \leq G$ to indicate that H is a subgroup of G and $H < G$ if H is a proper subgroup of G.

5.4.2 *Examples*

(i) $H = \{[123],[231],312]\}$ is a subgroup of S_3.

(ii) Every group G is a subgroup of itself and $\{e\}$, the set consisting of e alone is a subgroup of G, called the *trivial subgroup*.

(iii)The group of symmetries of a square is a subgroup of $GL(2,\mathbf{R})$ (See Example 5.2.4(x)).

(iv) $(\mathbf{Z},+)$ is a subgroup of $(\mathbf{R},+)$.

(v) All the groups of symmetries of regular polygons are subgroups of the group of symmetries of the circle (see Example 5.2.6(iv)). These groups are the so-called *dihedral groups*. We usually denote the group of symmetries of a regular n-sided polygon by D_n. Using the notation established in Example 5.2.6 we can list the elements of D_n quite easily. There are two cases to consider according to whether n is even or n is odd.

First suppose n is even, say $n = 2m$, then:

$$D_{2m} = \{R_{k\pi/m} \mid k = 0,1,2,...,2m-1\} \cup \{M_{k\pi/2m} \mid k = 0,1,2,...,2m-1\}.$$

If n is odd, say, $n = 2m+1$, then:

$$D_{2m+1} = \{R_{2k\pi/2m+1} \mid k = 0,1,2,...,2m\} \cup \{M_{2k\pi/2m+1} \mid k = 0,1,2,...,2m\}.$$

Observe that in either case, the order of the group of symmetries of a regular n-gon is $2n$. ∎

The next theorem tells us how to check that a given subset of a group is a subgroup. The theorem is expressed in multiplicative notation and we leave it up to the reader to formulate the theorem in additive notation. We first prove a lemma and corollary.

Recall that an element x of a semigroup is *idempotent* if $x^2 = x$.

5.4.3 *Lemma*

Let $(G,.)$ be a group. Then the only idempotent element of G is the identity e.

Proof. Suppose $g^2 = g$. Multiplying both sides of this equation by g^{-1} on the right, we obtain $g = e$. ∎

5.4.4 *Corollary*

If H is a subgroup of G , then e, the identity of G , is contained in H.

Proof. If f is the identity of H, then $f^2 = f$ and since f is an element of G, by the previous result, $f = e$. ∎

5.4.5 *Theorem*

A subset H of a group $(G,.)$ is a subgroup of G if, and only if, the following hold:

  (i) $H \neq \phi$.
  (ii) For all $a,b \in G$, if $a,b \in H$, then $ab \in H$.
  (iii) For all $a \in G$, if $a \in H$, then $a^{-1} \in H$.

Alternatively, H is a subgroup if, and only if,
  (i) $H \neq \phi$.
  (ii)' For all $a,b \in G$, if $a,b \in H$, then $ab^{-1} \in H$.

Proof. If H is a subgroup of G then clearly, by the very definition of a group $H \neq \phi$. Also, since H is a group under ., H is closed under . and so (ii) holds. Lastly, since the identity e of G is in H by Corollary 5.4.4 above, it follows that $a^{-1} \in H$ whenever $a \in H$.

   Conversely, assume H is a subset of G possessing Properties (i), (ii) and (iii).

   Associativity in H need not concern us since associativity holds in the whole of G and so, *a fortiori*, holds in H.

Since H is not empty by (i), pick $a \in H$. By (ii), $aH \subseteq H$ and $Ha \subseteq$ H. By (iii), $a^{-1} \in H$ and if $x \in H$, so is xa^{-1} by (ii) again. Thus

$$x = xa^{-1}a \in Ha$$

showing that $H \subseteq Ha$ for all $a \in H$. Similarly, $H \subseteq aH$ for all $a \in H$ thus proving that for all $a \in H$, $aH = Ha = H$. Therefore H is a group by Corollary 5.3.2.

We leave it to the reader to verify the alternative criterion. ∎

Remark. In additive notation, the criteria read as follows:

(i) $H \neq \phi$.

(ii) For all $a, b \in G$, if $a, b \in H$, then $a + b \in H$.

(iii) For all $a \in G$, if $a \in H$, then $-a \in H$.

(ii)' For all $a, b \in G$, if $a, b \in H$, then $a - b \in H$. ($a - b$ is short for $a + (-b)$.)

We apply the criterion established in the theorem just proved to prove that the intersection of any collection of subgroups of a group is again a subgroup.

5.4.6 *Theorem*

Let $\{H_i \,|\, i \in I\}$ be a collection of subgroups of a group G denoted multiplicatively. Then $\bigcap_{i \in I} H_i$ is a subgroup of G.

Proof. We check conditions (i), (ii) and (iii) of the previous theorem.

(i) $\bigcap_{i \in I} H_i \neq \phi$ since $e \in H_i$ for all $i \in I$.

(ii) Suppose $a, b \in \bigcap_{i \in I} H_i$. Then $a, b \in H_i$ for all $i \in I$. Since each H_i is a subgroup, it follows that $ab \in H_i$ for each $i \in I$. Therefore, $ab \in \bigcap_{i \in I} H_i$

(iii) Let $a \in \bigcap_{i \in I} H_i$. Then $a \in H_i$ for all $i \in I$. Since each H_i is a subgroup, $a^{-1} \in H_i$ for all $i \in I$. Therefore, $a^{-1} \in \bigcap_{i \in I} H_i$ and so $\bigcap_{i \in I} H_i$ is a subgroup. ∎

Among all the subgroups of a group, we single out those which are the simplest, namely, those *generated* by a single element according to the following definition.

5.4.7 *Definition*

Let T be a subset of a group G. The smallest subgroup of G containing T is called *the subgroup generated by T* and is denoted by $<T>$. The subset T is a *generating set* or a set of *generators* for $<T>$. If $T = \{g\}$, a singleton subset, the subgroup generated by g is denoted by $<g>$ rather than by $<\{g\}>$.

Remarks. By the smallest subgroup containing T we mean a subgroup H containing T and contained in any subgroup containing T. That such a subgroup exists follows from the fact that the intersection of all subgroups of G containing T is a subgroup by Theorem 5.4.6 and contains T. It is obviously the smallest one.

5.4.8 *Theorem*

Let g be an element of the group G. Then $<g> = \{g^n \mid n \in \mathbf{Z}\}$.

Proof. If H is a subgroup of G containing g, then clearly $g^n \in H$ for all $n \in \mathbf{Z}$. (Strictly speaking, an induction argument should be used to prove this). Hence,

$$\{g^n \mid n \in \mathbf{Z}\} \subseteq H .$$

If we can prove that $K = \{g^n \mid n \in \mathbf{Z}\}$ is a subgroup, then, by what we have just proved, this subgroup is contained in every subgroup containing g and so $K = <g>$. But, that K is a subgroup follows from:

(i) $K \neq \phi$ since $g \in K$.

(ii)' Let $a, b \in K$. Then $a = g^n, b = g^m$ for some $m, n \in \mathbf{Z}$. Thus

$$ab^{-1} = g^n g^{-m} = g^{n-m} \in K$$

since $n - m \in \mathbf{Z}$ and so K is a subgroup of G. Therefore $<g> = g^n \mid n \in Z\}$ ∎

The next concept is of fundamental importance in the theory of groups.

5.4.9 *Definition*

Let g be an element of the multiplicative group G. If there exists a positive integer n such that $g^n = e$, the identity of G, we say that g *is of finite order*. If no such positive n exists, we say that g *is of infinite order*.

If g is of finite order, the least positive integer m (it exists by the Well Ordering Principle) such that $g^m = e$ is called the order of g and is denoted by $|g|$. If g is of infinite order, we write $|g| = \infty$.

5.4.10 *Theorem*

(i) If g is of finite order m, then
$$< g > = \{ g^k \mid 0 \le k < m \}$$
and so $< g >$ contains exactly m elements.

(ii) If g is of infinite order, then
$$< g > = \{ ..., g^{-3}, g^{-2}, g^{-1}, e, g, g^2, g^3 ... \}$$
where $g^i \ne g^j$ for all integers $i \ne j$.

Proof. (i) We claim that the sequence
$$e = g^0, g^1 = g, g^2, ..., g^{m-1}$$
consists of distinct elements. For, suppose that $g^i = g^j$ where $0 \le i \le j < m$. Then $g^{j-i} = e$ and $0 \le j - i < m$. By the minimality of m, $j - i$ cannot be positive. Therefore, $i = j$ and so $e, g, g^2, g^3, ..., g^{m-1}$ are distinct.

Now let $a \in < g >$. Then $a = g^n$ for some $n \in \mathbf{Z}$. By the Division Algorithm, $n = qm + r$ where $0 \le r < m$. Thus
$$g^n = g^{qm+r} = (g^m)^q \, g^r = e^q g^r = g^r$$
since $g^m = e$. Therefore $g^n = g^r$ and so
$$< g > = \{ g^k \mid 0 \le k < m \}.$$
Hence $< g >$ contains exactly m elements, i.e, $|< g >| = |g|$.

(ii) Assume now that the order of g is infinite and suppose that $g^i = g^j$ for some integers i and j with $i \le j$. Then $g^{j-i} = e$ where $0 \le j - i$. But $j - i$ cannot be positive and so $i = j$. ∎

Remarks.

(i) Note that $|g|$ = the number of elements in $<g>$ even if g is of infinite order since we have set $|g|=\infty$ in that case. We denote the number of elements in a group G by $|G|$ and call this number the *order* of G (see Example 5.2.4(vi)).

It is a little confusing for the beginner to have the same word describing different ideas, but the context should make it clear whether we are talking about the size of the group generated by an element or the smallest positive power to which we must raise the element in order to get the identity.

(ii) In additive notation the order of an element g is the least positive m such that $mg = 0$. Note that mg does not mean m times g (which actually doesn't make sense), but rather g added to itself m times, that is, $g + g + ... + g$ m times if m is positive and $(-g) + (-g) + ... + (-g)$ $|m|$ times if m is negative.

If g is of infinite order, we just saw that $g^i \neq g^j$ if $i \neq j$. For an element g of finite order it is obvious that there will be integers i and j, with $i \neq j$ such that $g^i = g^j$. In the next theorem we determine precisely when this occurs.

5.4.11 *Theorem*

Let g be an element of finite order m of the group G. Then $g^i = g^j$ if, and only if, $m|(j-i)$. In particular, $g^n = e$ if, and only if, $m|n$.

Proof. Suppose that $g^i = g^j$ with $j \geq i$. By the Division Algorithm,
$$j - i = qm + r, \; 0 \leq r < m.$$
But $g^i = g^j$ implies $g^{j-i} = e$ and so
$$e = g^{j-i} = g^{qm+r} = (g^m)^q g^r = e^q g^r = eg^r = g^r \text{ since } g^m = e.$$
Thus $g^r = e$. But since r is a nonnegative integer less than m it follows that $r = 0$ and so $j - i = qm$.

Conversely, let $j - i = qm$. Then $g^{j-i} = (g^m)^q = e^q = e$ and multiplying both sides of the equation $g^{j-i} = e$ by g^i we get $g^i = g^j$. In particular, $g^n = e = g^0$ if, and only if, $m|n$. ∎

We now give a number of examples which will make the foregoing somewhat abstract results more palatable.

5.4.12 *Examples*

(i) The general linear group $GL(2,\mathbf{R})$ has elements of all orders. For example,

$$\begin{bmatrix} \cos\dfrac{2\pi}{n} & -\sin\dfrac{2\pi}{n} \\ \sin\dfrac{2\pi}{n} & \cos\dfrac{2\pi}{n} \end{bmatrix}$$

represents a rotation of $2\pi/n$ radians and so is an element of order n, while

$$\begin{bmatrix} 1 & 1 \\ 0 & 1 \end{bmatrix}$$

is an element of infinite order.

To prove the first matrix is of order n, the reader should use induction and the trigonometric identities

$$\cos(\alpha+\beta) = \cos\alpha\cos\beta - \sin\alpha\sin\beta$$
$$\sin(\alpha+\beta) = \sin\alpha\cos\beta + \sin\beta\cos\alpha .$$

(ii) $\lfloor 2341\rfloor \in S_4$ is of order 4.

We have

$$[2341]^2 = [3412], [2341]^3 = [4123], \ [2341]^4 = [1234].$$

Thus

$$<[2341]>=\{[1234],[2341],[3412],[4123]\}.$$

(iii) The order of an element of a group is 1 if, and only if, the element is the identity of the group.

(iv) The only element of finite order in $(\mathbf{Z},+)$ is the identity 0.

(v) In the multiplicative group of nonzero real numbers, the only elements of finite order are 1 and -1 whose orders are, respectively, 1 and 2. ∎

5.5 Cyclic Groups

Cyclic groups have a particularly simple structure which we investigate in this section. A subclass of the cyclic groups, namely cyclic groups of prime power order form the building blocks for all finite abelian groups, in the sense that every finite abelian group is a direct sum of such groups (see Section 6.5 for the definition of direct sum).

5.5.1 *Definition*

A group G is said to be *cyclic* if there exists $g \in G$ such that $G = <g>$. The element g is a *generator* of G.

5.5.2 *Examples*

(i) $(\mathbf{Z},+)$ is a cyclic group of infinite order. It has two generators, namely, 1 and -1.

(ii) The group $(U(\mathbf{Z}_7),.)$ of units of $\mathbf{Z}_7$ is a cyclic group of order 6 with two generators $\overline{3}$ and $\overline{5}$. We perform the computations for the generator $\overline{3}$.

$$\overline{3}^0 = \overline{1}, \ \overline{3}^1 = \overline{3}, \ \overline{3}^2 = \overline{2}, \ \overline{3}^3 = \overline{6}, \ \overline{3}^4 = \overline{4}, \ \overline{3}^5 = \overline{5} \ .$$

As theory predicts, $\overline{3}^6 = \overline{1}$ so that the order of $\overline{3}$ is 6.

Given an element g can we express the order of g^k in terms of $|g|$? The next theorem answers the question precisely. We shall see later on that this simple result is quite useful.

5.5.3 *Theorem*

If g is an element of finite order m, then

$$|g^k| = \frac{m}{d} \text{ where } d = \gcd(m,k).$$

Proof. If $|g^k| = s$, then $(g^k)^s = g^{ks} = e$ and so, by Theorem 5.4.11, $m \mid ks$. Therefore, $\frac{m}{d} \mid \frac{k}{d}s$. But $\gcd(\frac{m}{d}, \frac{k}{d}) = 1$ (why?) and so, by Theorem 4.4.2, $\frac{m}{d} \mid s$.

Conversely, we have $(g^k)^{\frac{m}{d}} = (g^m)^{\frac{k}{d}} = e^{\frac{k}{d}} = e$ since $g^m = e$. Thus, by Theorem 5.4.11, $s \mid \frac{m}{d}$ and so $s = |g^k| = \frac{m}{d}$. ∎

5.5.4 *Theorem*

A cyclic group is abelian.

Proof. Exercise. ∎

We observe that not all abelian groups are cyclic. For example, the real numbers under addition is an abelian group which is not cyclic.

It seems reasonable to expect that every subgroup of a cyclic group is also cyclic since cyclic groups are simple objects and it is unlikely that they would contain a more intricate structure.

5.5.5 *Theorem*

Every subgroup of a cyclic group is cyclic.

Proof. Let $G = <g>$ and suppose H is a subgroup of G. If $H = <e>$, H is trivially cyclic with generator e.

Assume therefore that $H \neq <e>$ and let $h \in H, h \neq e$. Then $h = g^k$ for some nonzero integer k. Since H is a subgroup,

$$(g^k)^{-1} = g^{-k} \in H.$$

Therefore, H contains positive powers of g. The Well Ordering Principle guarantees that there is a least positive power of g, say g^m, contained in H. We claim that

$$H = <g^m>.$$

Clearly $<g^m> \subseteq H$ since $g^m \in H$ and H, being a subgroup, is closed under the operation defined on G.

Conversely, let $h \in H$. Then $h = g^k$ for some integer k. By the Division Algorithm,

$$k = qm + r \text{ where } 0 \leq r < m.$$

Hence,

$$h = g^{qm+r} = (g^m)^q g^r \text{ and so } g^r = (g^m)^{-q} h.$$

But both $(g^m)^{-q}$ and h are in H so that their product, namely g^r, is in H. By the minimality of m and the fact that r is nonnegative and strictly less than m, it follows that $r = 0$. Therefore, $k = qm$ and so

$$h = (g^m)^q \in <g^m>,$$

proving that $H \subseteq <g^m>$. Therefore $H = <g^m>$. ∎

The next two theorems describe how to tell when one subgroup of a cyclic group is contained in another, providing us with a clear picture of the subgroup structure of the group. The subgroup structure of an arbitrary group is generally quite complicated.

5.5.6 *Theorem*

Let $G = <g>$ be an infinite cyclic group. Then
$$<g^m> \subseteq <g^n> \text{ if, and only if, } n \mid m.$$
Proof. If $<g^m> \subseteq <g^n>$, then $g^m \in <g^n>$ and so $g^m = (g^n)^q = g^{qn}$. By Theorem 5.4.10 $m = qn$, whence $n \mid m$.

If $m = qn$, then $g^m = (g^n)^q \in <g^n>$ and so $<g^m> \subseteq <g^n>$. ∎

5.5.7 *Corollary*

An infinite cyclic group $G = <g>$ has only two generators, namely, g and g^{-1}.

Proof. Let g^k be a generator. Then $<g^1> \subseteq <g^k>$ and so by the previous theorem, $k \mid 1$, proving that $k = \pm 1$. ∎

5.5.8 *Theorem*

Let $G = <g>$ be a finite cyclic group of order m. Then:

(i) For each divisor k of m there is exactly one subgroup of order k.
(ii) The order of any subgroup of G is a divisor of m.
(iii) $<g^s> \subseteq <g^t>$ iff $\gcd(t,m) \mid s$.
(iv) g^n is a generator of G if, and only if, $\gcd(m,n) = 1$.

Proof. (i) Suppose k divides m. Then $m = qk$ for some integer q and
$$|g^q| = \frac{m}{\gcd(m,q)} = \frac{qk}{q} = k$$
by Theorem 5.5.3. Hence $<g^q>$ is a subgroup of order k. We show that this is the only subgroup of order k.

Let H be another subgroup of order k. By Theorem 5.5.5, H is cyclic and so $H = <g^s>$ for some integer s. By Theorem 5.5.3,

$$k = |g^s| = \frac{m}{\gcd(m, s)} = \frac{qk}{\gcd(m, s)}$$

and so, cancelling k and rearranging, we get $\gcd(m, s) = q$. Therefore, $q \mid s$ and consequently $s = rq$. Thus $g^s = (g^q)^r \in <g^q>$ and from this it follows that $<g^s> \subseteq <g^q>$. But these two subgroups are of the same order by hypothesis and so $<g^s> = <g^q>$. Hence there is only one subgroup of order k.

(ii) Let H be a subgroup of G. By Theorem 5.5.5, $H = <g^s>$ and by Theorem 5.5.3,

$$|g^s| = \frac{m}{d} = |<g^s>| = |H| \text{ where } d = \gcd(m, s).$$

Therefore $d \mid H \mid = \mid G \mid$ and so the order of H divides the order of G.

(iii) The easiest way to prove this is to appeal to Theorem 4.6.4 which states that the equation $\overline{t}.\overline{x} = \overline{s}$ has a solution in $\mathbf{Z}_m$ if, and only if, $\gcd(t, m) \mid s$. The following series of equivalent statements proves the result:

$$<g^s> \subseteq <g^t> \text{ iff } g^s \in <g^t> \text{ iff } g^s = g^{kt}$$

for some integer k iff

$$m \mid (s - kt) \text{ (by Theorem 5.4.11 (i))},$$

iff

$$\overline{t}.\overline{x} = \overline{s} \text{ has a solution in } \mathbf{Z}_m \text{ iff } \gcd(t, m) \mid s.$$

(iv) Exercise. (Note that an element x generates a group of order n if, and only if, the order of x is n. Use this fact and Theorem 5.5.3). ∎

5.5.9 *Example*

Let us determine all the subgroups of a group $G = <g>$ of order 12. The divisors of 12 are 1, 2, 3, 4, 6, 12. Therefore G has six subgroups, including the trivial subgroup and the whole group. They are:

$$<e>, <g>, <g^2>, <g^3>, <g^4>, <g^6>.$$

Note that, for example, $|<g^3>| = 4, |<g^6>| = 2$.

What are the generators of G? By part (iv) of the previous theorem, they are g, g^5, g^7, g^{11}. What about the generators of $<g^2>$? Since this is a subgroup of order 6, we want an elements of G of order 6.

Suppose g^s is such an element. Then, by Theorem 5.5.3, we want

$6 = \dfrac{12}{\gcd(12, s)}$. The solutions of this equation are $s = 2$ and $s = 10$.

Therefore, g^2 and g^{10} generate the same subgroup of order 6 (remember that there is only one).

Is $< g^8 > \subseteq < g^{10} >$? By the theorem, since

$$\gcd(10,12) = 2 \text{ and } 2 \text{ divides } 8,$$

the containment is valid. Indeed, $g^8 = (g^{10})^2$. ∎

5.6 Exercises

1) Let $(G,*)$ and $(H,\bullet)$ be groups. On $G \times H$ define an operation denoted by juxtaposition by $(g_1, h_1)(g_2, h_2) = (g_1 * g_2, h_1 \bullet h_2)$. Prove that $(G \times H, .)$ is a group. It is called *the direct product of G and H*. Generalize this to the direct product of any finite number of groups.

2) Let $G = H = < -1 > \leq (\mathbf{R}^*, .)$ Construct the Cayley Table of $G \times H$. This is a group of order 4 called *the Klein 4-group* (see Example 5.2.6(ii)).

3) Construct the Cayley Tables of $(\mathbf{Z}_5, +)$ and $(\mathbf{Z}_7^*, .)$. Both of these groups are cyclic. Find all the generators in each case.

4) On the set of real numbers $\mathbf{R}$ define a relation $\sim$ by : $a \sim b$ if $a - b \in \mathbf{Z}$.

    (i) Prove that $\sim$ is an equivalence relation on $\mathbf{R}$ and denote the equivalence class to which a belongs by $\bar{a}$.

    (ii) Prove that if $a \sim b$ and $c \sim d$, then $a \pm c \sim b \pm d$.

    (iii) Prove that every equivalence class contains exactly one nonnegative real number strictly less than 1.

    (iv) Denote the set of all equivalence classes of $\sim$ by $\mathbf{R}/\mathbf{Z}$. On $\mathbf{R}/\mathbf{Z}$ define an operation $\oplus$ by $\bar{a} \oplus \bar{b} = \overline{a+b}$ and prove that it is well defined.

(v) Prove that $(\mathbf{R}\big/\mathbf{Z}, \oplus)$ is an abelian group which has elements of all orders.

5) Find the orders of the following elements:

(i) [234165] in S_6.

(ii) $\begin{bmatrix} \dfrac{1}{\sqrt{2}} & -\dfrac{1}{\sqrt{2}} \\[2mm] \dfrac{1}{\sqrt{2}} & \dfrac{1}{\sqrt{2}} \end{bmatrix}$

(iii) $\bar{8} \in (Z_{12}, +)$

(iv). $(\bar{\bar{2}}, \bar{\bar{3}})$ where $\bar{2} \in Z_6$ and $\bar{\bar{3}} \in Z_9$ and the operation is defined by $(\bar{\bar{a}}, \bar{\bar{b}}) + (\bar{\bar{c}}, \bar{\bar{d}}) = (\overline{a+c}, \overline{\overline{b+d}})$.

(v) $\begin{bmatrix} 1 & -1 \\ 1 & 1 \end{bmatrix}$

6) Prove that if a_i, $i = 1, 2, 3, ..., n$ are elements of a group, then $(a_1 a_2 a_3 ... a_n)^{-1} = a_n^{-1} a_{n-1}^{-1} ... a_3^{-1} a_2^{-1} a_1^{-1}$.

7) Prove that, if H and K are subgroups of a group, then $H \cup K$ is a subgroup if, and only if, either $H \subseteq K$ or $K \subseteq H$.

8) Let H be a finite subset of a group G such that
    (i) $H \neq \emptyset$;
    (ii) For all $a, b \in G$, $a, b \in H$ imply $ab \in H$.

Prove H is a subgroup of G. Is the result still true if H is infinite? Explain.

9) The *centralizer* of an element g of a group G is the set $C_G(g) = \{x \in G \mid gx = xg\}$. Prove that $C_G(g)$ is a subgroup of G.

10) Let G be a group and S the set of all subsets of G. On S define an operation by $XY = \{xy \mid x \in X, y \in Y\}$ for all $X, Y \in S$. Prove that $(S, \cdot)$

is a semigroup with identity whose group of units consists of the singleton subsets of G.

11) Let H and K be subgroups of an abelian group G. Prove that HK (see previous problem for the definition of HK) is a subgroup of G. Show by means of an example that if G is not abelian, HK is not necessarily a subgroup.

12) The *normalizer in G of a subgroup H of G* is the set $N_G(H) = \{g \in G \mid gHg^{-1} = H\}$. Prove this set is a subgroup of G containing H.

13) The *center of a group G* is the set $Z(G) = \{g \in G \mid gx = xg \ \forall \ x \in G\}$. Prove that $Z(G)$ is an abelian subgroup of G.

14) Prove that a group in which every non-identity element is of order 2 is abelian.

15) If a and b are elements of a group, prove that $|ab| = |ba|$.

16) Let Q denote the set of 8 matrices

$$\{\pm\begin{bmatrix}1&0\\0&1\end{bmatrix}, \pm\begin{bmatrix}0&-1\\1&0\end{bmatrix}, \pm\begin{bmatrix}0&i\\i&0\end{bmatrix}, \pm\begin{bmatrix}-i&0\\0&i\end{bmatrix}\}$$

over the complex numbers. (Here, $i = \sqrt{-1}$). Prove that Q is a group under ordinary matrix multiplication. This group Q is called the *quaternion group of order* 8. The notation can be considerably simplified by setting $1 = \begin{bmatrix}1&0\\0&1\end{bmatrix}$, $i = \begin{bmatrix}0&-1\\1&0\end{bmatrix}$, $j = \begin{bmatrix}0&i\\i&0\end{bmatrix}$, $k = \begin{bmatrix}-i&0\\0&i\end{bmatrix}$.
Then $Q = \{\pm 1, \pm i, \pm j, \pm k\}$ and multiplication is effected by the following rules:

$$i^2 = j^2 = k^2 = -1, ij = k = -ji, jk = i = -kj, ki = j = -ik.$$

Of course 1 acts like the identity. These rules are easily remembered as follows:

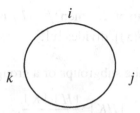

When multiplying two adjacent elements in the clockwise direction, the product is the third element. When multiplying two adjacent elements in the counter clockwise direction , the product is minus the third element.

17) List all the subgroups of the quaternion group above.

18) Fix an element g of the group G. Define a map $\alpha_g : G \to G$ by $\alpha_g(x) = gxg^{-1}$. Prove:

(i) α_g is a bijective map with the property that

$$\alpha_g(xy) = \alpha_g(x)\alpha_g(y) \text{ for all } x, y \in G.$$

(ii) $|x| = |\alpha_g(x)|$ for all $x \in G$.
(iii) If H is a subgroup of G, so is $\alpha_g(H)$.
(iv) $\alpha_x \circ \alpha_y = \alpha_{xy}$ where $\circ$ is composition of maps.
(v) $I(G) = \{\alpha_x \mid x \in G\}$ is a group under composition of maps.

$I(G)$ is called the *group of inner automorphisms* of G.

19) Let $\beta : (\mathbf{Z}, +) \to (\mathbf{Z}_n, +)$ be the map defined by $\beta(z) = \bar{z}$. Prove

(i) $\beta(x + y) = \beta(x) + \beta(y)$ for all $x, y \in Z$;
(ii) If H is a subgroup of $\mathbf{Z}$, then $\beta(H)$ is a subgroup of $\mathbf{Z}_n$;
(iii) If K is a subgroup of $\mathbf{Z}_n$, then $\beta^{-1}(K)$ is a subgroup of $\mathbf{Z}$.

20) Let $(G, .)$ and $(H, *)$ be groups and $\beta : G \to H$ a map with the property: $\beta(xy) = \beta(x) * \beta(y)$ for all $x, y \in G$. (Note that the map β of the previous question has this property). Prove:

(i) The image of the identity of G under β is the identity of H.
(ii) If K is a subgroup of G, then $\beta(K)$ is a subgroup of H.

(iii) If L is a subgroup of H, then $\beta^{-1}(L)$ is a subgroup of G.

(iv) For all $x \in G$, $|\beta(x)|$ divides $|x|$.

21) Let H and K be finite subgroups of a group G. Prove

$$|HK| = \frac{|H|.|K|}{|H \cap K|}.$$

(Hint: On $H \times K$ define a relation $\sim$ by: $(h_1, k_1) \sim (h_2, k_2)$ if, for some $x \in H \cap K$, $h_2 = h_1 x$, $k_2 = x^{-1}k_1$. Prove that $\sim$ is an equivalence relation and show that each equivalence class contains $|H \cap K|$ elements. Finally, prove that there is a bijective map between the equivalence classes of $\sim$ and the elements of HK.)

22) (i) Let $U = \{1,2\}$. Compute the Cayley Table of $(P(U), \Delta)$.

Recall that Δ denotes the symmetric difference. That is,

$$A \Delta B = (A \setminus B) \cup (B \setminus A).$$

Note that $(P(U), \Delta)$ is an abelian group of order 4.

(ii) Prove that if U is any set, then $(P(U), \Delta)$ is an abelian group. If U contains n elements, then the order of $(P(U), \Delta)$ is 2^n.

23) Compute the Cayley Table of the group of units of $\mathbf{Z}_{18}$ under multiplication.

24) In the following show, by computing the orders of the given elements and their products, that, in general, we cannot predict the order of the product of two elements knowing the orders of each of the factors:

(i) $[21453]$, $[21345]$ and $[21453].[21345]$ in S_5;

(ii) $\begin{bmatrix} 1 & 1 \\ 0 & 1 \end{bmatrix}$, $\begin{bmatrix} 0 & -1 \\ 1 & 0 \end{bmatrix}$ and $\begin{bmatrix} 1 & 1 \\ 0 & 1 \end{bmatrix}\begin{bmatrix} 0 & -1 \\ 1 & 0 \end{bmatrix}$

in $GL(2, \mathbf{R})$.

(iii) $\begin{bmatrix} 1 & -1 \\ 1 & 0 \end{bmatrix}, \begin{bmatrix} 0 & 1 \\ -1 & 0 \end{bmatrix}$ and $\begin{bmatrix} 1 & -1 \\ 1 & 0 \end{bmatrix}\begin{bmatrix} 0 & 1 \\ -1 & 0 \end{bmatrix}$

in $GL(2, \mathbf{R})$.

25) If the elements a and b of a group commute, that is, $ab = ba$, then for all $n \in \mathbf{Z}$, $(ab)^n = a^n b^n$).

26) Let a and b be commuting elements of a group of orders m and n respectively and let $\gcd(m, n) = d$.

Prove that $|ab|$ divides $\operatorname{lcm}(m, n)$ $(= \dfrac{mn}{d})$.

27) Show that the elements of the dihedral groups are as described. In Example 5.4.2(v)

28) (Compare with Theorem 3.4.11). Give an example of a semigroup S which is not a group and which satisfies the following;

    (i) S has a left identity e;
    (ii) for each $s \in S$ there exists s' such that $ss' = e$.

29) (See Example 5.2.6(iv)). Let C denote the group of symmetries of a circle. Prove that C contains elements of finite order n for all n and elements of infinite order.

30) Given a subgroup H of a group G and b an element of G such that b^m and b^n are in H, then $b \in H$ if $\gcd(m, n) = 1$.

Chapter 6

Further Properties of Groups

6.1 Introduction

We introduce two equivalence relations on a group, each of which is a generalization of the relation congruence modulo n defined on the integers in Chapter 4 and arrive quite naturally at the notion of a coset, the analogue of a congruence class. This then leads us to a fundamental theorem of group theory, known as Lagrange's Theorem, without which not much can be said about finite groups.

In the case of the integers, we turned the set of congruence classes into a group by defining an appropriate operation. We ask whether it is possible to do the same for cosets of a group. The situation is more complicated in an arbitrary group but under the right conditions we can indeed turn the set of cosets of a group G into a group called a factor group of G.

Closely related to factor groups are special maps from a group G to a group H called homomorphisms. It turns out that the image of G under a homomorphism is a subgroup of H which has, in general, a simpler structure than the group G but from which, nonetheless, we can infer properties of the original group G.

6.2 Cosets

Recall that in Chapter 4, we defined the relation $\equiv \bmod n$ on $\mathbf{Z}$ by: $a \equiv b \bmod n$ if $n \mid (a-b)$. We can couch this in the language of group theory as follows: $a \equiv b \bmod n$ if $a - b \in \ <n>$ where $<n> = \mathbf{Z}n$ is the subgroup generated by n. (Remember that since $(\mathbf{Z}, +)$ is cyclic, every subgroup is generated by some n.) This definition is, of course, given in

127

additive notation since the group in question is an additive group, namely, $(\mathbf{Z}, +)$.

Let us generalize to an arbitrary group G written multiplicatively. To $<n>$ corresponds some subgroup H of G. The expression $a - b$ is short for $a + (-b)$ and this, translated into multiplicative notation becomes ab^{-1}. We are therefore led to the following definition.

6.2.1 *Definition*

Let G be a group and H a subgroup of G. Define a relation $\sim_H$ on G by: $a \sim_H b$ if $ab^{-1} \in H$. We call this the *right congruence determined by H*. (We write $\sim$ rather than $\sim_H$ when the subgroup in question is clear).

The relation $\approx_H$ defined by $a \approx_H b$ if $a^{-1}b \in H$ is called the *leftt congruence determined by H*.

These definitions mimic precisely the definition of a congruence on $\mathbf{Z}$. In an abelian group the two congruences are identical. In a nonabelian group, they are, in general, different. They turn out to be identical when the subgroup H possesses a certain property called *normality*, to be discussed later on.

6.2.2 *Theorem*

Let $\sim$ (respectively, $\approx$) be the right (respectively, left) congruence determined by the subgroup H of G. Then $\sim$ (respectively, $\approx$) is an equivalence relation and the equivalence class of a is the set $Ha = \{ha \mid h \in H\}$ (respectively, $aH = \{ah \mid h \in H\}$).

Proof.

(i) $\sim$ is reflexive since, for all a in G, $aa^{-1} = e \in H$ given that H is a subgroup.

(ii) $\sim$ is symmetric. For, suppose $a \sim b$. Then $ab^{-1} \in H$ and since H is a subgroup, $ba^{-1} = (ab^{-1})^{-1} \in H$, and so $b \sim a$.

(iii) $\sim$ is transitive. Suppose $a \sim b$ and $b \sim c$. Then ab^{-1} and bc^{-1} are in H. Because H is a subgroup, their product. $ab^{-1}bc^{-1} = ac^{-1} \in H$. Therefore, $a \sim c$ and $\sim$ is transitive.

Hence, ~ is an equivalence relation and so G is the disjoint union of its classes.

Let $[a]$ denote the class of a and suppose $x \in [a]$. Then $x \sim a$ and so $xa^{-1} \in H$. Therefore, $x \in Ha$, and so $[a] \subseteq Ha$.

Conversely, if $x \in Ha$, then $x = ha$ for some h in H. Therefore, $xa^{-1} = h \in H$ and so $x \sim a$. Hence $Ha \subseteq [a]$ proving that $Ha = [a]$.

A similar proof establishes that $\approx$ is an equivalence relation whose classes are of the form aH. ∎

6.2.3 *Corollary*

$$Ha = Hb \text{ iff } ab^{-1} \in H; \ aH = bH \text{ iff } b^{-1}a \in H.$$

Proof. Exercise. ∎

6.2.4 *Definition*

If H is a subgroup of G and a is an element of G, the set Ha is a *right coset of H in G*. We also say that Ha is the right coset of a Similarly, aH is a *left coset of H in G* .

Remarks.

(i) For a given subgroup H, we have established that G is the disjoint union of the right (respectively, left) cosets of H in G . Note that $He = H = eH$ is both a right and a left coset. This is the only coset (right or left) which is a subgroup. All other cosets are merely *subsets* of G. (Compare with the congruence classes of **Z**.)

(ii) For the remainder of the chapter, we shall speak mainly of right cosets and leave it up to the reader to formulate the corresponding statements for left cosets.

6.2.5 *Theorem*

Let H be a subgroup of G. Then, for any given $a \in G$, the map $\alpha : H \rightarrow Ha$ defined by

$$\alpha(h) = ha \text{ for all } h \in H$$

is a bijection. In particular, if H is a finite subgroup of $G,$, all right cosets have the same number $|H|$ of elements.

Proof. Clearly α is onto. Suppose that $\alpha(h) = \alpha(k)$ for some $h, k \in H$. Then $ha = ka$, and so , cancelling a (Theorem 5.2.3 (iii)) we have $h = k$, proving that α is also $1:1$. ∎

It is now clear that all cosets, left or right, have the same number of elements. Are there the same number of right cosets of H in G as there are left cosets of H in G? The next theorem answers in the affirmative.

6.2.6 *Theorem*

Let Λ denote the set of left cosets and P the set of right cosets of a subgroup H in G. The map $\varphi : \Lambda \to P$ defined by

$$\varphi(aH) = Ha^{-1}$$

is a bijection. In particular, if there are a finite number n of left cosets of H in G, then there are n right cosets of H in G.

Proof. The map φ seems to be dependent on the representative of the coset involved and so, our first task is to ascertain that it is well defined; that is, that it depends only on the coset.

Hence, assume that $aH = bH$. Then $a \approx b$ and so $a^{-1}b \in H$. Therefore $a^{-1}(b^{-1})^{-1} \in H$ whence $a^{-1} \sim b^{-1}$. Therefore, $Ha^{-1} = Hb^{-1}$, proving that φ is well defined.

Suppose that $\varphi(aH) = \varphi(bH)$. Then $Ha^{-1} = Hb^{-1}$ and so $a^{-1}(b^{-1})^{-1} \in H$; that is, $a^{-1}b \in H$. Hence $a \approx b$ and so $aH = bH$. Thus φ is $1:1$.

For ontoness, let $Ha \in P$. Then $\varphi(a^{-1}H) = H(a^{-1})^{-1} = Ha$. ∎

6.2.7 *Definition*

The number of cosets (left or right) of H in G is called the *index of H in G* and is denoted by $|G:H|$. This can be finite or infinite.

The next theorem due to the French mathematician, Lagrange, has far-reaching consequences as we shall see.

6.2.8 *Corollary (Lagrange's Theorem)*

If G is a finite group and H a subgroup of G, then $|G| = |G:H|.|H|$ from which we deduce that the *order of a subgroup of a finite group divides the order of the group.*

Proof. There are $|G:H|$ cosets, each containing $|H|$ elements. These cosets do not overlap and cover the whole of G. Hence, $|G| = |G:H|.|H|$.
∎

We shall now derive a number of results directly from Lagrange's Theorem.

6.2.9 *Theorem*

If G is a finite group, then the order of any element of G divides the order of G.

Proof. By Theorem 5.4.11, $|<g>| = |g|$ and so, since $<g>$ is a subgroup, by Lagrange's Theorem, $|<g>|$ divides $|G|$. Hence, $|g|$ divides $|G|$. ∎

6.2.10 *Corollary*

If G is a finite group of order m, then for all $g \in G$, $g^m = e$.

Proof. By the previous theorem, $|g|$ divides $|G|$. and so $m = sk$ where $s = |g|$. Hence,

$$g^m = g^{sk} = (g^s)^k = e^k = e$$

since $g^s = e$. ∎

Group theory has many applications in Number Theory, the study of the properties of the positive integers. The next result was first proved by Pierre de Fermat who was actually not a mathematician by profession but a lawyer. He nevertheless made many discoveries in mathematics and tantalized the mathematical world for centuries by proclaiming that he had found a remarkable proof of the fact that the equation $x^n + y^n = z^n$ has no integral solutions if $n > 2$, but that, unfortunately, there was not enough room in the margins of his booklet to write out the proof! This infamous theorem goes by the name of "Fermat's Last Theorem" and was only proved in 1994.

6.2.11 *Corollary (Fermat's Little Theorem)*

If p is prime and a and p are relatively prime, then $a^{p-1} \equiv 1 \bmod p$.

Proof. The multiplicative group $\mathbf{Z}_p^*$ of nonzero classes $\bmod\, p$ is of order $p-1$ and contains $\overline{a}$ since a and p are relatively prime. Hence by the previous corollary, $\overline{a}^{p-1} = \overline{1}$, which, translated into the language of congruences, yields the result. ∎

6.2.12 *Corollary*

Every group of prime order is cyclic and every nonidentity element of the group is a generator.

Proof. Let $|G| = p$, a prime and suppose g is any nonidentity element of G. Then by Theorem 6.2.9, $|g|$ divides p. Since p is prime and $|g| \neq 1$, it follows that $|g| = p$. Therefore $|<g>| = p = |G|$ and so $<g> = G$. ∎

Using the previous results, we can now describe completely all groups whose orders are at most 5. The importance of this theorem does not lie so much in the result itself but rather in the method of proof which gives the reader a flavour of some of the proofs (s)he will encounter in group theory.

6.2.13 *Theorem*

All groups of order at most 5 are abelian. Moreover, there is basically one group each of orders $1, 2, 3, 5$ and two groups of order 4, one of which is cyclic.

Proof. There is clearly only one rather uninteresting group of order 1, namely $<e>$ where $e^2 = e$. Each of the groups of orders $2, 3$ and 5 is cyclic by the previous theorem and so abelian by Theorem 5.5.4.

Suppose now that G is of order 4, say $G = \{e,a,b,c\}$ where e is the identity. By Theorem 6.2.9, the orders of the elements of G divide 4, and so the non-identity elements must be either of order 4 or of order 2.

If G contains an element of order 4, then G is cyclic and so abelian.

Suppose that G is not cyclic. Then every nonidentity element of G is of order 2 and so every element of G is its own inverse. What are the possibilities for the product ab? They are:

$$ab = e, \ ab = a, \ ab = b \text{ and } ab = c.$$

(i) If $ab = e$, then by Theorem 5.2.3, $b = a^{-1}$, contradicting the fact that every element is its own inverse. Therefore $ab \neq e$.

(ii) If $ab = a$, then $ab = ae$ and cancelling a, we get $b = e$, a contradiction. Hence $ab \neq a$.

A similar argument shows that $ab \neq b$, proving that, if there is such a group of order 4, then $ab = c$. In a similar way, we can prove that the product of any two distinct nonidentity elements of G is the third nonidentity element of G. The Cayley Table of G is therefore as follows:

Table 6.1 Cayley Table of K_4

	e	a	b	c
e	e	a	b	c
a	a	e	c	b
b	b	c	e	a
c	c	b	a	e

Observe that the identity occurs along the main diagonal and that the table is symmetric about the same diagonal showing that the group is abelian.

Let us be clear about what we have established. We have proved that, *if a group of any one of the given orders exists, then it must be of the described form.* We should now establish that each group does indeed exist. For the cyclic groups this is easy. In fact, $(\mathbf{Z}_n, +)$ is a concrete example of a cyclic group of order n, and so we know that cyclic groups of orders 1, 2, 3, 4 and 5 exist.

Is there a concrete example of the alleged group of order 4 whose Cayley Table is given above? There is indeed and here it is: Let

$$K = \{[1234], [4321], [3412], [2143]\}$$

under composition of maps. This is a subgroup of S_4, but is, nevertheless, a group in its own right. Its Cayley Table follows.

Table 6.2 Cayley Table of K_4

$\circ$	[1234]	[4321]	[3412]	[2143]
[1234]	[1234]	[4321]	[3412]	[2143]
[4321]	[4321]	[1234]	[2143]	[3412]
[3412]	[3412]	[2143]	[1234]	[4321]
[2143]	[2143]	[3412]	[4321]	[1234]

The notation is too cumbersome to detect the similarity between this table and the preceding one, so let us simplify by writing: $1 = [1234]$, $2 = [2143]$, $3 = [3412]$, $4 = [4321]$ (Notice that in fact we have replaced the 4-tuple $[ijkl]$ by i). In terms of this new notation, the Cayley Table is:

Table 6.3 Cayley Table of K_4

	1	2	3	4
1	1	2	3	4
2	2	1	4	3
3	3	4	1	2
4	4	3	2	1

The pattern now is apparent. We can see that the identity occurs along the diagonal and that the product of any two nonidentity elements is the third nonidentity element, just as in Table 6.1. Indeed, if, in Table 6.3, we re-christen 1 as e, 2 as a, 3 as b, and 4 as c and draw up the table in the new notation, we get Table 6.1. Therefore, Table 6.3, which is the table of an honest-to-goodness concrete group is really Table 6.1 in disguise. Tables 6.1 to 6.3 and the group of symmetries of a rectangle which is not a square introduced in Chapter 5 (Table 5.4) represent the same group in different notation. This notion of apparently different groups being basically the same group, except for notation, gives rise to a fundamental concept in Abstract Algebra which we now discuss.

6.3 Isomorphisms and Homomorphisms

6.3.1 *Definition*

Let $(G,.)$ and $(H,*)$ be groups. An *isomorphism from G to H* is a bijection f from G to H with the following property:

$$\text{For all } x, y \in G, f(x.y) = f(x) * f(y).$$

If the isomorphism is from a group G to itself, then we call it an *automorphism of G.*

Remarks.

(i) The word "isomorphism" comes from Ancient Greek. "Isos" means "same" and "morphos' means "shape".

(ii) From an algebraic standpoint, if there is an isomorphism from G to H, then the groups are considered the same. In abstract algebra, we are not interested in the nature of the elements but rather in how these elements "behave" with respect to the operation(s) involved.

(iii) When we speak of isomorphisms in the abstract between groups, we generally denote the operation in each group by juxtaposition, so that the fundamental property above reads $f(xy) = f(x)f(y)$.

Since isomorphic groups are considered essentially the same, it is natural to put them all in one class, the way we put all elements equivalent to each other under some equivalence relation into one equivalence class.

6.3.2 *Theorem*

Let $f : G \rightarrow H$ and $s : H \rightarrow K$ be isomorphisms. Then:

(i) The inverse map f^{-1} (which exists since f is bijective) is also an isomorphism (from H to G, naturally);

(ii) The map $sf : G \rightarrow K$ is also an isomorphism.

Proof. (i) We must show

$$f^{-1}(ab) = f^{-1}(a)f^{-1}(b) \text{ for all } a,b \text{ in } H.$$

Since f is onto, there exist $x, y \in G$ such that $f(x) = a, f(y) = b$ so that

$$f^{-1}(a) = x, \ f^{-1}(b) = y$$

Hence

$$f^{-1}(ab) = f^{-1}[f(x)f(y)] = f^{-1}[f(xy)] = (f^{-1} \circ f)(xy) = xy$$
$$= f^{-1}(a)f^{-1}(b).$$

(The second equality follows from the fact that f is an isomorphism). Thus f^{-1} is an isomorphism.

(ii) Let x and y be elements of G. Then

$$sf(xy) = s[f(xy)] = s[f(x)f(y)] = s[f(x)]s[f(y)] = (sf)(x).(sf)(y).$$

The first equality follows by definition; the second because f is an isomorphism; the third because s is an isomorphism and the last by definition. ∎

6.3.3 *Notation*

On the family $\mathbb{F}$ of all groups, define a relation $\cong$ by $G \cong H$ if there is an isomorphism from G to H. We read $G \cong H$ as "G is isomorphic to H".

6.3.4 *Theorem*

$\cong$ is an equivalence relation on $\mathbb{F}$.

Proof. (i) For all $G \in \mathbb{F}$, $G \cong G$ since the identity map on G is trivially an isomorphism from G to G. Therefore, the relation is reflexive.

   (ii) Suppose that $G \cong H$. Then there exists an isomorphism f, say, from G to H. By Theorem 6.3.2, f^{-1} is an isomorphism from H to G, thereby establishing that $H \cong G$ and so $\cong$ is symmetric.

   (iii) Assume $G \cong H$ and $H \cong K$ via the isomorphisms f and s respectively. By Theorem 6.3.2, sf is an isomorphism from G to K. Therefore $G \cong K$, and so the relation is transitive. ∎

The equivalence classes of the relation $\cong$ are called *isomorphism classes of groups* and any two groups in the same isomorphism class are said to be *isomorphic*. From an algebraic standpoint, any two groups in the same isomorphism class are considered indistinguishable. Their algebraic properties, such as the number of subgroups, the orders of the elements, etc., are identical. The only difference is the nature of the elements.

6.3.5 *Examples*

   (i) The group $(\mathbf{R}^{+}, .)$ of positive real numbers under multiplication is isomorphic to the group $(\mathbf{R}, +)$ of real numbers under addition.

   Whenever we claim that two groups are isomorphic, we must produce an isomorphism between the two groups to establish our assertion. In this case, the natural logarithm from $(\mathbf{R}^{+}, .)$ to $(\mathbf{R}, +)$ works. Indeed, the natural logarithm, denoted by $\ln$, is a bijection (from Calculus days) and has the property:

$$\ln xy = \ln x + \ln y$$

for all positive real numbers x and y.

   (ii) Let G be a group and a a fixed element of G. The map $\alpha_a : G \rightarrow G$ defined by

$$\alpha_a(x) = a^{-1}xa$$

for all x in G is an automorphism of G. It is called *the inner automorphism of G induced by a*.

That α_a is bijective is left to the reader to prove. As for the fundamental property,

$$\alpha_a(xy) = a^{-1}xya = a^{-1}xeya = a^{-1}xaa^{-1}ya = (a^{-1}xa)(a^{-1}ya) = \alpha_a(x)\alpha_a(y)$$

for all x and y in G.   ∎

An indispensable tool in the study of groups is the notion of a homomorphism. This is a map between groups which, unlike isomorphisms, does not preserve all algebraic properties of the group. It does, however preserve enough of the structure of the group so that we can infer properties of the group from the image whose structure is in general more transparent.

6.3.6 *Definition*

A map f from a group $(G, .)$ to group $(H, *)$ is a *homomorphism* if
$$f(xy) = f(x) * f(y)$$
for all x and y in G. If f is onto, then f is said to be an *epimorphism*. If $G = H$, then f is an *endomorphism*.

Note. (i) The only difference between a homomorphism and an isomorphism is that we have dropped the requirement of bijectivity in the former.

(ii) For those familiar with linear transformations, a group homomorphism is nothing really new. In fact, if the groups are denoted additively, and f is a homomorphism, the defining property reads: $f(x + y) = f(x) + f(y)$. This should be reminiscent of one of the defining properties of a linear transformation.

6.3.7 *Examples*

(i) The map $f : (\mathbf{Z}, +) \rightarrow (\mathbf{Z}_n, +)$ defined by $f(a) = \bar{a}$ for all $a \in \mathbf{Z}$ is an epimorphism. The mapping f is clearly not 1:1 since $\mathbf{Z}$ is infinite while $\mathbf{Z}_n$ is finite.

(ii) The determinant map $\det : GL(2, \mathbf{R}) \rightarrow (\mathbf{R}^*, .)$ defined by
$$\det \begin{bmatrix} a & b \\ c & d \end{bmatrix} = ad - bc \text{ for all } \begin{bmatrix} a & b \\ c & d \end{bmatrix} \in GL(2, \mathbf{R})$$

is an epimorphism since $\det AB = \det A.\det B$ (see Appendix B) and, given any nonzero real number a, $\det \begin{bmatrix} a & 0 \\ 0 & 1 \end{bmatrix} = a.$ ■

We now gather together some of the elementary but important properties of homomorphisms.

6.3.8 *Theorem*

If $f: G \to H$ is a homomorphism, the following hold:

(i) If e is the identity of G, then $f(e)$ is the identity of H;

(ii) For all x in G and for all n in $\mathbf{Z}$, $f(x^n) = \{f(x)\}^n$.
In particular, $f(x^{-1}) = f(x)^{-1}$;

(iii) If K is a subgroup of G, then $f(K)$ is a subgroup of H;

(iv) If L is a subgroup of H, then $f^{-1}(L)$ is a subgroup of G;

(v) If $x \in G$ is of finite order, then $|f(x)|$ divides $|x|$.

(vi) $f(<x>) = <f(x)>$;

(vii) If G is abelian, so is $f(G)$.

Proof. (i) $f(e) = f(ee) = f(e)f(e)$ since f is a homomorphism. Therefore, $f(e)$ is an idempotent of H. By Theorem 5.4.3, the only idempotent of a group is the identity and so $f(e)$ is the identity of H.

(ii) We prove this by induction on n for nonnegative n.

(a) When $n = 0$, the statement reads: $f(x^0) = \{f(x)\}^0$, which, by virtue of (i), is true.

(b) Assume that for some nonnegative integer k, $f(x^k) = \{f(x)\}^k$
Then

$$f(x)f(x^k) = f(x)\{f(x)\}^k = \{f(x)\}^{k+1}.$$

But, since f is a homomorphism, we have

$$f(x)f(x^k) = f(xx^k) = f(x^{k+1})$$

and so

$$f(x^{k+1}) = \{f(x)\}^{k+1}.$$

By the First Principle of Mathematical Induction, the statement is true for all nonnegative integers.

To prove the result for negative integers, observe first that, since f is a homomorphism, we have $f(x^{-1})f(x) = f(x^{-1}x) = f(e) = \varepsilon$, (the identity of H) by (i) above. By Theorem 5.2.3(i) it follows that

$$f(x^{-1}) = f(x)^{-1}.$$

Since $x^{-n} = (x^{-1})^n$ (where n is a positive integer) we have

$$f(x^{-n}) = f((x^{-1})^n) = f(x^{-1})^n = [f(x)^{-1}]^n = f(x)^{-n}$$

where the second equality follows from the case where n is a positive integer.

(iii) Let K be a subgroup of G. We check the criteria established in Theorem 5.4.5.

(a) $f(K) \neq \phi$ since $K \neq \phi$.

(b) Let $x, y \in f(K)$. Then there exist $a, b \in K$ such that
$$f(a) = x, f(b) = y.$$

Now
$$xy = f(a)f(b) = f(ab)$$

since f is a homomorphism, and $ab \in K$ since K is a subgroup. Hence $xy \in f(K)$.

(c) If $x \in f(K)$, there exists $a \in K$ such that $x = f(a)$.
Hence, $x^{-1} = f(a)^{-1} = f(a^{-1})$ by (ii). But K is a subgroup and so $a^{-1} \in K$. Therefore, $x^{-1} \in f(K)$ and so $f(K)$ is a subgroup of H.

(iv) Let L be a subgroup of H. We proceed as in (iii).

(a) Since, by (i), $f(e) = \varepsilon$, the identity of H, it follows that $e \in f^{-1}(L)$. Therefore $f^{-1}(L) \neq \phi$.

(b) If $a, b \in f^{-1}(L)$, then $f(a), f(b) \in L$ and so $f(a)f(b) \in L$ since L is a subgroup. But, since f is a homomorphism, $f(a)f(b) = f(ab)$, proving that $f(ab) \in L$ and so $ab \in f^{-1}(L)$.

(c) If $a \in f^{-1}(L)$, then $f(a) \in L$ and so $f(a)^{-1} \in L$ since L is a subgroup. But $f(a)^{-1} = f(a^{-1})$ by (ii). Therefore, $f(a^{-1}) \in L$, which implies that $a^{-1} \in f^{-1}(L)$. Hence $f^{-1}(L)$ is a subgroup of G.

(v) If $x \in G$ and $|x| = m$, then $x^m = e$ and, applying f to both sides of this equation and bearing in mind (i), we get $f(x^m) = f(e) = \varepsilon$, the identity of H. By (ii), $f(x^m) = \{f(x)\}^m$ and so, $\{f(x)\}^m = \varepsilon$. By Theorem 5.4.11 (i), it follows that $|f(x)|$ divides $|x|$.

(vi)

$$f(<x>) = f\{x^n \mid n \in \mathbf{Z}\} = \{f(x^n) \mid n \in \mathbf{Z}\} = \{[f(x)]^n \mid n \in \mathbf{Z}\} = <f(x)>.$$

The third equality holds by virtue of (ii).

(vii) The proof of this is left as an exercise. ∎

6.3.9 *Corollary*

If $f : G \rightarrow H$ is an isomorphism, then, for all x in G, $|f(x)| = |x|$. Furthermore, G is abelian if, and only if, H is.

Proof. Since both f and f^{-1} are homomorphisms, it follows from (v) of the preceding theorem that the orders of x and $f(x)$ divide each other, and so they are equal. The second statement follows from (vii) ∎

Remark.To prove two groups are isomorphic, we must find an isomorphism from one to the other. This often proves to be a daunting task.

On the other hand, to prove two groups are not isomorphic, we need only find an algebraic property that one possesses but that the other one does not. For example, by the preceding corollary, if one group has m elements of a certain order while the other has n elements of that same order and m and n are different, then the groups are not isomorphic. A concrete example of this is a cyclic group of order 4 and the group whose Cayley Table is given in Example 6.2.13 (the *Klein 4-group*, K_4). These groups cannot be isomorphic since the cyclic group of order 4 has exactly one element of order 2 whereas the Klein 4-group has three.

In Theorem 6.3.8(iv) we saw that the inverse image of a subgroup under a homomorphism is a subgroup. The inverse image under a homomorhism of the trivial subgroup is therefore also a subgroup. This subgroup plays an important role in the theory of groups, much in the same way that the null space of a linear transformation plays a significant role in linear algebra.

6.3.10 *Definition*

Let $f : G \to H$ be a homomorphism and ε the identity of H. The subgroup $f^{-1}(<\varepsilon>)$ is called the *kernel of the homomorphism* and is denoted by $\ker f$.

6.3.11 *Examples*

(i) The map $f : (\mathbf{Z}, +) \to (\mathbf{Z}_n, +)$ defined by $f(a) = \bar{a}$ is a homomorphism (see Example 6.3.7). Its kernel is $f^{-1}(<\bar{0}>) = \mathbf{Z}n$.

(ii) Consider the dihedral group

$$D_4 = \left\{ \begin{bmatrix} 1 & 0 \\ 0 & 1 \end{bmatrix}, \begin{bmatrix} 0 & 1 \\ 1 & 0 \end{bmatrix}, \begin{bmatrix} 0 & -1 \\ 1 & 0 \end{bmatrix}, \begin{bmatrix} 0 & -1 \\ -1 & 0 \end{bmatrix}, \begin{bmatrix} -1 & 0 \\ 0 & 1 \end{bmatrix}, \begin{bmatrix} 1 & 0 \\ 0 & -1 \end{bmatrix}, \begin{bmatrix} -1 & 0 \\ 0 & -1 \end{bmatrix}, \begin{bmatrix} 0 & 1 \\ -1 & 0 \end{bmatrix} \right\}$$

of Example 5.4.2 (iii). Let $\det : D_4 \to (\mathbf{R}^*, .)$ be the ordinary determinant map which is a homomorphism (see Appendix B). The kernel of this homomorphism consists of those matrices in D_4 which have determinant equal to 1, the identity of $(\mathbf{R}^*, .)$. Thus

$$\ker(\det) = \{\begin{bmatrix} 1 & 0 \\ 0 & 1 \end{bmatrix}, \begin{bmatrix} 0 & -1 \\ 1 & 0 \end{bmatrix}, \begin{bmatrix} -1 & 0 \\ 0 & -1 \end{bmatrix}, \begin{bmatrix} 0 & 1 \\ -1 & 0 \end{bmatrix}\} = <\begin{bmatrix} 0 & -1 \\ 1 & 0 \end{bmatrix}>,$$

a cyclic group of order 4. ∎

Remark. The matrices occurring in D_4 can be described geometrically as follows, starting from the leftmost matrix:

(i) The identity linear transformation;
(ii) A reflection in the line $y = x$;
(iii) A rotation of 90 degrees in the counterclockwise direction;
(iv) A reflection in the line $y = -x$;
(v) A reflection in the $y-$axis;
(vi) A reflection in the $x-$axis;
(vii) A rotation of 180 degrees;
(viii) A rotation of 270 degrees in the counterclockwise direction.

We see from this that D_4 is in fact isomorphic to the group of symmetries of a square. The reader should establish an isomorphism between these two groups.

Yet another representation of this group can be found in Example 5.2.6(iv) and the reader should write out the elements in this notation.

Observe that the kernel of the determinant homomorphism consists of the rotations (consider the identity map as a rotation through 0 degrees).

6.4 Normal Subgroups and Factor Groups

There is a special property possessed by some subgroups and not by others. We motivate our investigation of such subgroups by an example.

6.4.1 *Example*

Consider first the subgroup $H = <[213]> = \{[123], [213]\}$ of S_3 and let us compute the left and right coset decompositions of S_3 relative to H. We list the left cosets of H in S_3 first:

$$[123]H = \{[123], [213]\} = [213]H = H;$$
$$[132]H = \{[132], [312]\} = [312]H;$$
$$[321]H = \{[321], [132]\} = [132]H.$$

Observe that S_3 is the disjoint union of these cosets, as expected. Now for the right cosets:

$$H[123] = \{[123], [213]\} = H[213] = H;$$
$$H[132] = \{[132], [231]\} = H[231];$$
$$H[321] = \{[321], [312]\} = H[312].$$

We see that, except for H, the left cosets are different from the right cosets, that is, $aH \neq Ha$.

In contrast to the example above, consider the quaternion group Q of order 8 introduced in Problem 16 of the exercises at the end of Chapter 5. Now

$$Q = \{\pm 1, \pm i, \pm j, \pm k\}$$

and the subgroup

$$H = <-1> = \{1, -1\}$$

has the property that its left cosets are identical to its right cosets as the following computations show:

$$1H = H1 = (-1)H = H(-1) = H;$$
$$iH = \{i, -i\} = Hi;$$
$$jH = \{j, -j\} = Hj;$$
$$kH = \{k, -k\} = Hk. \blacksquare$$

6.4.2 *Definition*

A subgroup H of G is said to be *normal in* G if, for all $a \in G$, the left coset of a is the same as the right coset of a, that is, $aH = Ha$. We write $H \triangleleft G$ to indicate that H is normal in G. If the group G is apparent, we just say that H is a normal subgroup.

6.4.3 *Examples*

(i) The trivial subgroup of G and G itself are normal subgroups of G.
(ii) In an abelian group all subgroups are normal.
(iii) Every subgroup of the quaternion group is normal. ∎

In the next theorem we give another criterion for normality. We first prove a lemma.

6.4.4 *Lemma*

On $P(G)$, the set of all nonempty subsets of the group G, define an operation denoted by juxtaposition by:
$$\{ XY = \{xy \mid x \in X, y \in Y\}.$$
Then $P(G)$ is a semigroup under this operation. Moreover, $P(G)$ has an identity, namely, the singleton set $\{e\}$, and the group of units of $P(G)$ consists of the singleton subsets of G.

Proof. We first need to show that for all
$$A,B,C \in P(G), (AB)C = A(BC).$$
If $x \in A(BC)$, then there exists $a \in A, b \in B$ and $c \in C$ such that $x = a(bc)$. Since the operation in G is associative, $a(bc) = (ab)c$ and so $x \in (AB)C$. Therefore $A(BC) \subseteq (AB)C$.

Similarly $(AB)C \subseteq A(BC)$ and so $(AB)C = A(BC)$.

To simplify notation, we write g for the singleton subset $\{g\}$.
If X is a subset of G, then $eX = \{ex \mid x \in X\} = \{x \mid x \in X\} = X$ and, similarly, $Xe = X$ so that e is the identity of $P(G)$.

Assume now that X is a unit. Then there exists Y such that $XY = e$. If X has more than one element, say $X = \{a,b,...\}$ where $a \neq b$, then $XY \supseteq \{ay, by \mid y \in Y\}$. Since $ay \neq by$, it follows that XY contains at least two elements and so cannot be equal to the singleton set $\{e\}$. Hence, if a set is a unit, it must be a singleton set.

The converse is obvious. ∎

6.4.5 *Theorem*

A subgroup H of a group G is normal in G if, and only if, $g^{-1}Hg = H$ for all $g \in G$.

Proof. By definition, $g^{-1}Hg = \{g^{-1}hg \,|\, h \in H\}$. Suppose first that H is normal in G. Then, for all g in G, $gH = Hg$. This is an equation in the semigroup $P(G)$ of the lemma. Multiplying both sides of this equation by g^{-1} on the left, we get

$$g^{-1}gH = g^{-1}Hg \text{ and so } eH = H = g^{-1}Hg.$$

Conversely, assume that $g^{-1}Hg = H$ for all $g \in G$. Again, we consider this as an equation in $P(G)$, and multiplying both sides of this equation on the left by g and appealing to the lemma, we obtain the result.∎

We establish some properties of the kernel of a homomorphism, among them, that it is always a normal subgroup.

6.4.6 *Theorem*

Let $f : G \to H$ be a homomorphism of groups and K the kernel of f. Then
   (i) $K = <e>$ if, and only if, f is 1:1.
   (ii) K is a normal subgroup of G.
   (iii) For all g in G, $f^{-1}\{f(g)\} = Kg$.

Proof. (i) Suppose $K = <e>$ and $f(x) = f(y)$. Then
$$f(x)\{f(y)\}^{-1} = \varepsilon$$
where ε is the identity of H.

Since f is a homomorphism, $\{f(y)\}^{-1} = f(y^{-1})$ by Theorem 6.3.8 (ii). Again, using the fact that f is a homomorphism,

$$f(x)f(y^{-1}) = f(xy^{-1}) \text{ and so } f(xy^{-1}) = \varepsilon.$$

Therefore, $xy^{-1} \in K = <e>$ and so $xy^{-1} = e$ from which it follows that $x = y$. Hence f is 1:1.

Conversely, assume that f is 1:1. We know that $f(e) = \varepsilon$ and since f is 1:1, no other element can map to ε. Therefore $K = <e>$.

(ii) We use the criterion established in Theorem 6.4.5 to prove the normality of K.

If g is an element of G, and k an element of K, then
$$f(g^{-1}kg) = f(g^{-1})f(k)f(g) = f(g)^{-1}\varepsilon f(g) = f(g)^{-1}f(g) = \varepsilon$$
where we have used the fact that f is a homomorphism repeatedly. Hence $g^{-1}kg \in K$ for all $k \in K$, proving that
$$g^{-1}Kg \subseteq K \text{ for all } g \in G.$$
Thus, in particular,
$$(g^{-1})^{-1}Kg^{-1} \subseteq K.$$
Multiplying this containment on the left by g^{-1} and on the right by, g we get $K \subseteq g^{-1}Kg$ which, together with the previous containment, proves that $g^{-1}Kg = K$ for all $g \in G$. Therefore, K is normal in G.

(iii) The following sequence of bi-implications establishes part (iii):
$$x \in f^{-1}(f(g)) \text{ iff } f(x) = f(g) \text{ iff } f(x)f(g)^{-1} = \varepsilon \text{ iff } f(x)f(g^{-1}) = \varepsilon$$
$$\text{iff } f(xg^{-1}) = \varepsilon \text{ iff } xg^{-1} \in K \text{ iff } x \in Kg.$$

The following diagram illustrates part (iii) of the theorem. ∎

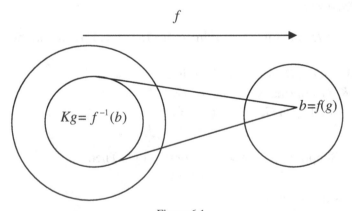

Figure 6.1

Remark. The size of the kernel of a homomorphism can be thought of as a measure of the "shrinkage" which results under the homomorphism. For example, if the kernel is of order 6, then the homomorphic image is one sixth the size of the group.

Our next objective is to prove a fundamental and powerful theorem about homomorphisms. Before doing this, however, we need to introduce the concept of a factor group which we have actually already encountered, in a more concrete form when we constructed the group $(\mathbf{Z}_n, +)$. We shall point out later on how $(\mathbf{Z}_n, +)$ can be considered as a factor group according to our definition below.

6.4.7 *Theorem*

Let K be a normal subgroup of the group G and denote the set of right (or left) cosets by G/K (read "G mod K"). On G/K define an operation "." by

$$Kg.Kh = Kgh.$$

Then "." is a well defined operation and $(G/K, .)$ is a group called the *factor group* G mod K. Furthermore, the map

$$v : G \to G/K \text{ defined by } v(g) = Kg$$

is an epimorphism, called the *natural homomorphism from* G *onto* G/K, whose kernel is K.

Proof. Observe two facts:

(i) First, since K is normal in G, the set of right cosets of K in G is the same as the set of left cosets;

(ii) Second, if we set $Kg = \overline{g}$, the operation is expressible as $\overline{g}.\overline{h} = \overline{gh}$ which, in words, says that the coset of g times the coset of h is the coset of gh. This should be reminiscent of the definition of addition and multiplication in $\mathbf{Z}_n$.

Once again we are faced with the problem of ensuring that the definition of "." depends only on the cosets involved and not on the representatives of these cosets. After all, the "rule" "." instructs us to do the following: to multiply two cosets together, pick a representative from each, multiply these representatives according to the operation defined on G and find the coset to which this product belongs. What if we choose different representatives? Will we end up with the same coset?

We show that, because of the normality of K in G, we do.

Let $x \in Kg, y \in Kh$. We must show that xy is in the same coset as gh, that is, we must show that. $xy \in Kgh$.

Since $x \in Kg, y \in Kh$, we have

$$xg^{-1} \in K, yh^{-1} \in K .$$

Now

$$xy(gh)^{-1} = xyh^{-1}g^{-1} = [x(yh^{-1})x^{-1}][xg^{-1}]$$

and since K is normal, $x(yh^{-1})x^{-1} \in K$. Therefore, $xy(gh)^{-1} \in K$ and so $xy \in Kgh$. Therefore the operation is well defined and we now proceed to show that $(G/K, .)$ is a group.

(i) We obviously have closure since the product of two cosets is again a coset.

(ii) The associativity of multiplication of cosets depends on the associativity of the operation defined on G as can be seen from the following:

$$(Ka.Kb)Kc = Kab.Kc = K(ab)c = Ka(bc) = Ka(Kbc) = Ka(Kb.Kc).$$

(iii) The identity of the factor group is the coset $Ke = K$ since

$$Ke.Ka = Kae = Ka \text{ and } Ka.Ke = Kae = Ka \text{ for all } Ka \in G/K.$$

(iv) Since $Ka.Ka^{-1} = Kaa^{-1} = Ke = Ka^{-1}a = Ka^{-1}.Ka$, it follows that

$$(Ka)^{-1} = Ka^{-1}.$$

(v) Finally we show that v is an epimorphism whose kernel is K.

(a) We have $v(ab) = Kab = Ka.Kb = v(a)v(b)$ for all $a, b \in G$.

(b) Given $Ka \in G/K$, $v(a) = Ka$ and so the natural map is onto.

Moreover,

$$\ker v = \{a \in G \mid v(a) = K\} = \{a \in G \mid Ka = K\} = \{a \in G \mid a \in K\} = K. \blacksquare$$

Remarks.

(i) When the group is denoted additively, right cosets and left cosets are of the form $K + a$ and $a + K$ respectively. The operation defined above in additive notation reads: $(K + a) + (K + b) = K + (a + b)$. Note that $K + a = \{k + a \mid k \in K\}$.

(ii) In the group $(\mathbf{Z}, +)$, all subgroups are normal since $(\mathbf{Z}, +)$ is abelian. Thus, for any fixed n, $\mathbf{Z}n = <n>$ is a normal subgroup. Let us compute $\mathbf{Z}/<n>$. As noted above, the cosets are of the form $<n> + a$ and $<n> + a = <n> + b$ if, and only if, $a - b \in <n>$, that is, if, and only if, $a \equiv b \bmod n$. Therefore, the cosets are precisely the equivalence classes $\bmod n$.

Addition of cosets is the same as the addition of classes $\bmod n$ defined in 4.5.5. Hence, $(\mathbf{Z}/_{< n >}, +)$ is none other than $(\mathbf{Z}_n, +)$.

(iii) If the subgroup K of G is not normal in G, the operation defined above is not well defined. Take as an example the subgroup $K = \{[123], [213]\}$ of S_3. The right cosets are:

$$K = K[123] = \{[123], [213]\} = K[213];$$
$$K[132] = \{[132], [231]\} = K[231];$$
$$K[321] = \{[321], [312]\} = K[312].$$

Suppose now we try to compute $K[132].K[321]$ according to our definition of the operation in a factor group. If we pick the representatives [132] and [321] from $K[132]$ and $K[321]$, respectively, multiply them in the group S_3 to get [231] and find the coset to which this element belongs, we get

$$K[132].K[321] = K[231].$$

On the other hand, if we choose representatives [231] and [312] from $K[132]$ and $K[321]$, respectively, multiply them to get $\lfloor 123 \rfloor$ we obtain

$$K[132].K[321] = K[123] = K \neq K[231]$$

and so the two computations do not lead to the same answer. The reason is that K is not normal.

(iv) In Lemma 6.4.4 we defined the product of two subsets X and Y of a group by $XY = \{xy \mid x \in X, y \in Y\}$. We could have defined the product of two elements Kg and Kh in the factor group as the product of the two sets Kg and Kh, that is, $(Kg)(Kh) = \{kgk'h \mid k, k' \in K\}$. In general, this set is not a coset. However, if K is normal, it is. Moreover, since gh is in this set, it is in fact the coset Kgh. Hence the alternative definition coincides with the original one.

We sketch the proof that, when K is normal, $(Kg)(Kh)$ as defined above is also a coset:

$$(Kg)(Kh) = K(gK)h = K(Kg)h = K^2 gh = Kgh \text{ since } K^2 = K.$$

Observe that normality was used at the point where we replaced gK by Kg.

The next theorem is of great importance in the theory of groups and, in slightly modified form, in other areas of abstract algebra.

6.4.8 Theorem. (First Isomorphism Theorem for Groups)

If $f : G \to H$ is an epimorphism of groups and K the kernel of f then $G\!\big/_{\!K} \cong H$.

Proof. By Theorem 6.4.6(ii), K is normal in G and so we can form the factor group G/K. Define a map

$$\varphi : G\!\big/_{\!K} \to H \text{ by } \varphi(Kg) = f(g).$$

Again, we must make sure that the map is well defined.

Suppose $Kx = Ky$. We must show that $f(x) = f(y)$. Indeed,

$$Kx = Ky \text{ iff } xy^{-1} \in K \text{ iff } f(xy^{-1}) = \varepsilon \text{ iff } f(x)f(y^{-1}) = \varepsilon$$

$$\text{iff } f(x)f(y)^{-1} = \varepsilon \text{ iff } f(x) = f(y)$$

and so φ is well defined.

Since f is onto, clearly φ is onto.

That φ is a homomorhism follows from:

$$\varphi(Kx.Ky) = \varphi(Kxy) = f(xy) = f(x)f(y) = \varphi(Kx)\varphi(Ky).$$

Finally,

$$\ker \varphi = \{Kx \mid \varphi.(Kx) = \varepsilon\} = \{Kx \mid f(x) = \varepsilon\} = \{Kx \mid x \in K\} = \{K\}.$$

Hence, by Theorem 6.4.6(i), φ is 1:1 and therefore an isomorphism. ∎

Remarks.

(i) There is a difference between the equations $\ker f = K$ and $\ker \varphi = \{K\}$. In the former, if K is big, then $\ker f$ is big. In the latter, no matter how big K is, $\ker \varphi$ consists of exactly one element, namely, the "bag" K, which is considered as an indivisible unit. Thus, in the last line of the previous theorem, $\ker \varphi$ contains only one element, namely, the identity K of the factor group $G\!\big/_{\!K}$ and so establishes that φ is 1:1.

(ii) Observe that, if G is finite, $\left|G\!\big/_{\!K}\right| = {|G|}\!\big/_{\!|K|}$ and so the bigger the kernel, the smaller the homomorphic image. Therefore, the size of the kernel is a measure of how much "collapsing" the homomorphism effects, a fact remarked on above.

We now apply the First Isomorphism Theorem to prove what is commonly known as the Second and Third Isomorphism Theorems. We start with a lemma.

6.4.9 *Lemma*

If H is a normal subgroup of G and K is any subgroup then HK is a subgroup of G.

Proof.

(i) Clearly $HK \neq \phi$ since $H \neq \phi$ and $K \neq \phi$.

(ii) Let x and y be elements of HK. Then $x = h_1 k_1$ and $y = h_2 k_2$ where the h's are in H and the k's are in K. Therefore

$$xy = h_1 k_1 h_2 k_2 = h_1 k_1 h_2 k_1^{-1} k_1 k_2 = h_1 h'_2 k_1 k_2$$

where $h'_2 = k_1 h_2 k_1^{-1}$ is an element in H since H is normal in G. Hence HK is closed under the operation.

(iii) Let $x = hk \in HK$. Then

$$x^{-1} = k^{-1} h^{-1} = k^{-1} h^{-1} k k^{-1} = h' k^{-1}$$

where h' is in H since H is normal in G. Therefore HK is closed under the taking of inverses and so HK is a subgroup. ∎

6.4.10 *Theorem (The Second Isomorphism Theorem)*

If H is a normal subgroup of G and K any subgroup of G, then

$$\frac{HK}{H} \cong \frac{K}{H \cap K}.$$

Proof. Define a map

$$\theta : K \to \frac{HK}{H} \text{ by } \theta(k) = \bar{k}$$

where $\bar{k} = kH$. We see that θ is a homomorphism since

$$\theta(k_1 k_2) = \overline{k_1 k_2} = \bar{k}_1 . \bar{k}_2 = \theta(\bar{k}_1)\theta(\bar{k}_2).$$

To show that θ is onto, observe that $\bar{h}.\bar{k} = \bar{k}$ since $\bar{h} = \bar{e}$ for all $h \in H$.

The kernel of θ is

$$\{k \in K \mid \bar{k} = \bar{e}\} = \{k \in K \mid k \in H\} = K \cap H.$$

Applying the First Isomorphism Theorem we obtain the result. ∎

6.4.11 *Theorem (The Third Isomorphism Theorem)*

If H and K are normal subgroups of G with $K \leq H$, then
$$(G/K)/(H/K) \cong G/H .$$
Proof. Define a map $\varphi : G/K \to G/H$ by $\varphi(gK) = gH$. φ is well defined since $xK = gK$ implies that $g^{-1}x \in K \leq H$ and so $xH = gH$.

Clearly φ is an epimorphism and
$$\ker \varphi = \{ gK \mid \varphi(gK) = H \} = \{ gK \mid gH = H \} = \{ gK \mid g \in H \} = H/K .$$
Applying the First Isomorphism Theorem yields the result. ∎

One more result of some importance is the following. Before giving a proof we give an example to motivate the theorem.

6.4.12 *Example*

Consider the quaternion group $Q = \{\pm 1, \pm i, \pm j, \pm k\}$ and the Klein 4-group K_4. Below are the diagrams of the posets of the subgroups of Q and the subgroups of K_4 (in its incarnation as the group of symmetries of a rectangle) under the partial order of containment (see 3.2.9).

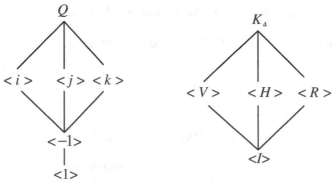

Fig. 6.2

The map $\varphi : Q \to K_4$ defined by
$$\varphi(1) = I = \varphi(-1), \varphi(i) = V = \varphi(-i), \varphi(j) = H = \varphi(-j), \varphi(k) = R = \varphi(-k)$$
is an epimorphism whose kernel is $<-1>$.

Observe that there is a 1:1 correspondence between the subgroups of Q containing the the kernel and all the subgroups of K_4 defined by
$$< -1 > \leftrightarrow < I >, < i > \leftrightarrow < V >, < j > \leftrightarrow < H >, < k > \leftrightarrow < R >, Q \leftrightarrow K_4.$$
This correspondence is effected by taking the inverse images under φ of the relevant subgroups of K_4.

It might help the reader to keep this example in mind as (s)he works through the proof of the following theorem.

6.4.13 *Theorem*

Let $f : G \to H$ be an epimorphism with kernel K. Then there is a one-one inclusion-preserving correspondence between all the subgroups of H and all the subgroups of G containing K. Moreover, in this correspondence normal subgroups correspond to normal subgroups.

Proof. Let $[G, K]$ and $[H, < \varepsilon >]$ denote the set of all subgroups of G containing K and the set of all subgroups of H containing the trivial subgroup $< \varepsilon >$, respectively. Define a map
$$\theta : [G, K] \to [K, < \varepsilon >] \text{ by } \theta(J) = f(J).$$
We see from Theorem 6.3.8 that $f(J)$ is a subgroup of H.

(i) We show that θ is inclusion-preserving: If $A \leq B$, then $x \in A$ implies $x \in B$ and so $f(A) \leq f(B)$.

(ii) Define a map $\varphi : [H, < \varepsilon >] \to [G, K]$ by $\varphi(L) = f^{-1}(L)$. Since the identity e of G is in L, it follows that K is contained in $f^{-1}(L)$. Moreover, by Theorem 6.3.8, $f^{-1}(L)$ is a subgroup of G.

(iii) We show that θ and φ are inverses of each other. To do this we need to show:

(a) $\varphi\theta(J) = J$ for all $J \in [G, K]$ and
(b) $\theta\varphi(L) = L$ for all $L \in [H, < \varepsilon >]$.

(a) $\varphi\theta(J) = f^{-1}\{f(J)\}$.
It is always the case that
$$J \leq f^{-1}\{f(J)\} \text{ (why?)}.$$

Conversely, if $x \in f^{-1}\{f(J)\}$, then $f(x) \in f(J)$ and so there exists $j \in J$ such that $f(x) = f(j)$. Since f is a homomorphism, $f(xj^{-1}) = \varepsilon$, the identity of H. Hence $xj^{-1} \in K \leq J$ and so $x \in J$. Therefore $\varphi\theta$ is the identity map.

(b) is proved similarly and is left as an exercise for the reader.

As for normality, suppose J is a normal subgroup of G containing K. If $y \in H$, then

$$y\theta(J)y^{-1} = yf(J)y^{-1}.$$

But since f is surjective, there exists $x \in G$ such that $f(x) = y$ and so

$$y\theta(J)y^{-1} = yf(J)y^{-1} = f(x)f(J)f(x^{-1}) = f(xJx^{-1}) = f(J) = \theta(J).$$

Conversely, if L is a normal subgroup of H, then

$$x\varphi(L)x^{-1} = xf^{-1}(L)x^{-1}.$$

We show $xf^{-1}(L)x^{-1} = f^{-1}(L)$.

Let $z \in xf^{-1}(L)x^{-1}$. Then $z = xyx^{-1}$ where $f(y) \in L$. Therefore

$$f(z) = f(x)f(y)f(x)^{-1} \in L$$

since L is normal in H. Hence $z \in f^{-1}(L)$ and so

$$xf^{-1}(L)x^{-1} \leq f^{-1}(L).$$

To prove the reverse inclusion, let $z \in f^{-1}(L)$. Then $f(z) \in L$ and, since L is normal,

$$f(x^{-1})f(z)f(x) = f(x)^{-1}f(z)f(x) \in L.$$

Thus $f(x^{-1}zx) \in L$ and so $x^{-1}zx \in f^{-1}(L)$. Therefore, $z \in xf^{-1}(L)x^{-1}$, and so $f^{-1}(L)$ is normal in G. ∎

6.5 Direct Products of Groups

Given a finite collection of groups there are a number of ways of "stitching" them together to form a new group. In this section we explore the simplest way of doing this by forming what we call the direct product of the groups as now defined:

6.5.1 *Definition*

Let $\{G_i \mid i = 1, 2, ..., n\}$ be a finite set of groups, each with operation denoted by juxtaposition. The *direct product* of these groups, denoted

$G_1 \otimes G_2 \otimes ... \otimes G_n$, consists of the set of all n-tuples $(g_1, g_2,...,g_n)$ where $g_i \in G_i$ with operation defined by

$$(g_1, g_2,...,g_n)(h_1, h_2,...,h_n) = (g_1 h_1, g_2 h_2,..., g_n h_n).$$

We often write $\bigotimes_{i=1}^{n} G_i$ rather than the more cumbersome $G_1 \otimes G_2 \otimes ... \otimes G_n$.

6.5.2 *Theorem*

If $G_i, i = 1, 2,..., n$ are groups, then $\bigotimes_{i=1}^{n} G_i$ is a group.

Proof. Closure is obvious. Associativity follows as a result of associativity in each of the groups G_i. The identity is $(e_1, e_2,..., e_n)$ and $(g_1, g_2,..., g_n)^{-1} = (g_1^{-1}, g_2^{-1},..., g_n^{-1})$. ∎

Remark. When the group is denoted additively, we call this construct the *direct sum* and write $G_1 \oplus G_2 \oplus ... \oplus G_n$ or more succinctly, $\bigoplus_{i=1}^{n} G_i$.

6.5.3 *Theorem*

The subset $\widehat{G_i} = \{(e_1, e_2,..., g_i,..., e_n) \mid g_i \in G_i\}$, where e_j is the identity of G_j, is a normal subgroup of $G_1 \otimes G_2 \otimes ... \otimes G_n$ and:

(i) $\bigotimes_{i=1}^{n} G_i = \widehat{G_1}\widehat{G_2}....\widehat{G_n} = \{\hat{g}_1 \hat{g}_2..., \hat{g}_n \mid \hat{g}_i \in \widehat{G_i}, i = 1, 2,..., n\}$;

(ii) $\widehat{G_i} \cap \widehat{G_1}\widehat{G_2}...\widehat{G_{i-1}}\widehat{G_{i+1}}...\widehat{G_n} = \{(e_1, e_2,..., e_n)\}$ for all $i, i = 1, 2,... n$;

(iii) Each element of the direct product is *uniquely* expressible as $\widehat{g_1}\widehat{g_2}....\widehat{g_n}$ where $\hat{g}_i = (e_1, e_2,..., g_i,..., e_n)$.

Proof. Exercise. ∎

The next result is the converse of the previous theorem.

6.5.4 *Theorem*

Let G be a group with normal subgroups $G_1, G_2, ..., G_n$ such that

(i) $G = G_1 G_2 .... G_n$.

(ii) $G_i \cap G_1 G_2 ... G_{i-1} G_{i+1} ... G_n = <e>$ for all $i, i = 1, 2, ..., n$.

Then G is isomorphic to the direct product of the G_i and each element of G is uniquely expressible as $g_1 g_2 ... g_n$ where $g_i \in G_i$.

Proof. Let us observe first that, by (ii), $G_i \cap G_j = <e>$ if $i \neq j$. It now follows, using normality, that if x_i and x_j are elements of G_i and and G_j, respectively, and $i \neq j$, then

$$(x_i^{-1} x_j^{-1} x_i) x_j = x_i^{-1} (x_j^{-1} x_i x_j) \in G_i \cap G_j = <e>$$

and so, $x_i x_j = x_j x_i$.

To prove each element of G is uniquely expressible as $\prod_{i=1}^{n} g_i$ where $g_i \in G_i$ for all $i, i = 1, 2, ..., n$, let

$$g = g_1 g_2 ... g_n = h_1 h_2 ... h_n$$

where g_i and h_i are in G_i. Then, since the elements of G_i commute with those of G_j whenever $i \neq j$, we find

$$h_i^{-1} g_i = \prod_{j \neq i} (h_j g_j^{-1}) \in G_i \cap \prod_{j \neq i} G_j = <e>$$

and so $h_i = g_i$ for all $i, i = 1, 2, ..., n$.

Define a map

$$f : G \to G_1 \otimes G_2 \otimes ... \otimes G_n$$

as follows: If $g = g_1 g_2 ... g_n$, then

$$f(g) = (g_1, g_2, ..., g_n) .$$

By the uniqueness of the representation of g as a product of the g_i's, the map is well defined and is clearly bijective. Moreover, if $g = g_1 g_2 ... g_n$ and $h = h_1 h_2 ... h_n$ where g_i and h_i are in G_i, then

$$f(gh) = f(\prod g_i \prod h_i) = f(\prod g_i h_i) = (g_1 h_1, g_2 h_2, ..., g_n h_n)$$
$$= (g_1, g_2, ..., g_n)(h_1, h_2, ..., h_n) = f(g) f(h)$$

and so f is an isomorphism. ∎

Remark. When G has normal subgroups with the properties set out in the theorem above, we sometimes say that G is the *internal direct product* of the subgroups. When G is constructed as in the definition above, we occasionally refer to G as the *external direct product* of the groups. In this case it is the internal direct product of the subgroups $\widehat{G_i}$ In general, though, we make no distinction and refer to both the internal and external direct products as merely the direct product since the groups are isomorphic.

6.5.5 *Example*

Consider the Klein 4-group as the group of symmetries of a rectangle which is not a square. Then using the notation established in Example 5.2.6 (iv),

$$K_4 = \{I, M_0, M_{\pi/2}, R_\pi\}.$$

Then the subgroups $G_1 = \langle M_0 \rangle$ and $G_2 = \langle M_{\pi/2} \rangle$ are normal subgroups

and $K_4 = G_1 \otimes G_2$ which can be thought of as an internal or an external direct product. In the latter case, the elements would consist of the ordered pairs

$$\{(I,I),(M_0,I),(I,M_{\pi/2}),(M_0,M_{\pi/2})\}.$$

The map $\varphi: K_4 \to G_1 \otimes G_2$ defined by

$$\varphi(I) = (I,I), \; \varphi(M_0) = (M_0,I), \; \varphi(M_{\pi/2}) = (I,M_{\pi/2}), \; \varphi(R_\pi) = (M_0,M_{\pi/2})$$

is an isomorphism. As an internal direct product, we would write $K_4 = \{I,M_0\}.\{I,M_{\pi/2}\}$. ∎

We end with a simple application of the direct product which will be useful when we discuss fields.

6.5.6 *Theorem*

If G is a finite abelian group in which every nonidentity element is of order a prime p, then G is a direct product of cyclic groups, each of order p. Moreover, G is of prime power order.

Proof. Observe first that if H is a subgroup and x is any element of G, then either $x \in H$ or $<x> \cap H = <e>$. For, if $<x> \cap H \neq <e>$, then $x^j \in H$ for some j, $1 \leq j < p$ and so $<x> = <x^j> \subseteq H$ since every nonidentity element of a cyclic group of prime order is a generator of the group.

Let $\{x_1, x_2, ..., x_n\}$ be a minimal generating set for G, that is, $G = <x_1, x_2, ..., x_n>$ but no proper subset of $\{x_1, x_2, ..., x_n\}$ generates G.

Suppose

$$<x_i> \cap (<x_1> . <x_2> ... <x_{i-1}> . <x_{i+1}> ... <x_n>) \neq <e>.$$

Then by the remark above, $x_i \in <x_1, x_2 ..., x_{i-1}, x_{i+1}, ..., x_n>$ and so

$$<x_1, x_2 ..., x_{i-1}, x_{i+1}, ..., x_n> = G$$

contradicting the minimality of $\{x_1, x_2, ..., x_n\}$. Therefore

$$<x_i> \cap <x_1> . <x_2> ... <x_{i-1}> . <x_{i+1}> ... <x_n> = <e>$$

for all i, $i = 1, 2, ..., n$. By Theorem 6.5.3,

$$G = <x_1> \otimes <x_2> \otimes ... \otimes <x_n>$$

and so $|G| = p^n$. ∎

6.6 Exercises

1) Prove that the dihedral group of order 8 is isomorphic to the group of symmetries of a square.

2) Let a and b be elements of a group and suppose $ab = ba$.

(i) Prove that if $\gcd(|a|, |b|) = 1$, then $|ab| = |a| \, \| \, b \|$.

(ii) Using (i) and induction, prove that if $a_i, i = 1, 2, 3, ..., n$ are elements of an abelian group whose orders are pairwise relatively prime, then the order of the product of the a_i is the product of their orders.

3) Prove that if H and K are finite subgroups of a group whose orders are relatively prime, then $H \cap K = <e>$.

4) Show by means of an example that, in general, if H and K are subgroups of a group, then $HK = \{hk \mid h \in H, k \in K\}$ is not necessarily a subgroup and that $HK \neq KH$ in general. Prove, however, that if at least one of H and K is normal, then HK *is* a subgroup and $HK = KH$. Furthermore, show that, in fact, HK is a subgroup if, and only if, $HK = KH$.

5) Prove that all subgroups of an abelian group are normal but that there is a nonabelian group all of whose subgroups are normal. (Hint: it's of order 8).

6) Show that S_3 has exactly one nontrivial proper normal subgroup.

7) Let $f : G \to H$ be a homomorphism of groups with kernel K. Prove that $f(G) \cong G/K$.

8) If G is a finite abelian group and m is the maximum order of the elements in G, prove that the order of any element in G divides m. Show by means of an example that this is not true if G is not abelian. [Hint: Assume that there exists $g \in G$ such that $|g|$ does not divide m and then, by using Problem 2 and the existence of an element whose order is m, construct an element whose order is greater than m.]

9) Let G be a finite abelian group and let $T = \{g \in G \mid g^2 = e\}$. Prove that

$$\prod_{g \in G} g = \prod_{g \in T} g.$$

Hence prove that if the order of G is odd, then the product of all the elements of G is the identity.

("$\prod$" means "product").

10) (i) Prove that in $(\mathbf{Z}_p^*, .)$, p prime, there is only one element of order 2.

(ii) Using Problem 9 and part (i), prove *Wilson's Theorem*: p is prime if, and only if, $(p-1)! \equiv -1 \bmod p$.

11) Prove that any two cyclic groups of the same order are isomorphic. [Hint: If $G = <g>$ is a group of order n, define the map $f : (\mathbf{Z}_n, +) \to G$ by $f(\overline{m}) = g^m$ and prove f is an epimorphism. Find the kernel of f and use the First Isomorphism Theorem].

12) Let n be a positive integer and let $\varphi(n)$ denote the number of positive integers less than n which are relatively prime to n. Prove that $\sum_{d|n} \varphi(d) = n$. [Hint: Consider a cyclic group of order n and count the number of generators of each subgroup].

13) Let D_n be the *dihedral group* of symmetries of a regular n-gon.
    (i) Prove that the order of D_n is $2n$;
    (ii) List the elements of the group D_3 of symmetries of an equilateral triangle and the group D_5 of symmetries of a regular pentagon.
    (iii) Prove that $D_3 \cong S_3$.

14) Prove that the dihedral group D_4 (of order 8) is not isomorphic to the quaternion group Q.

15) Let $K = <R_2>$ be the subgroup of D_4 of order 2 generated by a rotation of 180 degrees about the centre of the square. Prove
    (i) K is a normal subgroup of D_4.
    (ii) Prove that D_4 / K is isomorphic to the Klein 4-group.

16) Prove that if $K \le H \le G$ and $|H : K| = m, |G : H| = n$, then $|G : K| = mn$.

17) Prove that the number of homomorphisms from $(\mathbf{Z}_m, +)$ to $(\mathbf{Z}_n, +)$ is $\gcd(m, n)$.

18) If f is an endomorphism of $(\mathbf{Z}_{30}, +)$, such that $\ker f = \{\overline{0}, \overline{10}, \overline{20}\}$ and $f(\overline{23}) = \overline{6}$, determine all elements which map to $\overline{6}$ under f.

19) Prove that any homomorphic image of a cyclic group is cyclic.

20) Prove that if G and H are groups, then $G \otimes H$ is cyclic if, and only if, both G and H are finite and cyclic and $\gcd(|G|, |H|) = 1$.

21) Let $v : (\mathbf{Z}, +) \to (\mathbf{Z}/_{<12>}, +)$ be the natural homomorphism. Exhibit the correspondence established in Theorem 6.4.13.

22) Interpret Theorem 6.4.13 in case the homomorphism f is the natural homomorphism from G to $G/_K$ where K is some normal subgroup of G.

23) Suppose there is a homomorphism from a group G onto $(\mathbf{Z}/_{<10>}, +)$. Prove G has two normal subgroups, one of index 2, the other of index 5.

24) Suppose that f is a homomorphism from G onto a cyclic group of order 12 and that $\ker f$ is a subgroup of order 5. Prove G has normal subgroups of orders 5, 10, 15, 20, 30, 60.

25) Let H be a homomorphic image of a finite group G. Prove that if H has an element of order n, then so does G.

26) Suppose H and K are distinct subgroups of G of index 2. Prove that $H \cap K$ is a normal subgroup of G of index 4 and that $G/_{H \cap K} \cong K_4$, the Klein 4-group.

27) Let G be an abelian group, n a fixed integer and $f : G \to G$ the map defined by $f(x) = x^n$ for all $x \in G$. Prove that f is a homomorphism. If $|G| = m$, find an expression for the kernel of f and deduce that if $\gcd(m, n) = 1$, then f is an automorphism.

28) Prove that if the map $f : G \to G$ defined by $f(x) = x^2$ for all $x \in G$ is a homomorphism, then G must be abelian.

29) Let F be the set of all maps $f : Q \rightarrow Q$ of the form $f(x) = \dfrac{ax+b}{cx+d}$ where $a,b,c,d \in Q$ and $ad - bc \neq 0$. Prove that F under composition of maps, is a group.

If $\varphi : GL(2,Q) \rightarrow F$ is defined by $\varphi \left[\begin{pmatrix} a & b \\ c & d \end{pmatrix} \right] = f$ where f is the map $f(x) = \dfrac{ax+b}{cx+d}$, prove that φ is a homomorphism and find its kernel.

30) Let $G = < a >$ and $H = < b >$ be cyclic groups of orders 12, and 9 respectively. If $K = < (a^4, b^{-3}) >$,

   (i) What is the order of $(G \otimes H) \big/ K$?

   (ii) Is $(G \otimes H) \big/ K$ cyclic? Prove or show why it is not cyclic.

31) Let $C = < g >$ be a cyclic group of order mn. Find a generator for the subgroup H of order m, the subgroup K of order n and the subgroup $H \cap K$. Prove that if m and n are relatively prime, $C \cong H \otimes K$.

32) If $\varphi : G \rightarrow H$ is an epimorphism and $W \triangleleft H$, prove that $G \big/ \varphi^{-1}(W) \cong H \big/ W$.

Show how this is a generalization of the First Isomorphism Theorem.

33) Prove that the number of epimorphisms from $\mathbf{Z}_m$ to $\mathbf{Z}_n$ is $\delta \varphi(n)$ where φ is the Euler φ-function and

$$\delta = \begin{cases} 0 \text{ if } n \text{ does not divide } m \\ 1 \text{ if } n \text{ divides } m. \end{cases}$$

34) Prove that there is no nontrivial homomorphism from $\mathbf{Z}_8 \oplus \mathbf{Z}_2$ to $\mathbf{Z}_4 \oplus \mathbf{Z}_4$.

35) Let $T_{a,b} : \mathbf{R} \rightarrow \mathbf{R}$ where $a \neq 0$ be defined by

$$T_{a,b}(x) = ax + b.$$

(i) Prove that $G = \{T_{a,b} \mid a,b \in \mathbf{R}, a \neq 0\}$ is a group under composition of maps.

(ii) If $T_{a,b} \circ T_{c,d} = T_{p,q}$ find p and q in terms of a, b, c and d.

(iii) What is $T_{a,b}^{-1}$ in terms of a and b?

(iv) Compute the centre of G.

(v) Prove that $H = \{T_{1,b} \mid b \in \mathbf{R}\}$ is a normal subgroup of G and find G/H.

(vi) Given $T_{a,b}$, find all solutions to $T_{a,b} \circ X = X \circ T_{a,b}$.

36) (i) Let $G = <g>$ be a cyclic group of order m and a an element of G. Prove that $x^n = a$ has a solution in G iff $a^{\frac{m}{\gcd(m,n)}} = e$, the identity of G. Moreover, show that the subgroup of nth powers is of order $m/\gcd(m,n)$ and is generated by $g^{\gcd(m,n)}$.

(ii) Let $G = G_1 \otimes G_2 \otimes ... \otimes G_l$ where G_i is a cyclic group of order m_i, $i = 1,2,...,l$. Prove that an element $(a_1, a_2, ..., a_l)$ of G is an nth power iff $a_i^{\frac{m_i}{\gcd(m_i,n)}} = e$ for all $i, i = 1,2,...,l$. Moreover, show that the subgroup of nth powers is of order $\prod_{i=1}^{l} m_i / \gcd(m_i, n)$.

37) Let $f, g : G \to (\mathbf{R}^*, .)$ be nontrivial homomorphisms and suppose $L_f = \{x \in G \mid f(x) < 1\} \subseteq L_g = \{x \in G \mid g(x) < 1\}$. Prove that $L_f = L_g$ and $\ker f = \ker g$.

If there exist $x_0 \notin \ker f$ such that $f(x_0) = g(x_0)$, prove $f = g$.

Chapter 7

The Symmetric Groups

7.1 Introduction

In the investigation of finite groups the symmetric groups play an important role. Often we are able to achieve a better understanding of a group if we can think of it as a group of permutations on some set, often one with some geometrical structure. In this chapter we apply much of the theory we have developed so far to study the structure of S_n, the group of permutations on the set $\{1, 2, 3, ..., n\}$. We shall show that S_n can be partitioned into two sets, one of which is a subgroup of considerable importance. A partitioning of each of these two sets into conjugacy classes provides us with an important tool which we take advantage of in our proof that the converse of Lagrange's theorem is not valid.

7.2 The Cayley Representation Theorem

We begin with a well-known theorem which says that any abstract group is isomorphic to a subgroup of some symmetric group. Of course, knowing that a group is contained in some other group does not, in general, tell us much about the group itself, especially if the containing group is big compared with the contained group, as is the case in this representation. Nevertheless, what is important is the idea that an abstract group can be realized as a more concrete object, an idea that has led to a branch of group theory called Representation Theory.

7.2.1 Theorem (Cayley Representation Theorem)

Every group G is isomorphic to a subgroup of some symmetric group.

Proof. Let S_G be the group of permutations on the underlying set G of the group $(G,.)$. Define a map $f : G \to S_G$ as follows: $f(g) = \Pi_g$ where Π_g is the permutation of the set G which g induces by left multiplication; that is, for all $x \in G$, $\Pi_g(x) = gx$.

We must first show that Π_g is a bijective map on G which will establish that Π_g is in S_G.

Π_g is 1:1. Suppose $\Pi_g(x) = \Pi_g(y)$. Then $gx = gy$ and by cancelling g we get $x = y$, proving that Π_g is 1:1.

Π_g is onto G. Let $x \in G$ Then $\Pi_g(g^{-1}x) = gg^{-1}x = x$.

Next we show that f is a homomorphism. To do this we must prove that $f(gh) = f(g)f(h)$ for all $g,h \in G$ and this means showing that

$$\Pi_{gh} = \Pi_g \Pi_h.$$

Naturally, the operation in S_G is composition of maps. Now

$$\Pi_{gh}(x) = (gh)x = g(hx) = \Pi_g(hx) = \Pi_g \Pi_h(x)$$

for all $x \in G$ and so $\Pi_{gh} = \Pi_g \Pi_h$. Therefore, f is a homomorphism.

Lastly, we prove f is 1:1. If $g \in \ker f$, then Π_g is the identity map on G and so, in particular, $\Pi_g(e) = e$. Therefore $ge = e$, whence $g = e$ and so $\ker f = <e>$, thereby establishing that f is 1:1.

It now follows that $G \cong f(G)$ where $f(G)$ is a subgroup of S_G. ∎

Observe that the size of the isomorphic copy of the group G is, in general, very small compared to the size of S_G. For example, if the order of G is 6, then the order of S_G is $6! = 720$. The subgroup (of order 6) of S_G isomorphic to G is inside a group 120 times its size!

7.2.2 *Example*

We realize the Klein 4-group K_4 as a permutation group. Recall that $K_4 = \{e,a,b,c\}$ where the square of any element is e and the product of any two distinct nonidentity elements is the third element.
We have:

$$\Pi_e = I = \begin{pmatrix} e\ a\ b\ c \\ e\ a\ b\ c \end{pmatrix};\ \Pi_a = \begin{pmatrix} e\ a\ b\ c \\ a\ e\ c\ b \end{pmatrix};\ \Pi_b = \begin{pmatrix} e\ a\ b\ c \\ b\ c\ e\ a \end{pmatrix};\ \Pi_c = \begin{pmatrix} e\ a\ b\ c \\ c\ b\ a\ e \end{pmatrix}.$$

If we set $e = 1, a = 2, b = 3, c = 4$, the permutations are [1234] , [2143] , [3412], and [4321] form a subgroup of S_4. ∎

7.3 Permutations as Products of Disjoint Cycles

In order to study S_n in detail, we need to introduce notation for the elements of S_n which conveys at a glance the nature of the elements. The notation we have been using so far doesn't do this. For example, [1234] is a rather "tame" element of S_4 since it moves none of $1, 2, 3, 4$. On the other hand, [2341] is quite "dynamic" since every number is moved by it. The eye, however, does not immediately distinguish between the two.

7.3.1 *Definition and notation*

A permutation π such that: $\pi(a_i) = a_{i+1}, 1 \le i \le k - 1, \pi(a_k) = a_1$ is denoted by $(a_1\ a_2\ a_3....a_k)$ and is called an *n-cycle* or a cycle of length n. Two cycles without any common symbols are said to be *disjoint*. The identity permutation is written as (1) although, strictly speaking we should write it as $(1)(2)(3)...(n)$, but this is clearly too cumbersome.

Note. The *n*-cycles
$$(a_1\ a_2\ a_3...a_k),\ (a_2\ a_3\ a_4....a_k\ a_1),...,(a_k\ a_1\ a_2...a_{k-1})$$
all represent the same permutation.

7.3.2 *Example*

$(12)(345)(6)(7)$ is the permutation which maps 1 to 2, 2 to 1, 3 to 4, 4 to 5, 5 to 3 and leaves 6 and 7 fixed.

We would say that this permutation is a product of disjoint cycles. In general the 1-cycles are omitted. We would therefore write $(12)(345)$ when it is clear that the permutation is an element of S_7. In short, each omitted symbol is assumed to be mapped onto itself but we must indicate which S_n the permutation belongs to.

7.3.3 *Theorem*

Every permutation of S_n is a product of disjoint cycles.

Proof. Let $g \in S_n$ and for some $i \in \{1, 2, 3, ..., n\}$ consider the sequence

$$i = g^0(i), g^1(i) = g(i), g^2(i), ..., g^m(i),...$$

There must be a repetition in this sequence since each $g^m(i) \in \{1,2,3,...,n\}$ for all m. Let k be the first positive integer such that $g^k(i)$ is a repetition of some preceding term of the sequence, so that

$$i = g^0(i), g(i), g^2(i), ..., g^{k-1}(i)$$

are distinct while $g^k(i) = g^p(i)$ for some nonnegative $p \leq k$. Applying g^{-p} to each side of this equation, we get $g^{k-p}(i) = g^0(i) = i$. Now $0 \leq k - p \leq k$ and $g^{k-p}(i)$ is a repetition of a preceding term of the sequence, namely, i. Therefore, by the minimality of k, $k - p$ cannot be strictly less than k. Therefore, $p = 0$, showing that the first repetition is i, i.e., $g^k(i) = i$.

We can illustrate the foregoing by a diagram:

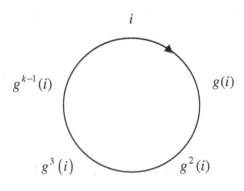

Fig. 7.1.

We claim that $\{g^m(i) \mid m \in \mathbf{Z}\} = \{i, g(i), g^2(i), ..., g^{k-1}(i)\}$. It is obvious that the right-hand side is contained in the left-hand side.

Conversely, for any integer m, $m = qk + r$ where $0 \le r < k$ by the Division Algorithm and so

$$g^m(i) = g^{kq+r}(i) = g^r[(g^k)^q(i)] = g^r(i)$$

since $(g^k)^q(i) = i$ (why? See the diagram above), and so our assertion is proved. (The subset $\{i, g(i), g^2(i), ..., g^{k-1}(i)\}$ of $\{1, 2, 3, ..., n\}$ is called the *orbit of i under g* and is denoted by $O_g(i)$.)

Next, pick any j in $\{1, 2, ..., n\}$ but not in $O_g(i)$ (if such a j exists). Proceed as above to get

$$\{j, g(j), g^2(j), g^3(j), ..., g^{t-1}(j)\}$$

where $g^t(j) = j$.

We claim that $O_g(i) \cap O_g(j) = \phi$.. Indeed, suppose $g^u(i) = g^v(j)$. Then $g^{u-v}(i) = j \in O_g(i)$, a contradiction since $j \notin O_g(i)$.

Continue producing orbits as above until all the elements of $\{1, 2, ..., n\}$ have been exhausted. We obtain a collection of non-overlapping subsets of $\{1, 2, ..., n\}$ whose union is $\{1, 2, ..., n\}$. It is now clear that

$$g = \left(i\ g(i)\ g^2(i)..g^{k-1}(i)\right)\left(j\ g(j)\ g^2(j)..g^{t-1}(j)\right)...\left(l\ g(l)\ g^2(l)..g^{v-1}(l)\right).$$

∎

Let us take a specific example of our discussion above.

7.3.4 *Example*

Consider $g = [21453789] \in S_9$. We compute the orbits of g. The orbit of 1 is $\{1, g(1) = 2\}$ since $g^2(1) = g(2) = 1$. Similarly, the orbits of 3 and 6 are, respectively, $\{3, 4, 5\}$ and $\{6, 7, 8, 9\}$. The following diagram exhibits quite clearly the behavior of g

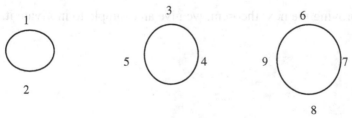

Figure 7.2

We can indicate the behavior of g in the first orbit by writing (12). We take this to mean that g maps 1 to 2 and 2 back to 1. As for the second orbit, we write (345) which conveys that g maps 3 to 4, 4 to 5 and 5 back to 3. For the last orbit, g maps 6 to 7, 7 to 8, 8 to 9 and 9 back to 6. Therefore we can write g as (12)(345)(6789). ∎

When computing with permutations expressed as products of disjoint cycles, the next theorem is useful.

7.3.5 *Theorem*

Disjoint cycles commute.

Proof. Suppose α, β are disjoint cycles. Then α fixes all symbols moved by β and vice-versa. Symbols which do not appear in either are fixed by both $\alpha\beta$ and $\beta\alpha$. Suppose now that j is a symbol appearing in α (and therefore not in β). Then $\alpha\beta(j) = \alpha(j)$ and, since $\alpha(j)$ does not appear as a symbol in β, it follows that $\beta[\alpha(j)] = \alpha(j)$. Hence $\alpha\beta(j) = \alpha(j) = \beta\alpha(j)$.

A similar argument shows that if k is a symbol appearing in β (and so not in α), we have $\alpha\beta(k) = \beta(k) = \beta\alpha(k)$. Therefore, $\alpha\beta = \beta\alpha$. ∎

7.3.6 *Corollary*

If $\alpha, \beta, ..., \pi$ are pairwise disjoint cycles, then, for all integers m, $(\alpha\beta....\pi)^m = \alpha^m \beta^m ....\pi^m$.

Proof. Exercise. (Use the previous theorem and induction). ∎

Before proving the next theorem, we give an example to motivate it.

7.3.7 *Example*

What is the order of the permutation $(12)(345)(6789)$?

Squaring, we get

$$[(12)(345)(6789)]^2 = (12)^2(345)^2(6789)^2 = (354)(68)(79).$$

Observe that $(12)^2 = (1)$ and so (12) disappears.

Cubing yields

$$(12)^3(345)^3(6789)^3 = (12)(6987).$$

This time (345) disappears since $(345)^3 = (1)$. This suggests that to "suppress" an m-cycle we must raise the cycle to a multiple of m. Therefore, to "suppress" all symbols occurring in the given permutation, we must raise it to the power 12 which is the least common multiple of the lengths of the cycles and this is the smallest such power. Therefore the order of $(12)(345)(6789)$ is 12. ∎

7.3.8 *Lemma*

The order of an m-cycle is m.

Proof. Suppose the m-cycle is $\alpha = (j_0 \ j_1 \ j_2 .... j_{m-1})$. Then $\alpha^k(j_s) = j_{s+k \bmod m}$ where for this proof only, $s + k \bmod m$ means the unique integer t such that $t \equiv s + k \bmod m, 0 \le t < m$. For α^k to be the identity map, it is necessary and sufficient for k to be the smallest positive integer for which $s + k \equiv s \bmod m$ for all s, $0 \le s < m$; that is, subtracting s from each side of the congruence, we want the smallest positive integer k such that $k \equiv 0 \bmod m$. Obviously, $k = m$ fits the bill and so the order of an m-cycle is m. ∎

7.3.9 *Theorem*

Suppose the permutation π is the product of the disjoint cycles $\alpha_i, i = 1, 2, ..., s$ where α_i is an m_i-cycle. Then the order of π is $\text{lcm}\{m_i \mid i = 1, 2, ..., s\}$.

Proof. Let $n = \text{lcm}\{m_i\{i=1,2,....,s\}$. By Corollary 7.3.6 and the preceding lemma, $\pi^n = \alpha_1{}^n \alpha_2{}^n .... \alpha_s{}^n = (1)$ and so $|\pi|$ divides n. . On the other hand, $\pi^{|\pi|} = (1)$ and so $\alpha_i^{|\pi|} = (1)$ for all i, $i = 1, 2, ..., s$ which implies that m_i divides $|\pi|$ for all i, $i = 1, 2, ..., s$ by Lemma 7.3.8. Therefore n divides $|\pi|$ and so $|\pi| = n$. ∎

7.3.10 *Example*

Express the permutation $\alpha = (134)(92631)(4735)(789)$ as a product of disjoint cycles and find its order. Note that the cycles in α are not disjoint.

To express α as a product of disjoint cycles we apply α to each of the symbols appearing in its cycles. We get $\alpha(7) = 8$, $\alpha(8) = 2$, $\alpha(2) = 6, \alpha(6) = 4, \alpha(4) = 7$ and the cycle is complete. Hence, one of the disjoint cycles is (78264). The symbol 1 does not appear in this cycle and so we apply α to 1 and proceed as above to obtain the cycle (1935). Therefore $\alpha = (78264)(1935)$ and so its order is 20, the lcm of 5 and 4, by Theorem 7.3.9. ∎

Remark. The decomposition of a permutation as a product of disjoint cycles enables us to write down its inverse immediately merely by writing the cycles backwards. For example, in the previous example, $\alpha^{-1} = (46287)(5391)$. The reader should provide a proof of the general statement.

7.4 Odd and Even Permutations

As mentioned in the Introduction, we show that the elements in S_n fall into two classes, the odd permutations and the even permutations. There are a number of proofs of this result which is quite tricky. We have chosen the one presented because it puts homomorphisms to use.

7.4.1 *Definition*

A permutation which interchanges two symbols and leaves all others fixed is called a *transposition*.

7.4.2 *Lemma*

Every m-cycle is a product of $m-1$ transpositions.

Proof. It is a simple matter to check that
$$(j_1 \, j_2 \, j_3 \cdots j_m) = (j_1 \, j_m)(j_1 \, j_{m-1})(j_1 \, j_{m-2}) \cdots (j_1 \, j_3)(j_1 \, j_2),$$
a product of $m-1$ transpositions. (Remember that the right-most map is applied first.)■

7.4.3 *Theorem.*

Every permutation is a product of transpositions.

Proof. Each permutation is a product of disjoint cycles and each cycle, by the lemma, is a product of transpositions. Therefore, each permutation is a product of transpositions.■

Remark. If a permutation is expressible as a product of disjoint cycles of lengths $r, s, t, \ldots, w$ then it can be expressed as a product of $(r-1)+(s-1)+(t-1)+\ldots+(w-1)$ transpositions since a cycle of length k is a product of $k-1$ transpositions, by Lemma 7.4.2.

7.4.4 *Example*

$$(123)(45)(6789) = (13)(12)(45)(69)(68)(67). ■$$

Remarks.
   (i) Observe that, in general, the transpositions are not disjoint.
   (ii The decomposition of a permutation as a product of transpositions is not unique. For example,
$$(123)(45)(6789) = (13)(12)(45)(69)(68)(67)$$

Abstract Algebra

$$= (12)(23)(45)(47)(46)(49)(48)(47).$$

Note that even the number of transpositions in the two decompositions is not the same. However, the number of transpositions in each decomposition is even.

(iii) As another example we have
$$(123)(4567) = (13)(12)(47)(46)(45)$$
$$= (23)(34)(14)(34)(47)(46)(45).$$

Again, although the number of transpositions in each of the two decompositions is different, each is odd.

(iv) We shall eventually prove that if a permutation decomposes as a product of an even (respectively, odd) number of transpositions, then any decomposition of that permutation will have an even (respectively, odd) number of transpositions. We shall then be able to classify a given permutation as odd or even.

To prove the assertion in (iv), we need to introduce some notation and prove two lemmas.

Notation. Let π be a permutation and let
$$S(\pi) = \prod_P \frac{\pi(i) - \pi(j)}{i - j}$$
where P is the set of all (unordered) pairs $\{i, j\} \subseteq \{1, 2, 3, \ldots, n\}$. We see that S is a map from S_n to the rational numbers.

Remark. Observe that the order of the pairs i, j is immaterial since
$$\frac{\pi(i) - \pi(j)}{i - j} = \frac{\pi(j) - \pi(i)}{j - i}.$$

7.4.5 *Example*

Let $\pi = (12)(34)$. Then there are 6 subsets of $\{1,2,3,4\}$ of size 2, namely, $\{1,2\}, \{1,3\}, \{1,4\}, \{2,3\}, \{2,4\}$ and $\{3,4\}$. Then
$$S(\pi) = \frac{[\pi(1) - \pi(2)]\ldots[\pi(2) - \pi(3)][\pi(2) - \pi(4)][\pi(3) - \pi(4)]}{[1-2] \times [1-3] \times [2-3] \times [2-4] \times [3-4]}$$

$$= \frac{[2-1][2-4][2-3][1-4][1-3][4-3]}{[1-2][1-3][1-4][2-3][2-4][3-4]} = 1$$

In practice, when computing $S(\pi)$, it is convenient to write π as follows:

$$\begin{pmatrix} 1 & 2 & 3 & 4 \dots \dots n \\ \pi(1) & \pi(2) & \pi(3) & \pi(4) \dots \pi(n) \end{pmatrix}.$$

Then the pairs can be listed as

$$\{1,2\}, \{1,3\}, \{1,4\},...,\{1,n\}; \{2,3\}, \{2,4\},...,\{2,n\}; ....;\{n-1,n\}$$

and $S(\pi)$ easily computed. In the listing of the pairs, we first list all pairs with 1 as the smaller number of the pair, then all pairs with 2 as the smaller number, all pairs with 3 ... and so on.

7.4.6 *Lemma*

If α and β are permutations, then

$$S(\alpha\beta) = S(\alpha)S(\beta).$$

Therefore, S is a homomorphism from S_n to the multiplicative group of nonzero rational numbers.

Proof. $S(\alpha\beta) = \prod_P \dfrac{\alpha\beta(i) - \alpha\beta(j)}{i-j}$ and $S(\beta) = \prod_P \dfrac{\beta(i) - \beta(j)}{i-j}$. We observe that α can be written as

$$\begin{pmatrix} \beta(1) & \beta(2) & \beta(3) & \beta(4)........................\beta(n) \\ \alpha\beta(1) & \alpha\beta(2) & \alpha\beta(3) & \alpha\beta(4).....................\alpha\beta(n) \end{pmatrix}$$

and so, we can write

$$S(\alpha) = \prod_P \frac{\alpha\beta(i) - \alpha\beta(j)}{\beta(i) - \beta(j)}.$$

Multiplying $S(\alpha)$ and $S(\beta)$ together, we get

$$S(\alpha)S(\beta) = \prod_P \frac{\alpha\beta(i) - \alpha\beta(j)}{\beta(i) - \beta(j)} \prod_P \frac{\beta(i) - \beta(j)}{i-j} = \prod_P \frac{\alpha\beta(i) - \alpha\beta(j)}{i-j}$$

$$= S(\alpha\beta)$$

since the $\beta(i) - \beta(j)$ cancel. Therefore, $S(\alpha\beta) = S(\alpha)S(\beta)$. ∎

7.4.7 Corollary

If $\alpha_1, \alpha_2, \alpha_3, ..., \alpha_k$ are permutations, then

$$S(\alpha_1\alpha_2\alpha_3.....\alpha_n) = S(\alpha_1)S(\alpha_2)S(\alpha_3)......S(\alpha_n).$$

Proof. Use the above result and induction. ∎

The final result we need is that the image of a transposition under S is -1.

7.4.8 Lemma

If τ is a transposition, then $S(\tau) = -1$.

Proof. Without loss of generality we may assume $\tau = (12)$. Then

$$S(\tau) = S\begin{pmatrix} 1\ 2\ 3\ 4...n \\ 2\ 1\ 3\ 4...n \end{pmatrix} =$$

∎

$$\frac{(2-1)(2-3)...(2-n)(1-3)...(1-n)(3-4)(3-5)...([n-1]-n)}{(1-2)(1-3)...(1-n)(2-3)...(2-n)(3-4)(3-5)...([n-1]-n)} = -1.$$

7.4.9 Theorem

If, $\alpha = \tau_1\tau_2\tau_3..... \tau_p = \sigma_1\sigma_2\sigma_3..... \sigma_q$ where the τ's and σ's are transpositions, then $(-1)^p = (-1)^q$, or, in other words, either both p and q are even or both odd.

Proof. Applying S to α, we have
$$S(\alpha) = S(\tau_1\tau_2...\tau_{p-1}\tau_p) = S(\sigma_1\sigma_2...\sigma_{q-1}\sigma_q).$$
By Corollary 7.4.7,
$$S(\tau_1)S(\tau_2)S(\tau_3)....S(\tau_p) = S(\sigma_1)S(\sigma_2)S(\sigma_3).....S(\sigma_q).$$
By Lemma 7.4.8,
$$S(\tau_i) = S(\sigma_j) = -1 \text{ for all } i, 1 \le i \le p \text{ and for all } j, 1 \le j \le q.$$

Therefore, $(-1)^p = (-1)^q$ from which we deduce that either both p and q are odd or both even. ∎

Remark. Note that we have incidentally proved that the map S is actually from the group to the multiplicative subgroup $\{1, -1\}$ of nonzero reals.

7.4.10 *Definition*

A permutation is said to be *even* if it is expressible as a product of an even number of transpositions. Otherwise, it is said to be *odd*. The property of being odd or even is the *parity* of the permutation.

7.4.11 *Corollary*

The set of all even permutations of $S_n, n > 1$, is a normal subgroup of S_n of order $n!/_2$ called the *alternating group of degree n* and denoted by A_n.

Proof. By the First Isomorphism Theorem, Lemma 7.4.6 and Theorem 7.4.9 above, $S_n/_{\ker S} \cong \{1, -1\}$. But $\ker S = A_n$ and so

$$S_n/_{A_n} \cong \{1, -1\}$$

and A_n is normal of index 2 in S_n.

Hence there are only two cosets of A_n in S_n, one consisting of all the even permutations and the other all the odd permutations. It now follows that $|A_n| = |S_n|/_2 = n!/_2$. ∎

7.4.12 *Example*

We list the elements of S_4 according to their *cycle structure*. We say that two permutations have the same cycle structure if in each decomposition as a product of disjoint cycles, the number of cycles is the same and the number of cycles of a given length is the same in each. For example, (123)(45)(6789) and (689)(2345)(17) have the same cycle structure.

In the table below the number of elements of a given cycle structure is calculated using fundamental principles of counting. For example, to compute the number of elements having the same cycle structure as (123) (considered as an element of S_4), we argue as follows: there are $24 = 4.3.2$ different ordered triples since there are 4 ways of choosing the first position, 3 ways of choosing the second and 2 ways of choosing the third. These 24 ordered triples, however, do not yield 24 different permutations. Indeed, the ordered triples 123, 231 and 312 give rise to the same permutation. If we collect together those ordered triples which yield the same permutation, each such collection will contain three ordered triples and so, there will be $24/3 = 8$ collections. Therefore, there are 8 permutations with the same cycle structure as (123) in S_4

Table 7.1. Cycle structure of S_4

Cycle Structure	Number	Order	Parity
(1)	1	1	Even
(12)	6	2	Odd
(123)	8	3	Even
(1234)	6	4	Odd
(12)(34)	3	2	Even

7.5 Conjugacy Classes of a Group

We take a brief excursion away from symmetric groups to develop some more theory which we need to further our study of the symmetric groups. We first define a relation called conjugacy and show that it is an equivalence relation whose classes we call the conjugacy classes of the group. This decomposition of a group makes its structure more transparent and we shall see later on that we shall make full use of it.

7.5.1 *Definition*

An element x of a group G is said to *be conjugate in G* to an element y if there exists an element g of the group such that $y = gxg^{-1}$. We write $x \simeq y$ and read: x is conjugate to y. To *conjugate x by g* means to compute $gxg^{-1} = {}^g x$ (some authors mean $x^{-1}gx = g^x$ for conjugation by x).

7.5.2 *Theorem*

Conjugacy is an equivalence relation on G whose equivalence classes are called the *conjugacy classes of G*. The conjugacy class of a is denoted by $K(a)$.

Proof. (i) (Reflexivity) For all $x \in G$, $x \simeq x$ since $x = exe^{-1}$.

(ii) (Symmetry) Suppose $x \simeq y$. Then there exists $g \in G$ such that $y = gxg^{-1}$. Multiplying this equation on the left by g^{-1} and on the right by g, we get $x = g^{-1}y(g^{-1})^{-1}$ and so $y \simeq x$.

(iii) (Transitivity) Suppose that $x \simeq y$ and $y \simeq z$. Then there exist g and h in G such that $y = gxg^{-1}$ and $z = hyh^{-1}$. Hence

$$z = hgxg^{-1}h^{-1} = (hg)x(hg)^{-1}$$

and so $x \simeq z$.

Therefore $\simeq$ is an equivalence relation. ∎

Remark. The conjugacy classes of an abelian group are all singleton sets so conjugacy turns out to be quite uninteresting in the abelian case.

When are two elements conjugate in S_n? The answer turns out to be quite esthetically pleasing. We begin with a lemma.

7.5.3 *Lemma*

Let π be a permutation. Then

$$(*)\ \pi(a_1 a_2 ... a_k)\pi^{-1} = (\pi(a_1)\ \pi(a_2)...\pi(a_k)).$$

Proof. Applying the left-hand side of equation above to $\pi(a_1)$, we get

$$\pi(a_1 a_2 ... a_k) \pi^{-1} [\pi(a_1)] = \pi(a_1 a_2 ... a_k)[a_1] = \pi(a_2).$$

In general, applying the left-hand side to $\pi(a_j)$, $j < k$, we get $\pi(a_{j+1})$ and when $j = k$,

$$\pi(a_1 a_2 ... a_k) \pi^{-1} [\pi(a_k)] = \pi(a_1 a_2 ... a_k)[a_k] = \pi(a_1).$$

Therefore, one of the nontrivial cycles of the permutation $\pi(a_1 a_2 ... a_k) \pi^{-1}$ is $(\pi(a_1) \, \pi(a_2) .... \pi(a_k))$.

The permutation on the right-hand side of (*) moves the symbols $\pi(a_j), 1 \le j \le k$ and no others. We need only show that the same is true of $\pi(a_1 a_2 ... a_k) \pi^{-1}$ and the result will be proved. Suppose

$$b \notin \{\pi(a_1), \pi(a_2), ..., \pi(a_k)\}.$$

Then $\pi^{-1}(b) \notin \{a_1, a_2, ..., a_k\}$ (why?). Hence $(a_1 a_2 ... a_k)[\pi^{-1}(b)] = \pi^{-1}(b)$ and so

$$\pi(a_1 a_2 .... a_k) \pi^{-1}[b] = \pi\{(a_1 a_2 .... a_k)[\pi^{-1}(b)]\} = \pi\pi^{-1}(b) = b. \quad \blacksquare$$

7.5.4 *Theorem*

Two permutations are conjugate in S_n if, and only if, they have the same cycle structure.

Proof. Suppose first that the two permutations λ and μ have the same cycle structure, say

$$\lambda = \tau_1 \tau_2 .... \tau_k, \quad \mu = \sigma_1 \sigma_2 .... \sigma_k$$

where $\tau_j = (a_{j1} \, a_{j2} \, a_{j3} .... a_{jt_j})$ and $\sigma_j = (b_{j1} \, b_{j2} \, b_{j3} .... b_{jt_j})$. The number of symbols appearing in each of λ and μ is $\displaystyle\sum_{j=1}^{k} t_j$ and so the number of symbols fixed by each is $n - \displaystyle\sum_{j=1}^{k} t_j$. Let ρ be any bijection from the symbols fixed by λ to the symbols fixed by μ and define a permutation π as follows:

$$\pi(x) = \begin{cases} b_{js} \text{ if } x = a_{js}; \\ \rho(x) \text{ otherwise.} \end{cases}$$

In the decomposition of λ as a product of disjoint cycles, insert $\pi^{-1}\pi$ between each pair of consecutive τ's and multiply λ by π on the left and by π^{-1} on the right. We get

$$\pi\lambda\pi^{-1} = (\pi\tau_1\pi^{-1})(\pi\tau_2\pi^{-1})(\pi\tau_3\pi^{-1}).....(\pi\tau_k\pi^{-1}).$$

By Lemma 7.5.3,

$$\pi\tau_j\pi^{-1} = \sigma_j \text{ for all } j, 1 \le j \le k,$$

proving that $\pi\lambda\pi^{-1} = \mu$.

Conversely, suppose λ and μ are conjugate, say $\mu = \gamma\lambda\gamma^{-1}$ and let the decomposition of λ as a product of disjoint cycles be as in the first part of the proof. Then

$$\mu = \gamma\lambda\gamma^{-1} = (\gamma\tau_1\gamma^{-1})(\gamma\tau_2\gamma^{-1})(\gamma\tau_3\gamma^{-1}).....(\gamma\tau_k\gamma^{-1}).$$

By the lemma, each $\gamma\tau_j\gamma^{-1}$ is a cycle of the same length as that of τ_j. Moreover, the cycles $\gamma\tau_j\gamma^{-1}, j = 1, 2, ..., k$ are disjoint (why?) and so λ and μ have the same cycle structure. ∎

7.5.5 Example

Find a permutation π such that $\mu = \pi\lambda\pi^{-1}$ where

$$\lambda = (1,2)(3,5,10)(12,4,6,7) \text{ and } \mu = (10,11)(2,4,3)(9,8,1,5).$$

The symbols fixed by λ are 8, 9 and 11; those fixed by μ are 6, 7 and 12. We define any bijection between these two sets of symbols, say

$$8 \to 6, 9 \to 7, 11 \to 12.$$

Next, place the decomposition of μ under the decomposition of λ so that cycles of the same length line up under each other. We get:

$$\begin{pmatrix} 1 & 2 & 3 & 5 & 10 & 12 & 4 & 6 & 7 \\ 10 & 11 & 2 & 4 & 3 & 9 & 8 & 1 & 5 \end{pmatrix}.$$

This array defines a map which sends a symbol in the top line to the symbol immediately below it. This map, together with the bijection defined above between the symbols fixed by the two permutations, yields the permutation $(1,10,3,11,12,9,7,5,4,8,6)$, which is the required permutation π.

Note that since we have a choice for the bijection between the elements fixed by each permutation, the solution above is one of six possible answers. There are in fact many more which we obtain by cyclically permuting the symbols in each cycle and following the recipe above. ∎

Lagrange's Theorem (Corollary 6.2.7) states that the order of a subgroup of a finite group divides the order of the group. However, given a divisor of the order of a group, there does not necessarily exist a subgroup of order the given divisor. We shall give an example of a group of order 12 which contains no subgroup of order 6. To do this, we first prove a couple of simple results.

7.5.6   *Theorem*

If H is a subgroup of G of index 2 in G, then H is normal in G.

Proof. We need to show that, for all $g \in G$, $gH = Hg$.

If g is an element of H, $gH = Hg = H$. Suppose g is not an element of H. Then, since H is of index 2, we have $G = H \cup gH = H \cup Hg$. Since this is a disjoint union, the complement of H in G is gH and also Hg. Hence, $gH = Hg$. ∎

7.5.7   *Theorem*

If H is a normal subgroup of G and a is an element of H, then the full conjugacy class of a in G is contained in H.

Proof. Every element of the conjugacy class of a is expressible as gag^{-1} for some g in G. But $a \in H$ implies $gag^{-1} \in gHg^{-1} = H$, the last equality being a consequence of the normality of H and Theorem 6.4.5. Therefore, the full conjugacy class of g is contained in H. ∎

7.5.8   *Theorem*

The alternating group A_4 of Example 7.4.12 has no subgroup of order 6.

Proof. Let us suppose that there exists such a subgroup H. Since $|A_4| = 12$, this hypothetical subgroup is of index 2 in A_4 and so, by 7.5.6 above, H is normal in G.

Now A_4 contains the identity, eight 3-cycles and the elements $(12)(34)$, $(13)(24)$ and $(14)(23)$. Hence our subgroup H must contain at least one 3-cycle, say, (abc). By 7.5.7, H must contain all conjugates of (abc) in A_4. Note that if we knew that H was normal in S_4, it would have to contain all eight 3-cycles by 7.5.7 and 7.5.4, which would be impossible since the order of H is only six. In this manner, we would have proved quite easily that H does not exist. However, we know only that H is normal in A_4, and so we have to work a little harder to force more elements in H.

We compute some of the conjugates of (abc), being careful to conjugate only by elements of A_4 only:

$$(abd)(abc)(abd)^{-1} = (bdc); .$$
$$(bad)(abc)(bad)^{-1} = (dac);$$
$$(bdc)(abc)(bdc)^{-1} = (adb).$$

Thus H must contain $\{(abc), (bdc), (dac), (adb)\}$ as well as the identity (1). Therefore, five elements are already accounted for. Also, since H is a subgroup, it must contain the inverse of each of its elements, and so must contain:

$$(abc)^{-1} = (acb),\ (bdc)^{-1} = (bcd),\ (dac)^{-1} = (dca),\ (adb)^{-1} = (abd).$$

But this means that H must contain at least nine elements, which is absurd since the order of H is only six. Therefore, H does not exist. ∎

(See Problem 19 for another proof).

7.6 Exercises

1) Follow the steps in the proof of Theorem 7.2.1 in the particular case that the group whose representation is being computed is a cyclic group of order 5.

2) Compute the order and parity of the following permutations:

    (i)   (347)(2315)(4735)(219847);

    (ii) (12)(153)(243)(36);

    (iii) (1357)(2463)(1234567);

    (iv) (123)(1342)(5461)(136).

3) Express each of the permutations in 2) as products of transpositions in two different ways.

4) Draw up a table for S_5 similar to the one in Example 7.4.12.

5) Let m and n be relatively prime positive integers.

    (i)   If $1 = rm + sn$, prove that $\gcd(r,n) = \gcd(s,n) = 1$.

    (ii) If b is an element of a group of order mn where m and n are as above, prove that $|b^{rm}| = n$ and $|b^{sn}| = m$. Hence prove that b is expressible as a product of an element of order m and an element of order n.

    (iii)Express (123456) as a product of an element of order 2 and an element of order 3 and write each of these elements as a product of disjoint cycles.

6) If π is a permutation such that

    (i) $\pi(1234)(56) = (3726)(14)$, express π as a product of disjoint cycles;

    (ii) $\pi(1234)(56)\pi^{-1} = (3726)(14)$, express π as a product of disjoint cycles. (There is more than one answer.)

7) In S_{13} find an element of order 42 and one which moves all symbols and is of order 20.

8) Prove that the smallest subgroup of S_n containing (12) and (123....n) is S_n itself. [Hint: Show that any subgroup containing the two given elements must contain all transpositions.]

9) If σ and τ are transpositions, show that $\sigma\tau$ can be expressed as a product of (not necessarily disjoint) 3-cycles.

10) Let G be a finite group and g an element of G. Recall (see Problem 9, Chapter 5) that the centralizer of g in G is the subgroup

$$C_G(g) = \{x \in G \mid xg = gx\}.$$

To simplify notation we shall denote this subgroup by C.

Prove that the number of elements in the conjugacy class $K(g)$ of g in G is $\left| G:C_G(g) \right|$, the index of $C_G(g)$ in G.

[Hint: Define a map φ from the set of left cosets of C in G to $K(g)$ by $\varphi(xC) = xgx^{-1}$. First show that φ is well defined. This has to be done since it appears that the image of a coset under φ depends on the coset representative. Then show that the map φ is bijective.]

11) What is the order of:

(i) $C_{S_5}((12)(34))$?

(ii) $C_{S_n}((123...n))$?

12) Prove that the conjugacy class of (12345) in A_5 is one half the size of the conjugacy class of (12345) in S_5 but that the conjugacy class of (123) in A_5 is of the same size as the conjugacy class of (123) in S_5.

13) Prove that if $n > 2$, then every element of A_n is expressible as a product of 3-cycles.

14) (Hard!) Prove that if H is a nontrivial normal subgroup of A_5, then $H = A_5$.

15) Prove $Z(S_n) = \langle (1) \rangle$. (See Problem 13, Chapter 5).

16) Prove that if τ_1, τ_2, τ_3 are transpositions, then $\tau_1 \tau_2 \tau_3 \neq (1)$.

17) Prove that if τ_1, τ_2 are distinct transpositions, then the order of $\tau_1 \tau_2$ is 2 or 3.

18) If the permutation $\begin{pmatrix} 1 & 2 & 3 & 4 & 5 & 6 & 7 & 8 & 9 \\ 3 & 1 & 2 & ? & ? & 7 & 8 & 9 & 6 \end{pmatrix}$ in S_9 is even, what must the images of 4 and 5 be?

19) Give another proof of the fact that A_4 has no subgroup of order 6 as follows:

    (i) If H is a subgroup of A_4 of order 6, then H is a normal subgroup of A_4;

    (ii) Show that for each $g \in A_4$, $g^2 \in H$;

    (iii) Prove then that H would have too many elements.

20) If $\alpha = (13579)(246)(8,10)$ and α^m is a 5-cycle, what can you say about m?

21) Describe the conjugacy classes of A_6 and compute the number of elements in each class.

22) Show that a permutation of odd order must be of even parity.

23) If $\alpha \in S_{13}$ and $\alpha^2 = (1,10,11,6,3,12,2,9,5,13,7,4,8)$ what is α?

24) Show that if H is a subgroup of S_n, then either every element of H is an even permutation or exactly half of them are even.

25) If σ and τ are permutations, prove that $\sigma\tau$ and $\tau\sigma$ have the same cycle structure.

27) Does the symmetric group S_7 contain an element of

    (i) order 5? order 10? order 15?

    (ii) What is the largest possible order of an element of S_7?

28) A deck of cards numbered 1 to $2n$ is in the order 1, 2,...,$2n$. A "perfect" shuffle is performed so that the cards end up in the order $n+1, 1, n+2, 2, n+3, 3,..., 2n, n$. How many perfect shuffles must be performed to return the cards to their initial order?

Chapter 8

Rings, Integral Domains and Fields

8.1 Rings

The set of all n-dimensional square matrices with real entries is a typical example of a ring. Let us distill the essence of this system.

We have a set of elements, namely, the matrices, with two operations which we call matrix addition and matrix multiplication, denoted by + and juxtaposition, respectively. Under addition, the matrices form an abelian group and under multiplication, a semigroup. Moreover, these two operations are linked by distributivity. These are the properties which we extract and call any system possessing them a ring. More formally, we have:

8.1.1 *Definition*

A *ring* is a set S with two binary operations, addition and multiplication, denoted by + and juxtaposition, respectively, and satisfying the following:

(i) $(S,+)$ is an abelian group with identity denoted by 0 and called the *zero* of the ring;

(ii) $(S,.)$ is a semigroup;

(iii) for all $a,b,c \in S$, $a(b+c) = ab + ac$ and $(b+c)a = ba + ca$.

These are called, respectively, the left and right distributivity laws.

Remarks.

(i) We have not insisted that a ring possess a multiplicative identity. When it does, we shall denote that identity by 1 and postulate $1 \neq 0$

187

(what if $1 = 0$?). We shall refer to the ring as a *ring with unity*. To distinguish the multiplicative identity from the additive identity, we call the former the *unity* of the ring.

(ii) $S = \{0\}$ with $0 + 0 = 0.0 = 0$ is a ring called the *trivial ring*.

(iii) When multiplication is commutative, we call the ring a *commutative ring*.

As in past chapters, we give a number of examples which the reader should work through and refer to whenever necessary.

8.1.2 *Examples*

(i) The integers under ordinary addition and multiplication form a commutative ring with unity which we denote by $(\mathbf{Z}, +, .)$.

(ii) The set $M_n(\mathbf{R})$ of all square matrices of dimension n over the reals under ordinary matrix addition and ordinary multiplication is a (noncommutative) ring with unity.

(iii) The set of even integers under ordinary addition and multiplication is a commutative ring (without unity).

(iv) The set $\mathbf{R}^\mathbf{R}$ of all maps from the reals to the reals with operations defined as follows:

Addition: for all f, $g \in \mathbf{R}^\mathbf{R}$ and for all $x \in \mathbf{R}$

$$(f + g)(x) = f(x) + g(x)$$

Multiplication: for all $f, g \in \mathbf{R}^\mathbf{R}$ and for all $x \in \mathbf{R}$

$$(fg)(x) = f(x)g(x)$$

This is a very large commutative ring with unity. The zero of this ring is the *zero map* $\mathbf{O}$ defined by $\mathbf{O}(x) = 0$ for all real numbers x and the unity is the map $\mathbf{1}$ which maps every real number to 1.

(v) The reals and the rationals under the ordinary operations of addition and multiplication form commutative rings with unity.

(vi) $(\mathbf{Z}_n, +, .)$ is a finite commutative ring with unity.

(vii) Let $(A, +)$ be a nontrivial abelian group and define a multiplication on A by $ab = 0$ for all a and b in A, where 0 is the identity of $(A, +)$. This is a ring with *trivial multiplication*. It clearly has no unity.

(viii) Let $(A,+)$ be an abelian group and let $\text{End}(A)$ denote the set of endomorphisms of A. On $\text{End}(A)$ define $+$ and $.$ as follows:

Addition: for all $f,g \in \text{End}(A)$ and for all $x \in A$

$$(f+g)(x) = f(x) + g(x).$$

Multiplication: For all $f,g \in \text{End}(A)$ and for all $x \in A$

$$(fg)(x) = (f \circ g)(x).$$

Notice that addition is defined as in (iv) but that multiplication is composition of maps in this case rather than "pointwise multiplication of maps" as in (iv). If in (iv) we had defined multiplication to be composition of maps, we would have constructed a system which would have satisfied all the axioms of a ring except for left distributivity.

The ring of endomorphisms of an abelian group is, in general, a noncommutative ring. It always has a unity, namely, the identity map.

(ix) (See Example 8.4.7). Given a set T, define on $P(T)$, the power set of T, the operations $+$ and $.$ by:

Addition: for all $X,Y \in P(T)$

$$X+Y = (X \cap Y') \cup (X' \cap Y).$$

Multiplication: for all $X,Y \in P(T)$

$$X.Y = X \cap Y.$$

This is a commutative ring with unity. Its identity is ϕ, its unity T.

(x) The set of all square upper triangular matrices of dimension n under the usual operations is a noncommutative ring with unity (a matrix (a_{ij}) is *upper triangular* if $a_{ij} = 0$ whenever $i > j$. If we require $a_{ij} = 0$ for all $i \geq j$,, then the matrix is *strictly upper triangular*). The set of strictly upper triangular matrices under the usual operations is also a ring, but without unity. ∎

Let us bear in mind that, since a ring is an abelian group under addition, any property we have already proved in the previous chapters about abelian groups will hold for the additive structure of the ring. The notation, however, is additive, whereas the theorems we proved in previous chapters are couched in multiplicative notation. Similarly, any theorem we have established for semigroups will be valid for the multiplicative structure of the ring.

We gather together some of the elementary properties of rings in the following.

8.1.3 *Theorem*

In a ring S the following hold:
   (i) For all $s \in S$, $s.0 = 0.s = 0$;
   (ii) For all $s, t \in S$, $s.(-t) = (-s).t = -(s.t)$;
   (iii) $-(-s) = s$;
   (iv) For all $s, t \in S$, $(-s)(-t) = st$.

Proof.
   (i) $s.0 = s.(0+0) = s.0 + s.0$ by left distributivity. Since the only idempotent in a group is the identity, it follows that $s.0 = 0$. Similarly, $0.s = 0$.
   (ii) $s.0 = s.(t + (-t)) = s.t + s.(-t)$, using distributivity. By (i), $s.0 = 0$ and so $s.t + s.(-t) = 0$. Adding $-(st)$ to each side of the last equation, we get the result.
   (iii) This is the additive equivalent of $(g^{-1})^{-1} = g$ (see Theorem 5.2.3(v)).
   (iv) By (ii) and (iii), $(-s)(-t) = -((-s).t) = -(-(s.t)) = st$.∎

Observe that (ii) and (iv) are the usual rules for multiplying signed real numbers and (i) is the "obvious" fact that zero times a real number is zero.

Remark. We shall write $a - b$ instead of $a + (-b)$ and call "—" "subtraction". It is an easy exercise to show that multiplication is distributive over subtraction. Note, however, that subtraction is not an associative operation.

From this point on, the reader will find many theorems and definitions reminiscent of corresponding theorems and definitions we have come across in our treatment of groups. For example, instead of subgroups, we shall be talking of subrings; instead of group homomorphisms, ring homomorphisms; instead of normal subgroups, ideals, etc., but fundamentally, the ideas and methods are very similar.

8.1.4 *Definition*

A subset T of a ring S is a *subring* of S if it is a ring in its own right under the operations inherited from S.

8.1.5 *Examples*

(i) $\{0\}$ and S are subrings of S.

(ii) The ring of integers is a subring of the ring of rational numbers, which, in turn, is a subring of the ring of real numbers.

(iii) The set of even integers is a subring of the ring of integers.

(iv) The ring $M_n(\mathbf{R})$ contains the subrings of upper triangular and strictly upper triangular matrices. ∎

The following theorem is the analogue of Theorem 5.4.5 for groups.

8.1.6 *Theorem*

A subset T of a ring S is a subring of S if, and only if,
  (i) $T \neq \varphi$;
  (ii) $s,t \in T$ implies $s - t \in T$;
  (iii) $s,t \in T$ implies $st \in T$.

Proof. If T is a subring, conditions (i) and (ii) hold since $(T,+)$ must be a subgroup of $(S,+)$ (see 5.4.5). Condition (iii) must also hold since T is closed under multiplication.

Conversely, by Theorem 5.4.5, conditions (i) and (ii) ensure that $(T,+)$ is a subgroup of $(S,+)$ and (iii) ensures that T is closed under multiplication, establishing that $(T,.)$ is a semigroup. As with groups, associativity of multiplication and all other properties of a ring hold in T since they hold in the big ring S. ∎

8.1.7 *Example*

Let us prove, using 8.1.6, that $T = \{m + n\sqrt{2} \mid m,n \in \mathbf{Z}\}$ is a subring of $(\mathbf{R},+,.)$.

  (i) $T \neq \varphi$ since $1 \in T$.

(ii) Suppose $m + n\sqrt{2}, \ p + q\sqrt{2} \in T$. Then
$$(m + n\sqrt{2}) - (p + q\sqrt{2}) = (m - p) + (n - q)\sqrt{2} \in T$$
since $m - p, n - q \in \mathbf{Z}$.

(iii) Suppose $m + n\sqrt{2}, \ p + q\sqrt{2} \in T$. Then
$$(m + n\sqrt{2})(p + q\sqrt{2}) = mp + 2nq + (mq + np)\sqrt{2} \in T$$
since $mp + 2nq, \ mq + np \in \mathbf{Z}$. ∎

In 4.5.13 we defined the concepts of zero divisors and units in $\mathbf{Z}_n$. We repeat the same definitions for arbitrary rings.

8.1.8 *Definition*

(i) Let $(S, +, .)$ be a ring with unity. An element s of S is a *unit* if s has a multiplicative inverse in S.

(ii) An element a of a *commutative* ring S is a *zero divisor* if there is a nonzero element b such that $ab = 0$.

Remark. We can define zero divisors in noncommutative rings, but then we would have to distinguish between left and right zero divisors, complicating matters beyond the usefulness of the concept in a first course in abstract algebra.

8.1.9 *Examples*

(i) The unity of a ring is always a unit. Sometimes it is the only one. For example, T is the only unit of the ring $(P(T), +, .)$ of 8.1.2(ix). In the same ring, every element except for T is a zero divisor.

(ii) The ring of integers has no nonzero zero divisors and the only units are $1, -1$.

(iii) In the ring $M_n(\mathbf{R})$ of matrices over $\mathbf{R}$, the set of units is the general linear group of dimension n. (See 5.2.4(v).)∎

8.1.10 *Theorem*

In a ring with unity, the set of units forms a group under the multiplicative operation of the ring. This group is called the *group of units of the ring*.

Proof. Let $U(S)$ denote the set of units of the ring S. Since multiplication is associative in a ring, we need only show that $U(S)$ is not empty and that it is closed under multiplication and the taking of inverses.

Clearly, $1 \in U(S)$ and so $U(S) \neq \phi$.

Suppose now that $a, b \in U(S)$. Since $abb^{-1}a^{-1} = 1 = b^{-1}a^{-1}ab$, it follows that $ab \in U(S)$.

Lastly, the equation $aa^{-1} = a^{-1}a = 1$ says that if a is a unit then so is a^{-1} with a as its inverse. ∎

8.1.11 *Example*

(i) Consider the ring T of 2×2 upper triangular matrices with real entries. The matrix $\begin{bmatrix} a & b \\ 0 & c \end{bmatrix}$ is a unit of this ring if, and only if, $ac \neq 0$.

Moreover, $\begin{bmatrix} a & b \\ 0 & c \end{bmatrix}^{-1} = \begin{bmatrix} a^{-1} & -a^{-1}c^{-1}b \\ 0 & c^{-1} \end{bmatrix}$.

(ii) The group of units of the ring $\mathbf{R}$ of real numbers consists of $\mathbf{R}^{*}$, the set of nonzero real numbers. ∎

In the ring of integers we can cancel nonzero integers, that is, if $ab = ac$, then we can deduce that $b = c$ provided $a \neq 0$. This property occurs frequently enough that we give it a name.

8.1.12 *Definition*

A commutative ring S is said to be *cancellative* if for all $a, b, c \in S$, $ab = ac$ and $a \neq 0$ imply $b = c$.

Remark. We have opted to define cancellativity only for commutative rings. We could have defined left and right cancellativity for noncommutative rings, and indeed, this is done in some texts. However, we shall only have occasion to use the concept in the context of commutative rings.

8.1.13 *Theorem*

A commutative ring S is cancellative if, and only if, S contains no nonzero zero divisors.

Proof. Suppose S is cancellative and let a be a zero divisor. Then there exists a nonzero element b in S such that $ab = 0$. Hence $ab = 0b$ and cancelling the nonzero element b, we get $a = 0$.

Conversely, let $ab = ac$ with $a \neq 0$. Then, $ab - ac = 0$ and so $a(b - c) = 0$. Since the only zero divisor in S is 0 and $a \neq 0$, it follows that $b - c = 0$. Therefore, $b = c$, proving that S is cancellative. ∎

8.2 Homomorphisms, Isomorphisms and Ideals

The next few definitions and theorems should give the reader a feeling of "déjà vu". They should call to mind analogous definitions and theorems (s)he has already come across in our treatment of groups.

8.2.1 *Definition*

Let S and T be rings. A mapping $f : S \to T$ is a *(ring) homomorphism* if:

(i) For all $x, y \in S$, $f(x + y) = f(x) + f(y)$.
(ii) For all $x, y \in S$, $f(xy) = f(x)f(y)$.
(iii) If both S and T are rings with unity, then $f(1_S) = 1_T$, where 1_S and 1_T are the unities of S and T, respectively.

A 1:1 ring homomorphism is a (ring) *monomorphism*; an onto ring homomorphism is a (ring) *epimorphism;* a 1:1 onto ring homomorphism is a (ring) *isomorphism*. As with groups, we write $S \cong T$ if there is an isomorphism between S and T and we say that the two rings are isomorphic.

Remarks.

(i) A ring homomorphism f is, in particular, a group homomorphism from $(S,+)$ to $(T,+)$. Therefore, all the theorems we have proved concerning group homomorphisms are valid for the additive structure of a ring. In particular, the kernel of f is an additive subgroup of S.

(ii) Although we have denoted the operations in the two rings by identical symbols, they are, in fact, different in general.

(iii) Under a ring homomorphism, the additive identity of S maps to the additive identity of T by the first remark above. It does not follow, however, from (i) and (ii) of the definition of a ring homomorphism, that the unity of S necessarily maps to the unity of T. Therefore, we stipulate in (iii) that if each ring has a unity, then f maps the one to the other.

(iv) We usually omit the qualifying adjective "ring" when speaking of homomorphisms, isomorphisms, etc. when it is clear that we are talking about rings.

(v) Just as with groups, $\cong$ is an equivalence relation on the class of all rings whose equivalence classes are called *isomorphism classes of rings*. From an algebraic standpoint, two rings from the same isomorphism class are indistinguishable.

Almost every theorem we proved for groups has a corresponding theorem for rings.We state the analogue of Theorem 6.3.8 and leave it up to the reader to supply proofs for most of the assertions.

8.2.2 *Theorem*

Let $f : S \to T$ be a (ring) homomorphism. Then:

(i) $f(0_S) = 0_T$ where 0_S and 0_T are the zeroes of S and T, respectively.

(ii) $f(nx) = nf(x)$ for all $x \in S, n \in \mathbf{Z}$.

(iii) If B is a subring of S, then $f(B)$ is a subring of S.

(iv) If C is a subring of T, then $f^{-1}(C)$ is a subring of S.

(v) $f(x^n) = f(x)^n$ for all $x \in S, n \in \mathbf{Z}^+$.

(vi) If both rings have unity and x is a unit in S, then $f(x)$ is a unit in T and $f(x^n) = f(x)^n$ for all $n \in \mathbf{Z}$.

Proof. We prove only (vi). Recall that, built into the definition of a ring homomorphism is the requirement that the unity of S map to the unity of T.

Suppose that x is a unit in S. Then.

$$f(x)f(x^{-1}) = f(xx^{-1}) = f(1_S)$$
$$= f(x^{-1}x) = f(x^{-1})f(x) = 1_T.$$

Therefore, $f(x)$ is a unit and $f(x)^{-1} = f(x^{-1})$.

To prove the second part of (vi), we observe that, because of (v) we need only show that $f(x^n) = f(x)^n$ for all $n \in \mathbf{Z}, n \leq 0$. For $n = -1$ the result was established above and for $n = 0$, the result is a consequence of the definition of x^0. If m is positive, then

$$f(x^{-m}) = f((x^{-1})^m) = f(x^{-1})^m$$

(the second equality by Part (v)). Since we have just proved that $f(x^{-1}) = f(x)^{-1}$, $f(x^{-m}) = [f(x)^{-1}]^m = f(x)^{-m}$. ∎

Normal subgroups are singled out among all subgroups of a group because they can be used to form factor groups which, by the First Isomorphism Theorem, turn out to be homomorphic images of the group.

In a similar manner, ideals are singled out among all subrings of a ring for the same reason.

8.2.3 *Definition*

A subring I of a ring S is an *ideal* (of S) if for all $s \in S$ and for all $x \in I$, $sx, xs \in I$.

Remarks. Since an ideal is, in particular, a subring, to show that I is an ideal of S we use the criteria established in Theorem 8.1.6, except that, instead of just showing closure under multiplication, we show the product of an element of I with any element of S is in I. Therefore to prove I is an ideal of S we show:

(i) $I \neq \phi$;

(ii) if $x, y \in I$, then $x - y \in I$;

(iii) if $s \in S$ and $x \in I$, then $sx \in I$ and $xs \in I$.

Think of an ideal as a "black hole" which draws into it any product of an element of the ideal with an element of the ring.

The following is the analogue of Theorem 6.4.6. Again, the proof (except for (i)) is left to the reader since it is entirely similar to that given in 6.4.6.

8.2.4 *Theorem*

Let $f : S \rightarrow T$ be a ring homomorphism. Then:

(i) $\ker f = f^{-1}(0_T)$ is an ideal of S.

(ii) f is 1:1 iff $\ker f = \{0_S\}$.

(iii) For all $s \in S$, $f^{-1}(f(s)) = s + \ker f$.

Proof. (i) We know from Theorem 6.4.6 that $\ker f$ is an additive subgroup of S. We therefore only need to prove the "black hole" property.

Let $s \in S, x \in \ker f$. Then $f(sx) = f(s)f(x) = f(s).0_T = 0_T$ and so $sx \in \ker f$. Similarly, $xs \in \ker f$. ∎

The analogue of Theorem 6.4.7 follows.

8.2.5 *Theorem*

Let S be a ring and I an ideal of S. Then I is a normal subgroup of $(S,+)$ since $(S,+)$ is abelian (see Example 6.4.3). Form the factor group $(S/_I, +)$ (see Theorem 6.4.7). On $S/_I$ define an operation denoted by . (or juxtaposition) as follows:

$$(a + I).(b + I) = ab + I \quad \text{for all } a + I, b + I \in S/_I.$$

Then $(S/_I, +, .)$ is a ring called the *factor ring of S mod I*. Moreover, the map $v: S \to S/_I$ defined by $v(s) = s + I$ is a (ring) epimorphism (called the *natural homomorphism*) whose kernel is I.

Proof. By Theorem 6.4.7, $(S/_I, +)$ is a group. That it is abelian follows from Theorem 6.3.8 and the fact that $(S/_I, +)$ is a homomorphic image of the abelian group $(S, +)$.

We now examine the multiplicative structure we have imposed on $S/_I$. First, we must show that the operation is well defined, since it appears that the multiplication of cosets depends on the representatives chosen. We show, therefore, that if
$$a + I = a' + I \text{ and } b + I = b' + I,$$
then $ab + I = a'b' + I$.

It is here that we have to use the special property of an ideal. We need to show that $ab - a'b' \in I$.

Since $\quad a + I = a' + I$ and $b + I = b' + I$, we have $a - a' \in I$ $\quad$ and $b - b' \in I$. Hence
$$ab - a'b' = ab - a'b + a'b - a'b'$$
$$= (a - a')b + a'(b - b') \in I$$
since $(a - a')b, a'(b - b') \in I$ because I is an ideal. Therefore, multiplication is well defined.

We simplify the notation by writing $\bar{x}$ instead $x + I$ and prove that $(S/_I, .)$ is a semigroup. Let $\bar{a}, \bar{b}, \bar{c} \in S/_I$. Then, by definition,

$$(\bar{a}\bar{b}).\bar{c} = \overline{ab}.\bar{c} = \overline{(ab)c} = \overline{a(bc)} = \bar{a}.\overline{(bc)} = \bar{a}.(\bar{b}.\bar{c}).$$

The third equality follows because multiplication is associative in S.

As for distributivity, we have

$$\bar{a}.(\bar{b} + \bar{c}) = \bar{a}.\overline{(b + c)} = \overline{a(b + c)} = \overline{ab + ac} = \bar{a}\bar{b} + \bar{a}.\bar{c}.$$

We have used left distributivity in S at the third equality. In a similar manner, we can prove right distributivity in $S/_I$ by appealing to right distributivity in S. Therefore, $(S/_I, +, .)$ is a ring.

That v is a homomorphism follows from

$$v(s+t) = \overline{s+t} = \overline{s} + \overline{t} = v(s) + v(t) \text{ and } v(st) = \overline{st} = \overline{s}.\overline{t} = v(s)v(t).$$

In case S has unity 1_S, then $v(1_S) = \overline{1_S}$ which is the unity of $S\!/_I$. Moreover, since for any given $\overline{s} \in S\!/_I$, we have $v(s) = \overline{s}$, , it follows that v is onto. ■

8.2.6 *Examples*

(i) In the ring $\mathbf{Z}$ of integers, all additive subgroups are of the form $\mathbf{Z}n$ for some nonnegative integer n. It so happens that in this ring every additive subgroup is an ideal (prove) and so we can form the ring $(\mathbf{Z}\!/_{\mathbf{Z}n}, +, .)$. By applying Theorem 8.3.1 below the reader should have little difficulty in proving that this ring is $(\mathbf{Z}_n, +, .)$.

(ii) In the ring T of upper triangular matrices of dimension 3, the set I of strictly upper triangular matrices is an ideal and $T\!/_I$ is a commutative ring although T itself is not commutative. ■

8.3 Isomorphism Theorems

The two theorems proved in this section are the analogues of Theorems 6.4.8 and 6.4.10.

8.3.1 *Theorem (First Isomorphism Theorem for Rings)*

Let $f : S \to T$ be a ring epimorphism with kernel J. Then $S\!/_J \cong T$ via the map $\tau : S\!/_J \to T$ defined by $\tau(s+I) = f(s)$.

Proof. The proofs that τ is well defined and a group isomorphism are, verbatim, the additive versions of 6.4.8. That τ preserves multiplication follows from the same property that f possesses. ■

8.3.2 *Corollary*

Let $f : S \to T$ be a homomorphism. Then $S\big/_{\ker f} \cong f(S)$.

Proof. The given map is onto $f(S)$. Apply the First Isomorphism Theorem. ∎

8.3.3 *Lemma*

If R is a ring, S a subring and J an ideal of R, then $S + J = \{s + j \mid s \in S \text{ and } j \in J\}$ is a subring of R. (compare with Lemma 6.4.9).

Proof. Closure under subtraction is easy to verify and is left to the reader. Indeed, closure under subtraction holds even in the case that J is merely a subring.

Closure under multiplication, however depends on J being an ideal as we can see from:

$$(s + j).(s' + j') = ss' + sj' + js' + jj'$$

where $ss' \in S$ by virtue of S being a subring and

$$sj' + js' + jj' \in J$$

since J is an ideal. ∎

8.3.4 *Theorem (Second Isomorphism Theorem)*

Let R, S and J be as in the lemma. Then

$$(S+J)\big/_J \cong S\big/_{S \cap J}.$$

(Compare with Theorem 6.4.10.)

Proof. Exactly as in the proof of Theorem 6.4.10, define $\theta : S \to (S+J)\big/_J$ by

$$\theta(s) = s + J = \bar{s}.$$

We have already proved that this is a group homomorphism in Theorem 6.4.10. This map also satisfies $\theta(st) = \overline{st} = \overline{s}.\overline{t}$ and so is a ring homomorphism. The kernel is $S \cap J$ and applying the First Isomorphism Theorem for rings, the result follows. ∎

The Third Isomorphism Theorem for rings is almost identical to the one for groups (see 6.4.11) and we leave the proof to the reader.

8.3.5 *Theorem (Third Isomorphism Theorem)*

If I and J are ideals of a ring R with $I \leq J$, then

$$(R/I)/(J/I) \cong R/J.$$

Proof. Exercise. ∎

8.4 Direct Sums of Rings

The reader should compare this section with Section 6.5 where (s)he will find the analogous definitions and theorems for groups.

8.4.1 *Definition*

Let $S_j, j = 1, 2, ..., n$ be rings. The *direct sum* of these rings, denoted $S_1 \oplus S_2 \oplus ..... \oplus S_n$, or more succinctly $\bigoplus_{i=1}^{n} S_i$, consists of the set of all ordered n-tuples $(s_1, s_2, ..., s_n)$ where $s_j \in S_j$, $j = 1, 2, ..., n$. The operations are defined by:

$$(s_1, s_2, ..., s_n) + (t_1, t_2, ..., t_n) = (s_1 + t_1, s_2 + t_2, ..., s_n + t_n);$$
$$(s_1, s_2, ..., s_n)(t_1, t_2, ..., t_n) = (s_1 t_1, s_2 t_2, ..., s_n t_n).$$

It is an easy exercise to verify that $\bigoplus_{i=1}^{n} S_i$ is a ring with $(0_{S_1}, 0_{S_2}, ..., 0_{S_n})$ as the zero of the ring and $(1_{S_1}, 1_{S_2}, ..., 1_{S_n})$ as the unity, provided that each S_i

has a unity 1_{S_i}. The other properties hold as a result of their holding in each S_i.

8.4.2 *Theorem*

If $S = \bigoplus_{i=1}^{n} S_i$ and $B = \{i_1, i_2, ..., i_k\}$ is a subset of $P = \{1, 2, ..., n\}$, then the map

$$\pi_B : S \to \bigoplus_{i \in B} S_i$$

defined by

$$\pi(s_1, ... s_{i_1}, ... s_{i_2}, ... s_{i_k}, ..., s_n) = (s_{i_1}, s_{i_2}, s_{i_3}, ..., s_{i_k})$$

is a homomorphism whose kernel consists of those n-tuples with 0 in the i_j coordinate position, $j = 1, 2, ..., k$. Moreover, by an abuse of notation, we have, by the First Isomorphism Theorem,

$$\bigoplus_{i=1}^{n} S_i \Big/ \bigoplus_{i \in B} S_i \cong \bigoplus_{i=1}^{t} S_{j_i}$$

where $\{j_1, j_2, ..., j_t\}$ is the complement of B.

Proof. Exercise. ∎

The notation gets in the way of this result, so here is a simple example which should clarify matters.

8.4.3 *Example*

Suppose $S = S_1 \oplus S_2 \oplus S_3 \oplus S_4$ and $B = \{1, 3\}$. Then $\pi_B(s_1, s_2\ s_3, s_4) = (s_1, s_3)$ and $\ker \pi_B = \{(0, s_2, 0, s_4) \mid s_2 \in S_2 \text{ and } s_4 \in S_4\}$. ∎

We have the analogue for rings of Theorems 6.5.2 and 6.5.3. We state the results for rings, leaving the details to the reader.

8.4.4 *Theorem*

If $S = \bigoplus_{i=1}^{n} S_i$ then $\hat{S}_i = \{\hat{s}_i = (0,0,...,s_i,0,0,...,0) \mid s_i \in S_i\}$ is an ideal of S.

Moreover, the following hold:

(i) $S = \hat{S}_1 + \hat{S}_2 + ... + \hat{S}_n = \{\hat{s}_1 + \hat{s}_2 + ... + \hat{s}_n \mid \hat{s}_i \in \hat{S}_i \text{ for all } i, 1 \le i \le n\}$.

(ii) Each element of S is uniquely expressible in the form $\hat{s}_1 + \hat{s}_2 + ... + \hat{s}_n$.

(iii) $\hat{S}_j \cap (\hat{S}_1 + \hat{S}_2 + ... \hat{S}_{j-1} + \hat{S}_{j+1} + ... + \hat{S}_n) = \{(0,0,...,0)\}$

for all $j, 1 \le j \le n$.

Proof. Exercise. ∎

Remark. The three conditions above are not independent since (i) and (ii) together imply (iii) and (i) and (iii) together imply (ii) (Proof?).

The next result is the converse of the preceding theorem in the sense that if a ring S has ideals S_i satisfying properties (i) and (ii) or (i) and (iii), then S is said to be the (internal) direct sum of the S_i. This ring is isomorphic to the (external) direct sum of the S_i. Because of this isomorphism, we do not distinguish between internal and external direct sums (see 6.5.4). Note that in the previous theorem, S is the (internal) direct sum of the $\hat{S}_i$.

In either case, we write $S = \bigoplus_{i=1}^{n} S_i$.

8.4.5 *Theorem*

Let S be a ring with ideals $S_i, 1 \le i \le n$ satisfying:

(i) $S = S_1 + S_2 + ... + S_n$;

(ii) $S_j \cap (S_1 + S_2 + ... + S_{j-1} + S_{j+1} + ... + S_n) = \{0\}$ for all $j, 1 \le j \le n$.

Then each element s of S is uniquely expressible in the form $s = s_1 + s_2 + ... + s_n$ where $s_i \in S_i$ for all $i, 1 \le i \le n$.

Proof. Suppose $s = s_1 + s_2 + ... + s_n = t_1 + t_2 + ... + t_n$. Then

$$s_j - t_j = \sum_{i \neq j} (t_i - s_i) \in S_j \cap \sum_{i \neq j} S_i = (0)$$

by (ii). Therefore $s_j = t_j$ for all relevant j, proving the uniqueness of the representation.

The isomorphism

$$(s_1, s_2, ..., s_n) \to s_1 + s_2 + ... + s_n$$

from the external direct sum of the S_i's to S establishes that the two rings are isomorphic. ∎

We now give a series of examples which the reader should work through carefully.

8.4.6 *Example*

Consider a direct sum of n copies of the ring $(\mathbf{Z}_2, +, .)$ and omit bars over the elements. To further simplify the notation, write $1001...1$ instead of $(1,0,0,1,...,1)$. We denote this ring by $\mathbf{Z}_2^n$.

Let us try to find all the ideals of this ring. Suppose $J \neq \{0\}$ is an ideal and let α be an element of J. Call the coordinate positions where the entry is 1 the support of the element, denoted $\mathrm{supp}(\alpha)$. For example, $\mathrm{supp}(11010011) = \{1, 2, 4, 7, 8\}$.

Let $\mathrm{supp}(\alpha) = B$. If ε_i is the element with 1 in the ith coordinate position and 0's elsewhere, then
$$\varepsilon_i \in J \text{ for all } i \in B \text{ since } \varepsilon_i.\alpha \in J \text{ for all } i.$$
Hence there is a subset C of $\{1, 2, 3, ..., n\}$ such that
$$J = \{\xi \mid \mathrm{supp}(\xi) \subseteq C\}.$$

If C is a subset of size k, then, as rings, $J \cong \mathbf{Z}_2^k$ and $\mathbf{Z}_2^n \big/ J \cong \mathbf{Z}_2^{n-k}$. Therefore every homomorphic image of $\mathbf{Z}_2^n$ is $\mathbf{Z}_2^m$ for some $m \leq n$. ∎

8.4.7 *Example*

(See Example 8.1.2(ix)) Let $T = \{1, 2, 3, ..., n\}$ and define two operations, $+$ and $.$ on the power set $P(T)$ as follows:

For all A and $B \in P(T)$, $A + B = A \cup B \setminus A \cap B$ and $A.B = A \cap B$.

Define a map $\varphi : \mathbf{Z}_2^n \to (P(T), +, .)$. by

$$\varphi(\alpha) = A \text{ where } A = \operatorname{supp}(\alpha).$$

We claim that φ has the following properties:

  (i) φ is bijective;

  (ii) $\varphi(\alpha + \beta) = \varphi(\alpha) + \varphi(\beta)$ for all α and β in $\mathbf{Z}_2^n$;

  (iii) $\varphi(\alpha.\beta) = \varphi(\alpha)\varphi(\beta)$;

  (iv) $\varphi(111...1) = T$.

(Observe that $111...1$ is the unity of $\mathbf{Z}_2^n$ and T is the unity of $(P(T), +, .)$.

The reader may wonder why we do not just say that φ is an isomorphism. If we did, then it would imply that we know that $(P(T), +, .)$ is a ring.

Let us assume that we don't know this. We show that once we establish (i), (ii) and (iii) above, we shall be able to claim that $(P(T), +, .)$ is a ring which is isomorphic to $\mathbf{Z}_2^n$.

This would be one way of proving that $(P(T), +, .)$ is a ring. There are other ways, but all, at some point, require us to prove the associativity of addition and this leads us to a messy computation with sets. Approaching the problem from this angle establishes rather neatly the associativity of addition. Of course, we are using a considerable amount of theory here.

Proof.

  (i) Suppose $\varphi(\alpha) = \varphi(\beta)$. Then $\operatorname{supp}(\alpha) = \operatorname{supp}(\beta)$. This means that α and β have 1's in exactly the same coordinate positions and therefore 0's in the same coordinate positions. Hence $\alpha = \beta$ and so φ is 1:1.

  (ii) $\operatorname{supp}(\alpha + \beta) = \{i \mid \text{the } i^{\text{th}} \text{ coordinate of } \alpha + \beta \text{ is } 1\}$. But the ith coordinate of $\alpha + \beta$ is 1

$$\text{iff } i \in \operatorname{supp}(\alpha) \setminus \operatorname{supp}(\beta) \text{ or } i \in \operatorname{supp}(\beta) \setminus \operatorname{supp}(\alpha)$$

$$= (\operatorname{supp}(\alpha) \cup \operatorname{supp}(\beta)) \setminus (\operatorname{supp}(\alpha) \cap \operatorname{supp}(\beta)).$$

Therefore,

$$\varphi(\alpha + \beta) = \varphi(\alpha) + \varphi(\beta) \text{ for all } \alpha \text{ and } \beta \text{ in } \mathbf{Z}_2^n;$$

  (iii) $\operatorname{supp}(\alpha.\beta) = \{i \mid i \in \operatorname{supp}(\alpha) \cap \operatorname{supp}(\beta)\}$ since we only get a 1 in the product when both coordinates are 1. Hence

$$\varphi(\alpha\beta) = \varphi(\alpha).\varphi(\beta).$$

(iv) This is left to the reader.

Finally, since $|\mathbf{Z}_2^n|=2^n=|P(T)|$, it follows that φ is onto.

We can now claim that $(P(T),+,.)$ is a ring. For example, if we want to prove associativity, of addition, let A,B and C be subsets of T and let α,β and γ be the pre-images of A,B and C, respectively. Then

$$A+(B+C)=\varphi(\alpha)+(\varphi(\beta)+\varphi(\gamma))=\varphi(\alpha)+\varphi(\beta+\gamma)=\varphi(\alpha+(\beta+\gamma))$$

since φ is a homomorphism. But $\mathbf{Z}_2^n$ is a ring and so $\alpha+(\beta+\gamma)=(\alpha+\beta)+\gamma$. Hence

$$A+(B+C)=\varphi((\alpha+\beta)+\gamma)=\varphi(\alpha+\beta)+\varphi(\gamma)$$
$$=(\varphi(\alpha)+\varphi(\beta))+\varphi(\gamma)=(A+B)+C.$$

The second and third equalities are a consequence of the fact that φ is a homomorphism. ∎

The reader should attempt the proof of the associativity of $+$ in $(P(T),+,.)$ using the definition of the operation.

8.5 Integral Domains and Fields

We specialize our investigation of rings to those which share many properties with some of the more familiar workaday rings such as the integers and the real numbers. We shall see that, in particular, principal ideal rings are reminiscent of the integers.

8.5.1 *Definition*

(i) An *integral domain* is a commutative, cancellative ring with unity.

(ii) A *field* is a commutative ring with unity in which every nonzero element has a multiplicative inverse.

Remark. (i) According to 8.1.13, we could have defined an integral domain as a commutative ring with unity, containing no nonzero zero divisors.

(ii) We could also define a field as a ring with unity in which the set of nonzero elements forms an abelian group under multiplication.

8.5.2 *Theorem*

Every field is an integral domain but there are integral domains which are not fields.

Proof. We dispose of the second statement first by offering the ring of integers as an example of an integral domain which is not a field.

To prove that a field F is an integral domain, it is sufficient to show that F is cancellative. Suppose, therefore, that $a,b,c \in F, a \neq 0$ and $ab = ac$. Since $a \neq 0$, a has a multiplicative inverse a^{-1} and so, multiplying the above equation by a^{-1} we get $a^{-1}ab = a^{-1}ac$. Therefore, $b = c$. ∎

As we saw above, not every integral domain is a field. If, however, we add the hypothesis of finiteness, then we have a field. Indeed, we don't even need to hypothesize the existence of a unity.

8.5.3 *Theorem*

A finite, commutative, cancellative ring D is a field.

Proof. By 8.1.13, D has no nonzero zero divisors. Hence, D^*, the set of nonzero elements of D, is a finite, cancellative semigroup under multiplication. Therefore $(D^*,.)$ is a commutative group by Theorem 5.3.3. By Remark (ii) above, D is a field. ∎

8.5.4 *Examples*

(i) The rational, the real and the complex numbers are all fields under the usual operations of addition and multiplication.

(ii) Let d be a fixed square-free positive integer (*square-free* means that there are no repeated factors in the factorization of d as a product of primes; thus 6 is square-free but 12 is not). Then
$$D = \{m + n\sqrt{d} \mid m, n \in \mathbf{Z}\}$$
is an integral domain (which is not a field) under the usual operations.

(iii) $F = \{r + s\sqrt{d} \mid r, s \in \mathbf{Q}\}$, where d is as in (ii) and $\mathbf{Q}$ is the field of rationals, is a field.

(iv) The factor ring $\mathbf{Z}/_{\mathbf{Z}n}$ is a field if, and only if, n is prime (see Example 8.2.6(i)). ∎

In the integral domain of integers, every ideal is of the form $\mathbf{Z}n$ for some fixed integer n. In an arbitrary integral domain D, the set $Da = \{xa \mid x \in D\}$ for some $a \in D$ is also an ideal containing a. In general, however, not all ideals are of this form, as we shall soon see.

8.5.5 *Definition*

(i) An ideal of the form Sa in a commutative ring S with unity is called a *principal ideal* and is denoted by (a). The ideal is said to be *generated by* a and a is called a *generator* of the ideal.

(ii) An integral domain is a *principal ideal domain* (PID for short) if every ideal is principal. Note that $\mathbf{Z}$ is a PID.

Let us first establish that if S is a commutative ring with unity, then Sa is indeed an ideal containing a.

8.5.6 *Theorem*

In any commutative ring S with unity, the set $Sa (= aS)$ for any a in S is an ideal of S containing a.

Proof.

(i) $Sa \neq \phi$ since $1.a = a \in Sa$.

(ii) Suppose $x, y \in Sa$. Then $x = sa, y = ta$ for some $s, t \in S$. Hence $sa - ta = (s - t)a \in Sa$.

(iii) Given $s \in S, x \in Sa$ we have $x = ta$ for some $t \in S$ and so $sx = (st)a \in Sa$. Also, $xs = (ta)s = s(ta) = (st)a \in Sa$. Therefore, Sa is an ideal of S. ∎

We showed above that a finite integral domain is a field. As we prove in the next theorem, a commutative ring with unity and with no ideals other than (0) and the whole ring is also a field.

8.5.7 *Theorem*

A commutative ring S with unity is a field if, and only if, (0) and S are the only ideals of S.

Proof. If S is a field and I is a nonzero ideal, pick $a \in I, a \neq 0$. Since S is a field, a has a multiplicative inverse a^{-1}. By the "black hole" property of an ideal, $aa^{-1} = 1 \in I$. But then, if x is any element of S, by the "black hole" property again $x = x1 \in I$ and so $I = S$.

Conversely, suppose S is a ring with no ideals other than (0) and S and let $a \in S, a \neq 0$. Then $(a) \neq (0)$ and so $(a) = S$. Therefore, since $1 \in S$, there exists $x \in D$ such that $xa = 1$ and so every nonzero element of D has a multiplicative inverse. ∎

Following the lemma below is an example of an integral domain which is not a *PID*. The reader is urged to work through the details which are quite interesting in themselves.

8.5.8 *Lemma*

If n is a positive integer which is not a perfect square, then $\sqrt{n}$ is irrational.

Proof. Assume the contrary and let $\sqrt{n} = \frac{p}{q}$ where p and q are relatively prime. Then $n = \frac{p^2}{q^2}$ and so $q^2 n = p^2$.

Since p and q are relatively prime, it follows that p^2 and q^2 are also relatively prime. Hence p^2 divides n and so $n = kp^2$ for some integer k. But then, substituting in $q^2 n = p^2$ we find that, after cancelling p^2 from both sides, $q^2 k = 1$, from which we deduce that k and q are both 1. This implies that $n = p^2$, contradicting our hypothesis that n is not a perfect square. Therefore $\sqrt{n}$ is irrational. ∎

8.5.9 *Example*

If D is any integral domain and a and b are fixed elements of D, then the set $I = \{xa + yb \mid x, y \in D\}$ is easily seen to be an ideal of D denoted (a,b).

Let $D = \{m + n\sqrt{10} \mid m, n \in \mathbf{Z}\}$ (see 8.5.4(ii)) and let
$$I = (3, 1 + \sqrt{10}).$$
We claim that I is not a principal ideal.

Proof.

(i)   We   first   show   that   if $m + n\sqrt{10} = p + q\sqrt{10}$,   then $m = p$ and $n = q$.

Suppose equality holds. Then $m - p = (q - n)\sqrt{10}$. If $q \neq n$ we get $\sqrt{10} = (m - p) \big/ (q - n)$, proving that $\sqrt{10}$ is rational, a contradiction by the lemma above. Therefore, $q = n$ and, consequently, $m = p$.

(ii) Next we show that $I \neq D$.

Suppose $I = D$. Then, in particular, $1 \in I$ and   so   there   exist $x + y\sqrt{10}$, $p + q\sqrt{10} \in D$ such that
$$1 = 3(x + y\sqrt{10}) + (1 + \sqrt{10})(p + q\sqrt{10}).$$
Hence, by (i),
$$(*)\, 1 = 3x + p + 10q \text{ and } (**)\, 0 = 3y + p + q.$$
But equation (*) can be written as $1 = 3x + p + q + 9q$ and substituting $-3y$ for $p + q$ from (**), we get
$$1 = 3x - 3y + 9q.$$
Since 3 divides the right-hand side but not the left, we arrive at a contradiction. Therefore $I \neq D$.

(iii) Suppose now, by way of contradiction, that the ideal I is principal and let $I = (a + b\sqrt{10})$ where a and b are fixed integers. Then there exist integers x, y, p and q such that

$$3 = (a + b\sqrt{10})(x + y\sqrt{10}) \text{ and } 1 + \sqrt{10} = (a + b\sqrt{10})(p + q\sqrt{10}).$$

From these equations and (i), we obtain the following:

$$(\text{I}) \quad \begin{array}{l} ax + 10by = 3 \\ bx + ay = 0 \end{array}$$

$$(\text{II}) \quad \begin{array}{l} ap + 10bq = 1 \\ bp + aq = 1 \end{array}$$

By the lemma, $a^2 - 10b^2 \neq 0$ and so we can apply Cramer's rule to the first of these systems of equations to get

$$x = \frac{\det \begin{bmatrix} 3 & 10b \\ 0 & a \end{bmatrix}}{\det \begin{bmatrix} a & 10b \\ b & a \end{bmatrix}} = \frac{3a}{a^2 - 10b^2}, \quad y = \frac{\det \begin{bmatrix} a & 3 \\ b & 0 \end{bmatrix}}{\det \begin{bmatrix} a & 10b \\ b & a \end{bmatrix}} = \frac{-3b}{a^2 - 10b^2}.$$

Since x and y are integers, it follows that $(a^2 - 10b^2) | 3a$, $(a^2 - 10b^2) | 3b$ Hence,

$$3a = r(a^2 - 10b^2) \text{ and } 3b = s(a^2 - 10b^2),$$

where r and s are integers.

Multiplying the second equation of (II) by 3, we get

$$3bp + 3aq = 3,$$

which, on substituting for $3a$ and $3b$, yields

$$s(a^2 - 10b^2)p + r(a^2 - 10b^2)q = 3.$$

It follows that $(a^2 - 10b^2) | 3$ and so either

$$a^2 - 10b^2 = \pm 1$$

or

$$a^2 - 10b^2 = \pm 3.$$

If $a^2 - 10b^2 = 1$, then

$$(a - b\sqrt{10})(a + b\sqrt{10}) = a^2 - 10b^2 = 1$$

and so $1 \in I$ since $a + b\sqrt{10} \in I$. But $1 \in I$ implies that $I = D$, contradicting (ii). Similarly, if $a^2 - 10b^2 = -1$, we arrive again at the same contradiction. Hence

$$a^2 - 10b^2 = \pm 3.$$

(iv) In the ring $\mathbf{Z}_5$ this last equation becomes $\overline{a}^2 = \overline{\pm 3}$. Hence, either $\overline{a}^2 = \overline{3}$ or $\overline{a}^2 = \overline{2}$; that is, either $\overline{2}$ or $\overline{3}$ is a square in $\mathbf{Z}_5$. But $\mathbf{Z}_5 = \{\overline{0}, \overline{1}, \overline{2}, \overline{3}, \overline{4}\}$ and the squares are

$$\overline{0}, \overline{1}, \overline{2}^2 = \overline{4}, \overline{3}^2 = \overline{4}, \overline{4}^2 = \overline{1}$$

and so $\overline{2}$ and $\overline{3}$ are not squares. This contradiction proves that I is not principal. ∎

A PID has many of the properties that the domain of integers has. For example, we can define divisibility and the notions of primeness and greatest common divisor. Moreover, it can be proved that in a PID every nonzero nonunit element can be (essentially) uniquely expressed as a product of irreducibles (the analogue of "prime" in the integers).

An integral domain in which each nonzero element is uniquely expressible as a product of irreducibles is a *Unique Factorization Domain*. The theory is not that hard to develop and is left as an exercise for the reader (see Exercise 13 at the end of this chapter).

8.6 Embedding an Integral Domain in a Field

The domain of integers is contained in the field of rational numbers. In fact, from a strictly logical point of view, they are not really contained in the field of rationals since a rational number, by definition, is a quotient of integers whereas an integer is not. However, we agree to call a rational number with denominator 1 an integer and write it omitting its denominator. Nothing untoward happens with this convention, since the domain of integers is isomorphic to the subdomain of the rationals consisting of those rational numbers with 1 in the denominator.

In what follows, given any integral domain, we shall attempt to construct a field containing the domain in such a way that the elements of the field are "quotients" of the elements of the domain, just as each rational number is a quotient of integers.

8.6.1 *Definition*

We say that a ring S is *embedded* in a ring T if there exists a ring monomorphism $f : S \rightarrow T$. f is called an *embedding map*.

Remark.

(i) If S is embedded in T with embedding map f, then $S \cong f(S)$.

(ii) The map $f(z) = \frac{z}{1}$ for all integers z embeds the integers in the rationals.

8.6.2 *Theorem*

Let D be an integral domain and form the Cartesian product $D \times D^* = \{(a,b) \mid a,b \in D, b \neq 0\}$. On $D \times D^*$ define a relation $\sim$ by
$$(a,b) \sim (c,d) \text{ if } ad = bc.$$
Then $\sim$ is an equivalence relation.

Denote the equivalence class of (a,b) by $[(a,b)]$ and the set of equivalence classes by $Q(D)$. On $Q(D)$ define two operations $+$ and $.$ as follows:
$$[(a,b)] + [(c,d)] = [(ad + bc, bd)]$$
$$[(a,b)].[(c,d)] = [(ac, bd)].$$
Then $(Q(D), +, .)$ is a field called the *field of quotients* of the integral domain D and the map $f : D \rightarrow Q(D)$ defined by $f(a) = [(a,1)]$ is a monomorphism which embeds D in its field of quotients.

Proof. We outline the proof leaving the details to the reader. We first prove that $\sim$ is an equivalence relation.

(i) Reflexivity:
$$(a,b) \sim (a,b) \text{ since } ab = ba;$$
(ii) Symmetry:
$$(a,b) \sim (c,d) \text{ iff } ad = bc \text{ iff } cb = da \text{ iff } (c,d) \sim (a,b);$$
(iii) Transitivity:

Suppose
$$(a,b) \sim (c,d) \text{ and } (c,d) \sim (e,f).$$
Then
$$ad = bc \text{ and } cf = de.$$
Multiplying the first equation by f on both sides, we get
$$adf = bcf.$$
Substituting for cf from the second equation gives
$$afd = bed.$$
Cancelling d (which is not zero, by hypothesis), we get $af = be$ and so $(a,b) \sim (e,f)$.

Once again, the definitions of the two operations appear to depend on the representatives of the equivalence classes involved. To show the operations are well defined we must prove that, no matter which representatives we choose, we arrive at the same answer. Therefore, suppose
$$(a,b) \sim (a',b') \text{ and. } (c,d) \sim (c',d').$$
We must show

(a) $(ad + bc, bd) \sim (a'd' + b'c', b'd')$;

(b) $(ac, bd) \sim (a'c', b'd')$.

We prove (a) only, leaving (b) to the reader.

To prove (a) we must show
$$(ad + bc)b'd' = bd(a'd' + b'c').$$
Multiplying out, we must prove
$$(*)\, adb'd' + bcb'd' = bda'd' + bdb'c'.$$
Now
$$(a,b) \sim (a',b') \text{ implies } ab' = ba'$$
and
$$(c,d) \sim (c',d') \text{ implies } cd' = dc'.$$
Substituting ba' for ab' and dc' for cd' on the left-hand side of $(*)$, we get
$$adb'd' + bcb'd' = ba'dd' + bc'db' = \text{ right-hand side.}$$
Therefore, addition is well defined.

In a similar manner, we can prove that multiplication is well defined.

Our next task is to prove that $(Q(D), +, .)$ is a field. This is quite straightforward but tedious and is left as an exercise.

Note that the multiplicative inverse of $[(a,b)]$ is $[(b,a)]$ provided $a \neq 0$; the additive identity is $[(0,b)]$ for any $b \neq 0$ and the unity is $[(a,a)]$ for any $a \neq 0$. When in doubt in the proof that $(Q(D),+,.)$ is a field, the reader should think in terms of the rationals.

To remind us of the rationals, we usually denote the class of (a,b) by $\frac{a}{b}$. In this notation,

$$\frac{a}{b} = \frac{c}{d} \text{ iff } ad = bc$$

which is the usual rule for determining when two rationals are equal. Also, in this notation, the rules for addition and multiplication are precisely those for ordinary addition and multiplication of rational numbers.

The map $f : D \to Q(D)$ defined by $f(a) = \frac{a}{1}$ is easily shown to be a monomorphism. ∎

Remarks.

(i) Just as in the case of the integers and the rationals, we say that D is contained in its field of quotients $Q(D)$ and write a instead of $\frac{a}{1}$.

(ii) $Q(D)$ is the "smallest" field containing D in the sense that, if F is another field containing D, then there exists a monomorphism embedding $Q(D)$ in F; that is, F contains an isomorphic copy of $Q(D)$ and so is possibly "bigger" than $Q(D)$.

8.7 The Characteristic of an Integral Domain

Since the additive structure of an integral domain is an abelian group, we can speak of the additive order of an element. We ask: What are the possible values of the additive order of an element? Clearly, if the integral domain is finite, the orders are all finite. If the integral domain is infinite, the orders of the elements can be infinite or finite, as we shall see in Chapter 9.

Can we be more precise? We can indeed. It turns out that the additive order of any nonzero element of an integral domain is either infinite or a prime. We give two simple examples before proving this assertion.

8.7.1 *Examples*

In the domain of integers, every nonzero integer is of infinite additive order whereas in the domain $\mathbf{Z}_p$, every nonzero element is of order p.

8.7.2 *Definition*

Let 1 be the unity of the integral domain D. If 1 is of infinite additive order we say that the integral domain is of *characteristic zero* and write $\text{char}(D) = 0$. If 1 is of finite additive order m, we say that the *characteristic of D is m* and write $\text{char}(D) = m$.

8.7.3 *Theorem*

If D is an integral domain and $\text{char}(D) \neq 0$, then $\text{char}(D) = p, p$ a prime. Moreover, $pa = 0$ for all $a \in D$.

Proof. (To avoid confusion, we set $1 = e$). Recall that, built into the definition of a ring with unity is that the unity is not equal to the additive identity. Therefore, the characteristic of an integral domain is not 1.

Let $n = \text{char}(D)$ and suppose that n is not prime. Then there exist positive integers r and s such that
$$n = rs, \ r < n \text{ and } s < n.$$
But
$$0 = ne = (rs)e = (re)(se).$$
Since there are no nonzero zero divisors in an integral domain, it follows that either $re = 0$ or $se = 0$. Either of these possibilities would contradict the minimality of n. Hence n is a prime p.

Also, $p.a = (pe).a = 0.a = 0$ for all $a \in D$, that is, every element is of additive order p. ∎

We end this chapter with a result which says that any field contains either a copy of the rationals or of $\mathbf{Z}_p$.

8.7.4 *Theorem*

(i) A field F of characteristic 0 contains a subfield isomorphic to the field of rationals.

(ii) A field F of characteristic p contains a subfield isomorphic to $\mathbf{Z}_p$. In each case, the subfield isomorphic to $\mathbf{Q}$ or $\mathbf{Z}_p$ is the smallest subfield of the respective fields and is called the *prime subfield of F*.

Proof. As above, to avoid confusion, we denote the unity of the field F by e.

(i) Since $ne \neq 0$ for all nonzero integers n, $(ne)^{-1}$ exists in F. Define a map $f : \mathbf{Q} \rightarrow F$ by

$$f\left(\frac{r}{s}\right) = re(se)^{-1}.$$

Since there are many representatives of a given rational number, we must first prove that f is well defined.

Suppose therefore that $\frac{r}{s} = \frac{u}{v}$. Then $rv = us$ and so $(rv)e = (us)e$. But

$$(rv)e = (re)(ve) \text{ and } (us)e = (ue)(se).$$

Therefore,

$$(re)(ve) = (ue)(se).$$

Multiplying both sides of this last equation by $(ve)^{-1}(se)^{-1}$ we see that

$$(re)(se)^{-1} = (ue)(ve)^{-1}$$

and so $f\left(\frac{r}{s}\right) = f\left(\frac{u}{v}\right)$.

Next, we prove that f is 1:1. Suppose $f\left(\frac{r}{s}\right) = f\left(\frac{u}{v}\right)$. Then

$$re(se)^{-1} = ue(ve)^{-1}$$

and consequently,

$$(re)(ve) = (ue)(se).$$

Therefore

$$(rv)e = (us)e \text{ and so } (rv - us)e = 0.$$

But the characteristic of F is 0 and so we conclude that $rv - us = 0$. Hence, $rv = us$, and so $\frac{r}{s} = \frac{u}{v}$ and f is 1:1.

To prove f is a homomorphism, we must show:

(a) $f(\frac{r}{s} + \frac{u}{v}) = f(\frac{r}{s}) + f(\frac{u}{v})$;

(b) $f(\frac{r}{s} \cdot \frac{u}{v}) = f(\frac{r}{s})f(\frac{u}{v})$..

We prove (a) only, leaving (b) as an exercise for the reader.

(a)

$$f\left(\frac{r}{s} + \frac{u}{v}\right) = f\left(\frac{(rv+us)}{sv}\right) = (rv+us)e.[(sv)e]^{-1}$$

$$= (rv)e[(se)(ve)]^{-1} + (us)e[(se)(ve)]^{-1}$$

$$= (re)(ve)(ve)^{-1}(se)^{-1} + (ue)(se)(se)^{-1}(ve)^{-1} = (re)(se)^{-1} + (ue)(ve)^{-1}$$

$$= f\left(\frac{r}{s}\right) + f\left(\frac{u}{v}\right).$$

The image of f is $\hat{Q} = \{re(se)^{-1} \mid r,s \in \mathbf{Z}, s \neq 0\}$ and $\hat{Q}$ is the smallest subfield contained in F since any subfield of F must contain ne for all integers n and their multiplicative inverses when $n \neq 0$.

(ii) Suppose now that F is a field of characteristic p. Then it is easily shown that, since $pe = 0$, the set $P = \{0, e, 2e, 3e, \dots, (p-1)e\}$ is a subfield of F isomorphic to the field Z_p under the map $f(\bar{s}) = se$. Moreover, as above, P is the smallest subfield of F. We leave the details to the reader. ∎

8.8 Exercises

1) Prove that if in a ring $S, 1 = 0$, then $S = (0)$. (This is the reason for stipulating $1 \neq 0$ in the definition of a ring with unity.)

2) Prove that all the examples given in 8.1.2 are rings.

3) Give an example of a map f from a ring S to a ring T satisfying (i) and (ii) of 8.2.1 but which fails to map the unity of S to the unity of T.

4) Prove (ii) and (iii) of 8.2.4.

5) Prove that the direct sum of rings is indeed a ring.

6) In 8.2.6 (ii), prove that $\frac{T}{I} \cong \mathbf{R} \oplus \mathbf{R} \oplus \mathbf{R}$ where $\mathbf{R}$ denotes the field of real numbers.

7) Prove in detail the First Isomorphism for Rings. (Theorem 8.3.1).

8) If S is a commutative ring with unity and a and b elements of S, prove that $I = \{xa + yb \mid x, y \in S\}$ is an ideal containing a and b.

9) Prove 8.6.2 in detail.

10) In 8.3.3 we showed that if S is a ring, T a subring of S and I an ideal of S, then $T + I = \{t + k \mid t \in T, k \in I\}$ is a subring of S. Give an example to show that if we merely assume I to be a subring, then $T + I$ is not necessarily a subring of S. Show further that if both T and I are ideals, then $T + I$ is also an ideal.

11) Prove that the ring $M_n(\mathbf{R})$ of $n \times n$ matrices over the reals has no ideals other than the zero ideal and itself. (A ring with no ideals other than the obvious ones is called *simple*.)

12) If in an integral domain $a^m = b^m$ and $a^n = b^n$ where m and n are relatively prime positive integers, then $a = b$.
[Hint: Embed the integral domain in its field of quotients.]

13) The following sequence of results aims to establish that a Principal Ideal Domain is a Unique Factorization Domain
   (i) Let $I_1 \subseteq I_2 \subseteq I_3 \subseteq .... \subseteq I_k \subseteq ...$ be ideals of a ring S. Prove that

$$J = \bigcup_{k=1}^{\infty} I_k \text{ is an ideal.}$$

   (ii) In a commutative ring S we say a divides b and write $a \mid b$ if $b = sa$ some $s \in S$. Prove that $(a) \subseteq (b)$ iff $b \mid a$. Prove also that, in an integral domain D the following are equivalent:

(a) $a\,|\,b$ and $b\,|\,a$;

(b) $(a) = (b)$;

(c) $b = ua$ for some unit u in D.

(iii) An ideal I of a ring S is said to be *maximal* if $I \neq S$ and there are no ideals other than S strictly containing I; that is, $I \subset B \subseteq S$ and B an ideal of S imply $B = I$ or $B = S$. Prove that in $\mathbf{Z}$ an ideal (n) is maximal if, and only if, n is prime. (Use part (ii).)

(iv) Prove that, in a commutative ring D with unity, an ideal I is maximal if, and only if, $D\!\!\big/_{\!I}$ is a field.

(v) Prove that, in a commutative ring with unity, an element is a unit if, and only if, it divides every element of the ring.

(vi) On an integral domain D define a relation $\sim$ by $a \sim b$ if $a\,|\,b$ and $b\,|\,a$. $a \sim b$ if $a\,|\,b$ and $b\,|\,a$. Prove that $\sim$ is an equivalence relation. Two elements in the same equivalence class are said to be *associates*.

(vii) A nonzero nonunit element a of an integral domain is said to be *irreducible* if the only divisors of a are units or associates of a. (Note: In $\mathbf{Z}$ a and b are associates if, and only if, $b = \pm a$ since ± 1 are the only units).

(viii) In an integral domain D, prove that $(a) = D$ if, and only if, a is a unit.

(ix) Prove that, in a PID D, every ideal different from D is contained in a maximal ideal.

Sketch of proof:

Suppose this is not the case. Then there exists an ideal $I \neq D$ which is not contained in any maximal ideal. In particular, I itself cannot be maximal and so there exists an ideal I_1 such that $I = I_0 \subset I_1 \neq D$. Again I_1 cannot be maximal (otherwise I would be contained in a maximal ideal, contrary to hypothesis) and thus there exists an ideal I_2 such that $I = I_0 \subset I_1 \subset I_2 \neq D$. Continue in this manner, eventually constructing an infinite sequence I_j, $j = 1, 2, 3, \ldots$ of ideals such that $I_j \subset I_{j+1}$ and $I_j \neq D$ for all j, $j = 0, 1, 2, \ldots$ Let $J = \bigcup_{k=1}^{\infty} I_k$. By (i), J is an ideal and since D is a PID, there exists $a \in D$ such that $(a) = J$. Hence $a \in J$ and so $a \in I_k$ for some positive k But then $J = (a) \subseteq I_k \subseteq J$ which implies that $I_k = J$. This is a contradiction since $I_k \subset I_{k+1} \subseteq J$ (and thus $I_k \neq J$). Therefore, it is the case that every ideal different from D is contained in a maximal ideal.

(x) Prove that in a PID D, an ideal I is maximal if, and only if, $I = (p)$ for some irreducible element p.

(xi) Prove that in a PID every nonzero nonunit element is divisible by an irreducible element.

(xii) Prove that in an integral domain any associate of an irreducible element is also irreducible.

(xiii) Prove that in a PID every nonzero nonunit is expressible as a product of a finite number of irreducible elements.

Sketch of proof:

Let a be a nonzero nonunit of the PID D. By (ix) and (x), $(a) \subseteq (p_1)$ where p_1 is irreducible. Hence $a = p_1 b_1$ for some $b_1 \in D$. If b_1 is a unit, we are done since, by (xii), $p_1 b_1$ is also irreducible. Thus assume that b_1 is not a unit. By the same argument as above, replacing a by b_1 we conclude that $b_1 = p_2 b_2$ where p_2 is irreducible. Thus $a = p_1 p_2 b_2$. If b_2 is a unit, we are done. If not, $b_2 = p_3 b_3$ for some irreducible element p_3. Note that $(a) \subset (b_1) \subset (b_2) \neq D$. Continue this process to get

$$a = p_1 b_1 = p_1 p_2 b_2 = p_1 p_2 p_3 b_3 = ... = p_1 p_2 p_3 .... p_n b_n$$

where the p_j are irreducible. If in this procedure we never arrive at a b which is a unit, we get an infinite chain of ideals, namely

$$(a) \subset (b_1) \subset (b_2) \subset (b_3) \subset ... \subset (b_n) \subset .. \neq D .$$

Taking the union of these ideals and using an argument similar to the one given in (ix), we arrive at a contradiction. Therefore, after a finite number of steps, say m, we arrive at a b_m which is a unit and so

$$a = p_1 p_2 p_3 ... p_m b_m = p_1 p_2 p_3 ... p'_m$$

where we have set $p'_m = p_m b_m$. Therefore, a is expressible as a product of a finite number of irreducible.

(xiv) In a PID D define a gcd of two elements a and b (not both 0) as an element d such that

    (a) d divides both a and b;

    (b) if g divides both a and b, then g divides d.

Prove that in a PID D, if d_1 and d_2 are gcd's of a and b, then d_1 and d_2 are associates.

Furthermore, show that gcd's exist and that if d is a gcd of a and b, then $d = xa + yb$ for some x and y in D; (Although two elements of a PID can have infinitely many gcd's, any two of them are associates and so one is a unit multiple of the other. Therefore, we often speak of *the* gcd of the two elements although this is not strictly correct.)

(xv) Let a and b be elements of a PID. If p is irreducible and p divides ab, prove that either p divides a or p divides b. Furthermore, using induction, prove that if p divides any finite product, then p divides one of the factors.

(xvi) Let a be a nonzero nonunit element of the PID D. Then by (xiii), a is expressible as a finite product of irreducible elements of D. Suppose that $a = p_1 p_2 \ldots p_n = q_1 q_2 \ldots q_m$ are two decompositions of a as a product of irreducible elements. Prove that $m = n$ and that, after renumbering if necessary, $p_j \sim q_j$ for all j, $j = 1, 2, \ldots, n$.

14) Let S be a ring with the property that $a^2 = a$ for all a in S. Prove that S is a commutative ring.

15) Let $(S, +, .)$ be a ring with unity. On the set S define new operations $\oplus$ and $\circ$ in terms of the original ones as follows:

$$a \oplus b = a + b + 1; \quad a \circ b = ab + a + b.$$

Prove that $(S, \oplus, \circ)$ is a ring isomorphic to $(S, +, .)$.

16) Give an example of a ring with unity which contains a subring with a unity which is different from the unity of the "big" ring.

17) Let $+$ and $.$ be operations defined on S satisfying all the axioms of a ring with unity except possibly commutativity of addition. Prove that in

fact $(S,+,.)$ is a ring. (Hint: Expand $(a+b)(1+1)$ in two ways). Does it matter if $1+1=0$?

18) Prove that the intersection of ideals is an ideal.

If $J_1, J_2,..., J_n$ are ideals of a ring S, prove that $S \Big/ \bigcap\limits_{i=1}^{n} J_i$ is isomorphic to a subring of the (external) direct sum $S\big/_{J_1} \oplus S\big/_{J_2} \oplus ... \oplus S\big/_{J_n}$. [Hint: define a homomorphism $\varphi : S \to S\big/_{J_1} \oplus S\big/_{J_2} \oplus ... \oplus S\big/_{J_n}$ in the obvious way and apply the First Isomorphism Theorem].

19) (i) An ideal P of a commutative ring S with unity is said to be *prime* if for all $a,b \in S$, $ab \in P$ implies $a \in P$ or $b \in P$. Prove that an ideal I of S is prime if, and only if, $S\big/_I$ is an integral domain.

(ii) What are the prime ideals of $\mathbf{Z}$?

20) (i) If A and B are ideals of a commutative ring S, let AB denote the set of all finite sums of products of elements from A and B, i.e.,

$$AB = \{ \sum_{k \in P} a_k b_k \mid a_k \in A,\ b_k \in B \text{ and } P \text{ is a finite subset of } \mathbf{Z}^+ \}.$$

Prove that AB is an ideal.

(ii) If $J_1, J_2,..., J_n$ are ideals of a commutative ring S with unity and $J_k + J_l = S$ for all $k \neq l$, prove that $\bigcap\limits_{i=1}^{n} J_i = \prod\limits_{i=1}^{n} J_i$.

[Hint: Show first that $S = J_i + \prod\limits_{k \neq i} J_k$. Choose $a_i^{(k)} \in J_i$ and $a_s \in J_s$ such that

$$1 = a_i^{(1)} + a_1 = a_i^{(2)} + a_2 = ... = a_i^{(i-1)} + a_{i-1} = a_i^{(i+1)} + a_{i+1} = ... = a_i^{(n)} + a_n.$$

Then

$$1 = (a_i^{(1)} + a_1)(a_i^{(2)} + a_2)...(a_i^{(i-1)} + a_{i-1})(a_i^{(i+1)} + a_{i+1})...(a_i^{(n)} + a_n)$$

and so $1 \in J_i + \prod\limits_{k \neq i} J_k$]. Now use induction on n to prove $\bigcap\limits_{i=1}^{n} J_i = \prod\limits_{i=1}^{n} J_i$.]

(iii) Suppose that S is a PID with ideals $J_1, J_2,..., J_n$ such that $J_k + J_l = S$ for all $k \neq l$. Write $x \equiv y \bmod J_k$ if $x - y \in J_k$. Given elements $a_i \in S$, $i = 1, 2,..., n$, prove that there is an $x \in S$ such that $x \equiv a_i \bmod J_i$, $i = 1, 2,..., n$. $x \equiv a_i \bmod J_i$, $i = 1, 2,..., n$. (This result is known as the Chinese Remainder Theorem). [Hint: Set $J_i = (m_i)$. From (ii) above,

$$1 = s_i m_i + t_i m_1 m_2 ... m_{i-1} m_{i+1} ... m_n$$

for some s_i and $t_i \in S$ and let $x = \sum_i^n a_i t_i m_1 m_2 ... m_{i-1} m_{i+1} ... m_n$].

21) If R is a ring, define the centre $Z(R)$ of R by

$$Z(R) = \{a \in R \mid ax = xa \text{ for all } x \in R\}.$$

Prove that $Z(R)$ is a subring of R. Find the centre of the ring of 3×3 matrices over $\mathbf{Z}_7$.

22) Let R be a ring with unity and e a central idempotent, that is, $e^2 = e$ and $e \in Z(R)$. Prove that $f = 1 - e$ is also an idempotent and that Re and Rf are ideals of R such that $Re \cap Rf = (0)$ and $R = Re \oplus Rf$.

23) An element b of a ring is said to be *nilpotent* if $b^n = 0$ for some positive integer n. Let R be a commutative ring with unity and b a nilpotent element of R.
(i) Prove that $1 + b$ is a unit.
(ii) Prove that if xy is a unit so are x and y.
(iii) Prove that if a is a unit and b a nilpotent element, then $a + b$ is a unit.

24) (i) Prove that in a commutative ring the set of all nilpotent elements forms an ideal.
(ii) If R is a commutative ring and J is the ideal of nilpotent elements, prove that R/J has no nonzero nilpotent elements.
(iii) Find the ideal J of nilpotent elements of the ring $\mathbf{Z}_{108}$. What familiar ring is $\mathbf{Z}_{108}/J$ isomorphic to?

(iv) For what values of n does $\mathbf{Z}_n$ have no nonzero nilpotent elements?

25) Let R be the ring of 2×2 matrices with entries from the field Z_p. How many elements does R have? How many units are there in R?

26) How many rings with unity are there of order a prime p? Justify your answer.

27) Give an example of a noncommutative ring with 16 elements.

28) Let $F = \{0,2,4,6,8\} \subset Z_{10}$. Prove that F is a subfield of Z_{10}.

29) Let $\varphi : R \to S$ be a ring epimorphism and J an ideal of S. Prove that

$$\frac{R}{\varphi^{-1}(J)} \cong \frac{S}{J}.$$

30) Let $R = \{a + bi \mid a, b \in \mathbf{Z}, i = \sqrt{-1}\}$.

(i) Prove that R is a subring of the field of complex numbers.

(ii) Prove that $I = (2 + 2i)$ is not a prime ideal of R. What is $\left|\frac{R}{I}\right|$?

What is char $\frac{R}{I}$? (See Problem 19 for the definition of prime ideal.)

(iii) Prove that $\frac{R}{(1-i)}$ is a field and find its order.

31) An ideal M of a ring R is *maximal* if M is a proper ideal of R which is not properly contained in any ideal other than R. Prove that M is a maximal ideal of a ring R with unity if, and only if, $\frac{R}{M}$ is a field.

32) If p is prime, show that $J = \{(px, y) \mid x, y \in \mathbf{Z}\}$ is a maximal ideal of $\mathbf{Z} \oplus \mathbf{Z}$. What is $\frac{(\mathbf{Z} \oplus \mathbf{Z})}{J}$?

33) Let $(A, +)$ be an abelian group and let End(A) be the set of endomorphisms of A. On End(A) define two operations $+$ and $.$ by:

(i) $(\alpha+\beta)(a)=\alpha(a)+\beta(a)$ for all $\alpha,\beta\in \text{End}(A)$ and all $a\in A$.

(ii) $(\alpha\beta)(a)=\alpha[\beta(a)]$ for all $\alpha,\beta\in \text{End}(A)$ and all $a\in A$.

Prove that $(\text{End}(A),+,.)$ is a ring with unity.

34) Determine all the ring endomorphisms of Z_6.

35) (i) If p is prime, how many idempotents does the ring $\mathbf{Z}_{p^n}$ have?

(ii) Prove that the number of idempotents in the ring $\mathbf{Z}_m$ is 2^k where k is the number of primes dividing m.

[Hint: If $m=\prod_{i=1}^{k} p_i^{a_i}$, then $Z_m\cong \bigoplus_{i=1}^{k} Z_{p_i^{a_i}}$ as rings.]

36) Show that $E=\{a+b\sqrt{2}\,|\,a,b\in \mathbf{Q}\}$ and $\{a+b\sqrt{3}\,|\,a,b\in \mathbf{Q}\}$ are nonisomorphic fields.

37) Let R be a ring and $\pi_1,\pi_2,...,\pi_n$ endomorphisms of R such that:

(i) $\pi_i\pi_j=0$, the zero endomorphism if $i\neq j$.

(ii) $\pi_i^2=\pi_i$ for all $i, 1\le i\le n$.

(iii) $\pi_1+\pi_2+...+\pi_n=$ the identity endomorphism of R.

(iv) If $R_i=\pi_i(R)$, for all $i, 1\le i\le n$, prove that $R=R_1\oplus R_2\oplus...\oplus R_n$.

38) (i) Let F be a finite field. Prove that if $\text{char}(F)=2$, every element of F is a square whereas if $\text{char}(F)\neq 2$, exactly half of the nonzero elements are squares. [Hint: Consider the map $\varphi(a)=a^2$.]

(ii) Use (i) to prove that in a finite field every element is the sum of two squares. The two squares are not necessarily distinct and one could be 0. Thus $0=0^2+0^2$ is an allowable representation as a sum of two squares.[Hint: Let S denote the set of squares (including 0) and for a fixed $a\in F$, consider $a-S=\{a-s\,|\,s\in S\}$. Show that $(a-S)\cap S\neq\phi$.]

39) Determine all ring homomorphisms from $\mathbf{Z}_{15}$ to $\mathbf{Z}_{25}$.

40) If I and J are ideals of a commutative ring R, let $IJ^{-1}=\{x\in R\,|\,xJ\subseteq I\}$. Prove that IJ^{-1} is an ideal of R containing I.

41) Let R be a commutative ring with unity. A subset M of R is said to be *multiplicatively closed* if

(a) $1 \in M$;

(b) If $a, b \in M$, then $ab \in M$.

(i) If P is a prime ideal of R, prove that $R \backslash P$ is multiplicatively closed (see Problem 19 for the definition of prime).

(ii) Let M be a multiplicatively closed subset of R. On $R \times M$ define a relation $\sim$ by $(a, m) \sim (b, n)$ if there exists $u \in M$ such that $(an - bm)u = 0$.

Prove that $\sim$ is an equivalence relation. Denote the equivalence class of (a, m) by a/m and show that the set of equivalence classes can be turned into a ring (denoted $M^{-1}R$) by defining addition and multiplication by:

Addition: $a/m + b/n = (an + bm)/mn$;

Multiplication: $(a/m)(b/n) = ab/mn$.

(iii) Prove that an element of the form u/v where u and v are in M is a unit.

(iv) If $M = R \backslash P$ for some prime ideal P, prove that

$$J = \{a/m \mid a \in P\}$$

is the unique maximal ideal in $M^{-1}R$, that is:

(a) $J \neq M^{-1}R$;

(b) If K is any proper ideal, then $K \subseteq J$.

In this particular case, the ring $M^{-1}R$ is denoted by R_P and the process of obtaining R_P is called *localization* at P since a ring with exactly one maximal ideal is called a *local ring*. Local rings arise quite naturally in algebraic geometry.

(v) Prove that if M is any multiplicatively closed set, the map $f : R \rightarrow M^{-1}R$ defined by

$$f(a) = a/1$$

is a homomorphism (which, in general is not injective).

Chapter 9

Polynomial Rings

9.1 Introduction

To the student of elementary algebra, a polynomial is an expression of the form $a_0 + a_1x + a_2x^2 + ... + a_{n-1}x^{n-1} + a_nx^n$ where the a's are real numbers called coefficients of the polynomial and x is thought of as a variable real number. In this chapter we shall view polynomials in a somewhat different light although their form will be identical to the polynomials of elementary algebra.

We shall show that the set of polynomials with coefficients from some commutative ring forms a ring under operations which, formally, are defined in exactly the same way the sum and product of high school days.

For most of the chapter we shall be concerned with polynomial rings over fields. As we shall see, these rings have many properties in common with the integers. In particular, in such rings we have a modified version of the Euclidean Algorithm which is used to compute the greatest common divisor of two polynomials in much the same way it was in the integers and to show that polynomial rings over fields are principal ideal domains. Furthermore, irreducible polynomials play the same role as primes in the integers and the Fundamental Theorem of Arithmetic has its counterpart in polynomial rings over fields. Finally, by factoring out appropriate ideals, we are able to construct, in a way analogous to the construction of $\mathbb{Z}/_{(n)}$, a large variety of fields, both finite and infinite.

9.2 Definitions and Elementary Properties

9.2.1 *Definition and notation*

(i) Let S be a commutative ring with unity. A *polynomial in x over S* is an expression of the form

$$a_0 + a_1 x + a_2 x^2 + \ldots + a_n x^n$$

where the a_j are elements of S and x is a symbol called an *indeterminate*.

(ii) The a_j are the *coefficients* of the polynomial and a_j is the *coefficient of x^j*.

(iii) If $a_n \neq 0$, the term $a_n x^n$ is the *leading term* of the polynomial and a_n is the *leading coefficient*.

(iv) The *degree* of a polynomial $p(x)$, denoted $\partial p(x)$, is the highest power of x with nonzero coefficient. If all the coefficients of the polynomial are zero, we define its degree to be $-\infty$.

(v) If the leading coefficient is 1, the polynomial is said to be *monic*.

(vi) A polynomial of the form ax^s is a *monomial*.

(vii) If it is not necessary to know the precise form of a polynomial, we write $p(x)$ for short and $p(x)_j$ for the coefficient of x^j. We shall also write

$$\sum_{j=0}^{n} a_j x^j \text{ for } a_0 + a_1 x + a_2 x^2 + \ldots + a_n x^n.$$

(viii) The set of all polynomials over S in the indeterminate x is denoted by $S[x]$.

(ix) Two polynomials $p(x)$ and $q(x)$ are said to be *equal* and we write $p(x) = q(x)$ if, for all relevant j, $p(x)_j = q(x)_j$.

Remarks. (i) The symbol x is to be viewed as a *place holder* and the plus signs do not, strictly speaking, denote addition. In fact, in some books, the polynomial $a_0 + a_1 x + a_2 x^2 + \ldots + a_n x^n$ is denoted by the infinite tuple $(a_0, a_1, a_2, \ldots, a_n, 0, 0, 0, \ldots)$, thus dispensing with the symbols x and $+$. In this notation, $(0, 1, 0, 0, \ldots, 0 \ldots)$ stands for the polynomial $1x$ which is usually written as just x in the notation we have adopted.

(ii) To conform with the optional notation mentioned in (i), we should, strictly speaking, write

$$a_0 + a_1 x + a_2 x^2 + ... + a_n x^n + 0 x^{n+1} + 0 x^{n+2} + ...$$

instead of $a_0 + a_1 x + a_2 x^2 + ... + a_n x^n$ but we usually leave out those terms with zero coefficient.

(iii) a_0 is the *constant term* of the polynomial and any polynomial all of whose coefficients are zero except possibly for the constant term is a *constant polynomial*. Note that a_0 is the coefficient of x^0 although we rarely write the constant term as $a_0 x^0$.

(iv) We adopt the following conventions regarding $-\infty$:
(a finite integer) $+ -\infty = -\infty = -\infty + $ (a finite integer) ; $-\infty + -\infty = -\infty$; $-\infty <$ any finite integer.

We shall now turn $S[x]$ into a ring by defining appropriate operations. Formally, these will be the same as the operations of addition and multiplication of polynomials in elementary algebra. From a logical standpoint, however, they are quite different.

9.2.2 *Definition*

Define $+$ and . (juxtaposition) on $S[x]$ as follows:

$$\sum_{j=0}^{m} a_j x^j + \sum_{j=0}^{n} b_j x^j = \sum_{j=0}^{\max(m,n)} (a_j + b_j) x^j ;$$

$$(\sum_{j=0}^{m} a_j x^j)(\sum_{j=0}^{n} b_j x^j) = \sum_{j=0}^{m+n} c_j x^j \text{ where } c_j = \sum_{k+l=j} a_k b_l .$$

Remarks. (i) In the definition of addition, the plus sign does triple duty. On the left-hand side, the plus signs occurring between the terms of each polynomial serve to separate the terms; the plus sign between the polynomials is the operation being defined. On the right-hand side, the plus sign between the a's and b's is the addition in the ring S.

(ii) The reader can easily verify that the definition of multiplication of polynomials can be arrived at by pretending that x is an element of S, using distributivity and collecting terms with the same power of x; in short, formally, multiplication of polynomials is the same as in elementary algebra.

9.2.3 *Theorem*

$(S[x], +, .)$ is a commutative ring called the *polynomial ring in the indeterminate x over S.*

Proof.

(i) Closure under + and . is clear.

(ii) Associativity of addition in $S[x]$ is an immediate consequence of associativity of the additive structure of the ring S.

(iii) The zero of the ring is the polynomial all of whose coefficients are zero, called the zero polynomial and denoted 0.

(iv) The additive inverse of a polynomial is that polynomial whose coefficients are the negatives (additive inverses) of the coefficients of the given polynomial.

(v) That addition and multiplication in $S[x]$ are commutative follows from the commutativity of addition and multiplication in S.

Other than the associativity of multiplication, the other properties of the multiplicative structure of $S[x]$ and distributivity are easily proved and are left as an exercise for the reader. As for associativity of multiplication, we proceed as follows:

Let $p(x)$, $q(x)$ and $r(x)$ be polynomials over S. If $p(x)$ is a monomial, i.e., $p(x) = ax^n$ associativity is easy to establish. In particular, associativity holds if $\partial p(x) = 0$.

Assume inductively that associativity holds for all polynomials $p(x)$, $q(x)$ and $r(x)$ provided $\partial p(x) < n$ for some $n, 1 \le n$ and suppose $\partial \hat{p}(x) = n$. Then, setting $\hat{p}(x) = s(x) + a_n x^n$ where $\partial s(x) < n$ we have

$$\hat{p}(x)[q(x)r(x)]$$
$$= [s(x) + a_n x^n][q(x)r(x)] = s(x)[q(x)r(x)] + a_n x^n [q(x)r(x)]$$

using distributivity. By the induction hypothesis

$$s(x)[q(x)r(x)] = [s(x)q(x)]r(x) \text{ and } a_n x^n[q(x)r(x)] = [a_n x^n q(x)]r(x).$$

Therefore,

$$\hat{p}(x)[q(x)r(x)] = [s(x)q(x)]r(x) + [a_n x^n q(x)]r(x)$$

$$= [s(x)q(x) + a_n x^n q(x)]r(x)$$
$$= \{[s(x) + a_n x^n]q(x)\}r(x) = [\hat{p}(x)q(x)]r(x).$$

It now follows by induction that multiplication is associative in $S[x]$. ■

9.2.4 *Example*

Let $\overline{2} + \overline{4}x + \overline{3}x^2, \overline{5} + \overline{2}x + \overline{4}x^2 + \overline{2}x^3$ be polynomials over $\mathbf{Z}_6$. Then their sum is

$$(\overline{2} + \overline{5}) + (\overline{4} + \overline{2})x + (\overline{3} + \overline{4})x^2 + (\overline{0} + \overline{2})x^3 = \overline{1} + x^2 + \overline{2}x^3.$$

Notice that when the coefficient of some power of x is the unity of the ring, we omit it. Also we leave out terms with coefficient the zero of the ring.

The product of these polynomials is
$$\overline{2.5} + \overline{24}x + \overline{31}x^2 + \overline{26}x^3 + \overline{20}x^4 + \overline{3.2}x^5$$
$$= \overline{4} + \overline{0}x + \overline{1}x^2 + \overline{2}x^3 + \overline{2}x^4 + \overline{0}x^5 = \overline{4} + x^2 + \overline{2}x^3 + \overline{2}x^4.$$ ■

9.2.5 *Theorem*

If $p(x)$, $q(x)$ are polynomials over the integral domain D, then
$$\partial(p(x)q(x)) = \partial(p(x)) + \partial(q(x)).$$

Proof. If one of the polynomials is the zero polynomial, then $p(x)q(x) = 0$ and so $\partial(p(x)q(x)) = -\infty$. On the other hand, since one of $p(x)$ and $q(x)$ is the zero polynomial, by definition, one of $\partial(p(x))$ and $\partial(q(x))$ is $-\infty$ and so, as postulated above, $\partial(p(x)) + \partial(q(x)) = -\infty$. Therefore, in this case, $\partial(p(x)q(x)) = \partial(p(x)) + \partial(q(x))$.

Suppose that neither $p(x)$ nor $q(x)$ is zero, say

$$p(x) = \sum_{j=0}^{n} a_j x^j \text{ and } q(x) = \sum_{j=0}^{m} b_j x^j$$

where $a_n \neq 0 \neq b_m$ Then, in the product, the term with the highest power of x is $a_n b_m x^{n+m}$. Since D is an integral domain $a_n b_m \neq 0$. Therefore

$$\partial(p(x)q(x)) = n + m = \partial(p(x)) + \partial(q(x)).$$ ■

Remark. Observe that in Example 9.2.4, the degree of the product of the two polynomials is 4 whereas the sum of the degrees is 5. Therefore, over an arbitrary ring we can only claim $\partial(p(x)q(x)) \leq \partial(p(x)) + \partial(q(x))$.

9.2.6 *Corollary*

If D is an integral domain, so is $D[x]$.

Proof. Let $p(x)$ and $q(x)$ be nonzero polynomials over D. Then $\partial(p(x)) \neq -\infty$ and $\partial(q(x)) \neq -\infty$. By the preceding theorem, $\partial(p(x)q(x)) = \partial(p(x)) + \partial(q(x))$ and, since each of the summands is a finite integer, so is the sum. Therefore, $p(x)q(x) \neq 0$, and so $D[x]$ has no nonzero zero divisors proving that $D[x]$ is an integral domain. ∎

Remark. If R is a commutative ring we can form the polynomial ring in two or more indeterminates. For example, we can first form the polynomial ring $R[x]$ and then the polynomial ring $(R[x])[y]$, consisting of polynomials in the indeterminate y with coefficients from $R[x]$. A typical element from this ring is of the form $\sum a_{ij} x^i y^j$ where the sum is over a finite number of terms and the a_{ij} are in R. This ring is denoted by $R[x, y]$. On account of 9.2.6, $R[x, y]$ is an integral domain if R is.

9.3 The Division Algorithm and Applications

We shall focus our investigations on polynomial rings over fields. It turns out that these algebraic structures have a great deal in common with the integers and we devote much of the rest of this chapter to establishing results which are analogues of corresponding results we have proved for the integers. The reader will notice that even the proofs have the same flavour.

Notation. From now on F will denote an arbitrary field and D an arbitrary integral domain.

9.3.1 *Lemma*

Let $b(x)$ and $r(x)$ be polynomials over the field F with $b(x) \neq 0$. If $\partial(r(x)) \geq \partial(b(x))$, then there exists $q(x) \in F[x]$ such that

$$\partial[r(x) - b(x)q(x)] < \partial(r(x)).$$

Proof. Let $\partial(r(x)) = v \geq n = \partial(b(x))$, $r(x)_j = r_j, b(x)_j = b_j$. Then $r_v \neq 0$ and $b_n \neq 0$. Since $b_n \neq 0$ and the coefficients are elements of a field, b_n^{-1} exists.

Let $q(x) = r_v b_n^{-1} x^{v-n}$ (Observe that $v - n \geq 0$ so that x^{v-n} makes sense.) Then $b(x)q(x)$ is a polynomial of degree v with $r_v x^v$ as leading term and so, $\partial[r(x) - b(x)q(x)] < v = \partial(r(x))$. ∎

9.3.2 *Theorem (Division Algorithm (for Polynomials))*

Let $F[x]$ be the polynomial ring over the field F and let $a(x), b(x) \in F[x], b(x) \neq 0$. Then there exist unique polynomials $q(x)$ and $r(x)$ in $F[x]$ such that:

(i) $a(x) = b(x)q(x) + r(x)$;

(ii) $\partial(r(x)) < \partial(b(x))$.($q(x)$ is the *quotient* and $r(x)$ the *remainder* upon dividing $a(x)$ by $q(x)$.)

Proof. Let $T = \{a(x) - b(x)t(x)\,|\, t(x) \in F[x]\}$.

(i) If $0 \in T$, then, for some $q(x) \in F[x]$ we have $a(x) = b(x)q(x)$ and so, in this case, the remainder is 0. By convention, $-\infty$ is less than any (finite) integer. Therefore,

$$a(x) = b(x)q(x) + r(x) \text{ where } \partial(r(x)) = \partial(0) = -\infty < \partial(b(x)).$$

(ii) If $0 \notin T$, then the set K of degrees of the polynomials in T is a subset of the nonnegative integers. By the Well Ordering Principle, K contains a least element, say v and there exists $q(x) \in F[x]$ such that $r(x) = a(x) - b(x)q(x)$ is of degree v. If $v \geq n = \partial(b(x))$, by the Lemma, there exists $w(x) \in F[x]$ such that $\partial[r(x) - b(x)w(x)] < v = \partial(r(x))$. But then, setting $u(x) = r(x) - b(x)w(x)$ we get

$$u(x) = r(x) - b(x)w(x) = a(x) - b(x)q(x) - b(x)w(x)$$
$$= a(x) - b(x)[q(x) + w(x)].$$

This contradicts the minimality of the degree of $r(x)$ since $u(x)$ is in T and so establishes the existence of the quotient and the remainder with the given properties.

Suppose now that

$$a(x) = b(x)q(x) + r(x) = b(x)q_1(x) + r_1(x)$$

where

$$\partial(r(x)) < \partial(b(x)) \text{ and } \partial(r_1(x)) < \partial(b(x)).$$

Then

$$(*)\, b(x)[q(x) - q_1(x)] = r_1(x) - r(x)$$

and so,

$$\partial(b(x)[q(x) - q_1(x)]) = \partial(b(x)) + \partial(q(x) - q_1(x)) = \partial(r_1(x) - r(x)).$$

Hence,

$$\partial(q(x) - q_1(x)) = \partial(r_1(x) - r(x)) - \partial(b(x)) < 0$$

since $\partial(r_1(x) - r(x)) < \partial(b(x))$. Therefore, $\partial(q(x) - q_1(x)) = -\infty$, from which it follows that $q(x) - q_1(x) = 0$ and so $q(x) = q_1(x)$. From $(*)$ we deduce $r(x) = r_1(x)$. Uniqueness is thus established. ∎

In the ring of integers there are only two units, namely, ± 1. The situation is quite different in polynomial rings over fields.

9.3.3　*Theorem*

The polynomial $p(x)$ is a unit in $F[x]$ if, and only if, $p(x)$ is a nonzero constant polynomial.

Proof. If $p(x)$ is a unit, then there exists $q(x)$ such $p(x)q(x) = 1$. Taking degrees of both sides, we get

$$\partial(p(x)q(x)) = \partial(1) = 0$$

and so, by Theorem 9.2.6, $\partial(p(x)) + \partial(q(x)) = 0$. It follows that $\partial(p(x)) = 0 = \partial(q(x))$, and so $p(x)$ is a nonzero constant polynomial.

Conversely, any nonzero constant polynomial a is a unit whose inverse is a^{-1}. ∎

9.3.4 Definition

The polynomial $a(x) \neq 0$ *divides* the polynomial $b(x)$ in $F[x]$ if there exists a polynomial $q(x)$ in $F[x]$ such that $b(x) = a(x)q(x)$. We write $a(x) \mid b(x)$. The polynomial $a(x)$ is said to be a *divisor* of $b(x)$.

The following facts are analogous to those found in Theorem 4.3.1.

9.3.5 Theorem

Let $a(x), b(x), c(x) \in F[x]$. Then:

(i) if $a(x) \mid b(x)$ and $b(x) \mid a(x)$, then $a(x) = ub(x)$ where u is a unit, that is, u is a nonzero element of F;

(ii) 0 is the only polynomial divisible by all nonzero polynomials;

(iii) if $a(x) \mid b(x)$ and $b(x) \mid c(x)$, then $a(x) \mid c(x)$;

(iv) if $a(x) \mid b(x)$, then $ua(x) \mid vb(x)$ for all units u and v;

(v) if $a(x) \mid b(x)$ and $a(x) \mid c(x)$, then $a(x) \mid (\pm b(x) \pm c(x))$;

(vi) if $a(x) \mid b(x)$ and $a(x) \mid (\pm b(x) \pm c(x))$, then $a(x) \mid c(x)$.

Proof. The proofs of these facts are so similar to those of 4.3.1 that we only prove (i) and (iv).

(i) We have $b(x) = a(x)q(x)$ and $a(x) = b(x)s(x)$. Hence

$$b(x) = b(x)s(x)q(x).$$

Since $b(x) \neq 0$ and $F[x]$ is an integral domain, we can cancel $b(x)$ to get $s(x)q(x) = 1$. Therefore, both $s(x)$ and $q(x)$ are units, say $s(x) = u$ and $q(x) = v$. Thus $a(x) = ub(x)$, that is, $a(x)$ is a constant multiple of $b(x)$.

(iv) Since $a(x) \mid b(x)$ we have $b(x) = a(x)q(x)$ and so $vb(x) = ua(x)[u^{-1}vq(x)]$, proving that $ua(x) \mid vb(x)$. ∎

9.3.6 *Corollary*

If $p(x)|q(x), q(x)|p(x)$ and both $p(x)$ and $q(x)$ are monic, then $p(x) = q(x)$.

Proof. By 9.3.5 (i), $p(x) = uq(x)$ for some constant u. Since the leading coefficient of $uq(x)$ is u because $q(x)$ is monic and that of $p(x)$ is 1, it follows that $u = 1$, proving that $p(x) = q(x)$. ∎

Note. Every nonzero polynomial can be expressed as a unit times a monic polynomial. Indeed, if $p(x) = \sum_{j=0}^{n} a_j x^j$ with $a_n \neq 0$, then

$$p(x) = a_n (a_n^{-1} a_0 + a_n^{-1} a_1 x + a_n^{-1} x^2 + \ldots + a_n^{-1} a_{n-1} x^{n-1} + x^n).$$

As a consequence of the Division Algorithm, we show in the next theorem that $F[x]$ is a PID. The proof follows closely the proof given for the corresponding result for integers (Theorem 4.2.10).

9.3.7 *Theorem*

If F is a field, then $F[x]$ is a PID. Moreover, if I is a nonzero ideal, then there exists a *unique* monic polynomial $m(x)$ such that $I = (m(x))$.

Proof. Let I be an ideal of $F[x]$. If $I = (0)$, there is nothing to prove. Assume therefore that $I \neq (0)$ and let $p(x)$ be a nonzero polynomial in I of least degree. Clearly

$$F[x]p(x) = (p(x)) \subseteq I$$

since I is an ideal of $F[x]$.

Conversely, let $a(x) \in I$. By the Division Algorithm, for some $q(x)$ and $r(x)$, $a(x) = p(x)q(x) + r(x)$ where $\partial(r(x)) < \partial(p(x))$. Rearranging the last equation, we have

$$r(x) = a(x) - p(x)q(x)$$

and so $r(x) \in I$ since $a(x), p(x)q(x) \in I$. But $p(x)$ is of minimal nonnegative degree in I and since $\partial(r(x)) < \partial(p(x))$ it follows that

$r(x) = 0$. Hence $a(x) = p(x)q(x) \in (p(x))$, proving that $I \subseteq (p(x))$. Therefore, $I = (p(x))$ and so $F[x]$ is a PID.

Suppose now that $p(x) = \sum_{j=0}^{n} a_j x^j$ with $a_n \neq 0$. Since I is an ideal, $m(x) = a_n^{-1} p(x) \in I$. But $m(x)$ is monic and since $p(x) = a_n m(x)$, it follows that $(m(x)) = I$.

To prove uniqueness of $m(x)$, assume $I = (q(x))$ where $q(x)$ is monic. By virtue of the fact that $q(x) \in (m(x))$, we have $m(x) | q(x)$.

Similarly, $m(x) \in (q(x))$ implies $q(x) | m(x)$. By Corollary 9.3.6, $q(x) = m(x)$. ∎

9.3.8 *Definition*

(See 4.3.2). Let $a(x)$ and $b(x)$ be polynomials over F, not both zero. A *monic* polynomial $d(x)$ is a *greatest common divisor of* $a(x)$ and $b(x)$ if:
  (i) $d(x) | a(x)$ and $d(x) | b(x)$;
  (ii) If $s(x) | a(x)$ and $s(x) | b(x)$, then $s(x) | d(x)$.

As in Chapter 4, we set out to prove the existence and uniqueness of greatest common divisors. The methods of proof are so similar to those for the integers that we merely sketch them for polynomials, leaving the details to the reader.

9.3.9 *Theorem*

If $a(x)$ and $b(x)$ are polynomials over F, not both zero, then the gcd of $a(x)$ and $b(x)$ exists and is unique. Moreover,

$$\gcd(a(x), b(x)) = p(x)a(x) + q(x)b(x)$$

for some polynomials $p(x)$ and $q(x)$.

Proof. The set $I = \{s(x)a(x) + t(x)b(x) | s(x), t(x) \in F[x]\}$ is a nonzero ideal of $F[x]$ and so, by Theorem 9.3.7, there exists a unique monic polynomial $d(x)$ such that $I = (d(x))$. Since $a(x), b(x) \in I$, it follows that $d(x) | a(x)$ and $d(x) | b(x)$ proving (i) of Definition 9.3.8.

Now $d(x) \in I$ implies

$$(*)\, d(x) = p(x)a(x) + q(x)b(x)$$

for some polynomials $p(x)$ and $q(x)$. Suppose that $s(x) \mid a(x)$ and $s(x) \mid b(x)$. Then

$$a(x) = u(x)s(x),\, b(x) = v(x)s(x).$$

Substituting for $a(x)$ and $b(x)$ in (*), we get

$$d(x) = p(x)u(x)s(x) + q(x)v(x)s(x) = [p(x)u(x) + q(x)v(x)]s(x)$$

and so $s(x) \mid d(x)$. Therefore, $d(x)$ is a greatest common divisor of $a(x)$ and $b(x)$.

If $d_1(x)$ is another gcd, then, as in the proof for integers, each of $d(x)$ and $d_1(x)$ divides the other. By Corollary 9.3.6, $d(x) = d_1(x)$ since each polynomial is monic, thus establishing uniqueness of the gcd. ∎

The proof we have just given is an existence proof which gives us no indication of how to compute the gcd of two polynomials. The Euclidean Algorithm for polynomials, which is entirely analogous to the Euclidean Algorithm for integers, provides us with an efficient way of computing gcd's. We give an example to show how the algorithm works, leaving it up to the reader to prove that the procedure does indeed yield the gcd in the general case.

9.3.10 *Example*

Find the gcd of

$$p(x) = 3x^3 + 5x^2 + 6x \text{ and } q(x) = 4x^4 + 2x^3 + 6x^2 + 4x + 5$$

over $\mathbf{Z}_7$ and express it as a linear combination of $p(x)$ and $q(x)$. (Note that we have omitted the bars over the elements of $\mathbf{Z}_7$.)

We apply the Division Algorithm repeatedly to obtain the following:

$$q(x) = p(x)(6x) + (5x^2 + 4x + 5)$$
$$p(x) = (5x^2 + 4x + 5)(2x + 5) + (4x + 3)$$
$$5x^2 + 4x + 5 = (4x + 3)(3x + 4). \blacksquare$$

As with the integers, the last nonzero remainder, if monic, is the gcd. Therefore
$$\gcd(p(x), q(x)) = 4^{-1}(4x+3) = x+6.$$
To find the linear combination which yields the gcd, we work from the bottom up to obtain:
$$4x+3 = p(x)(5x^2 + 2x+1) + q(x)(5x+2).$$
Multiplying both sides by $4^{-1} = 2$, we obtain
$$\gcd(p(x), q(x)) = p(x)(3x^2 + 4x+2) + q(x)(3x+4).$$

9.4 Irreducibility and Factorization of Polynomials

In this section we prove the analogue of the Fundamental Theorem of Arithmetic (Theorem 4.4.4). The reader will find that the proofs leading up to the main theorem are almost identical to the ones given for the integers and (s)he should compare the two.

9.4.1 *Definition*

The polynomial $p(x)$ over the field F is said to be *irreducible over F* if $p(x)$ is of positive degree and cannot be factored as a product of two polynomials over F, *each of positive degree*. If $p(x)$ is of positive degree and is not irreducible, then it is said to be *reducible*.

9.4.2 *Example*

(i) $x^2 + 1$ is irreducible over $\mathbf{R}$, the field of real numbers. It is, however, reducible over $\mathbf{C}$, the field of complex numbers since
$$x^2 + 1 = (x+i)(x-i).$$

(ii) $x^2 - 2$ is irreducible over $\mathbf{Q}$, the field of rational numbers. It is reducible over $\mathbf{R}$ since
$$x^2 - 2 = (x+\sqrt{2})(x-\sqrt{2}).$$
This is not a factorization of $x^2 - 2$ over $\mathbf{Q}$ since $\pm\sqrt{2} \notin \mathbf{Q}$.

(iii) Leaving out bars over the elements of $\mathbf{Z}_2$, $x^2 + 1$ is reducible over $\mathbf{Z}_2$ since $x^2 + 1 = (x+1)^2$.

(iv) Any polynomial of degree 1 is irreducible over any field. ∎

The examples above clearly show that, in general, irreducibility of a polynomial depends on the field over which the polynomial is considered.

Remark. It can be argued that $\frac{1}{2}(2x^2 + 2)$ is a factorization of $x^2 + 1$ over the real numbers. It is, however, a "bogus" factorization. To obtain a "real" factorization of a polynomial, *both* factors must have positive degree. This is equivalent to saying that the degree of each factor is strictly less than the degree of the given polynomial.

9.4.3 Definition

Two polynomials $p(x)$ and $q(x)$ are said to be *relatively prime* if their gcd is 1.

9.4.4 Theorem

(Compare with 4.4.2). Let $a(x), b(x), c(x)$ be polynomials over the field F. If $a(x)\big| b(x)c(x)$ and $a(x)$ is relatively prime to $b(x)$, then $a(x)|c(x)$.

Proof. The proof is entirely similar to that given in 4.4.2 and so we only give a brief sketch. By Theorem 9.3.9 and the hypotheses

$$\gcd(a(x), b(x)) = 1 = p(x)a(x) + q(x)b(x)$$

where $p(x)$ and $q(x)$ are polynomials over F. Thus

$$c(x) = c(x)p(x)a(x) + b(x)c(x)q(x)$$

and so $a(x)\big| c(x)$ since $a(x)\big| b(x)c(x)$ by assumption. ∎

9.4.5 Corollary

(Compare with 4.4.3). If $p(x)$ is irreducible over the field F and $p(x)\big| b(x)c(x)$, then either $p(x)\big| b(x)$ or $p(x)\big| c(x)$.

Proof. Same as the proof given in 4.4.3 with obvious changes in the wording. ∎

Remark. An easy induction argument establishes that if an irreducible polynomial divides any finite product of polynomials, then it must divide one of the factors.

9.4.6 *Theorem*

Every polynomial of positive degree over the field F is uniquely expressible as a unit times a product of monic polynomials, each irreducible over F.

Proof. *Existence of factorization.* We prove this portion by induction on the degree of the polynomial.

Assume that $a(x)$ is a polynomial over F of positive degree.

(i) If $\partial(a(x)) = 1$, then $a(x) = bx + c$ where b and c are elements of F, $b \neq 0$. Therefore $a(x) = b(x + b^{-1}c)$. The existence of the alleged factorization is established in this case since $x + b^{-1}c$ is monic and irreducible.

(ii) Assume that every polynomial of degree less than n is factorable as claimed and let $a(x)$ be a polynomial of degree n. If

$$a(x) = a_0 + a_1 x + a_2 x^2 + ... + a_n x^n$$

is irreducible, then

$$a(x) = a_n(x^n + a_n^{-1}a_{n-1}x^{n-1} + ... + a_n^{-1}a_0)$$

where $x^n + a_n^{-1}a_{n-1}x^{n-1} + ... + a_n^{-1}a_0$ is monic and irreducible.

Assume therefore that $a(x)$ is reducible, say $a(x) = b(x)c(x)$ where $\partial(b(x)) > 0$, $\partial(c(x)) > 0$: It follows that $\partial(b(x)) < n$, $\partial(c(x)) < n$ and so, by the induction hypothesis,

$$b(x) = up_1(x)p_2(x)...p_k(x), \quad c(x) = vq_1(x)q_2(x)...q_l(x)$$

where u and v are units (and thus elements of F) and the $p_j(x), q_j(x)$ are monic irreducible polynomials. Hence

$$a(x) = uvp_1(x)p_2(x)...p_k(x)q_1(x)q_2(x)...q_l(x)$$

and the factorization is of the required form. By the Second Principle of Mathematical Induction, the existence of the factorization is proved.

Uniqueness of factorization. For polynomials of degree 1, uniqueness is obvious.

Assume inductively that we have uniqueness of factorization for polynomials $p(x)$ of degree strictly less than $n, n \geq 2$ and let $\partial(a(x)) = n$. If $a(x)$ is irreducible, there is nothing to prove. From what we have just proved, $a(x)$ is factorable in the stipulated form. Suppose

$$a(x) = up_1(x)p_2(x)...p_k(x) = vq_1(x)q_2(x)...q_l(x)$$

where u and v are units, $k > 1$ and $l > 1$ and the p's and q's monic irreducible polynomials.

Now $p_1(x) | a(x)$ and so, by the Remark following Corollary 9.4.5, $p_1(x)$ divides one of the q's, say $q_1(x)$, without loss of generality. Since both $p_1(x)$ and $q_1(x)$ are monic irreducible, it follows that $p_1(x) = q_1(x)$. Cancelling $p_1(x)$, we get

$$b(x) = up_2(x)p_3(x)...p_k(x) = vq_2(x)q_3(x)...q_l(x) = b(x).$$

Since both k and l are greater than 1, $0 < \partial(b(x)) < \partial(a(x))$ and so by the induction hypothesis,

$$u = v \text{ and } p_j(x) = q_j(x), \ j = 2,3,...,k.$$

But $p_1(x) = q_1(x)$ also and so the two factorizations of $a(x)$ are identical (up to order) and uniqueness is established by induction. ∎

To determine whether a given polynomial over an arbitrary field is irreducible or not is, in general, very hard. There are, however, easily applicable criteria when the degree of the polynomial is 2 or 3 and there is one sufficient condition for irreducibility over the rational numbers. Also, we shall prove that over the real numbers, all irreducible polynomials are of degree 1 or 2 and over the complex numbers, only polynomials of degree 1 are irreducible (a result usually proved in a course in complex variables). Other than these cases, one has to use ad hoc methods to establish irreducibility.

9.4.7 *Definition*

Let S be a commutative ring with unity and b an element of S. Then the map $\sigma_b : S[x] \to S$ defined by

$$\sigma_b(\sum_{j=1}^{n} a_j x^j) = \sum_{j=1}^{n} a_j b^j$$

is called the *substitution map* associated with b. If $a(x) = \sum_{j=1}^{n} a_j x^j$ we

denote $\sum_{j=1}^{n} a_j b^j$ by $a(b)$.

9.4.8 *Theorem*

Any substitution map is a homomorphism.

Proof.
$$\sigma_b(p(x) + q(x)) = p(b) + q(b) = \sigma_b(p(x)) + \sigma_b(q(x));$$
$$\sigma_b(p(x)q(x)) = p(b)q(b) = \sigma_b(p(x))\sigma_b(q(x)).$$

These equalities hold since addition and multiplication of polynomials are defined as though x were an element of the ring. ∎

9.4.9 *Definition*

An element b of a commutative ring S is a *root* of the polynomial $p(x)$ over S if $p(b) = 0$. We also say that b is a *zero* of the polynomial.

We prove a simple but useful theorem with which the reader is probably familiar from elementary algebra.

9.4.10 *Theorem (The Remainder and Factor Theorem)*

If $p(x)$ is a polynomial over the field F and b is an element of F, then the remainder on dividing $p(x)$ by $x - b$ is $p(b)$. In particular, b is a root of $p(x)$ if, and only if, $x - b$ is a factor of $p(x)$.

Proof. By the Division Algorithm,

$$p(x) = (x - b)q(x) + r(x)$$

where $\partial(r(x)) < 1$. Therefore, $r(x)$ is a constant polynomial, say c.

Applying the substitution map associated with b to both sides of the equation above, we get

$$p(b) = (b - b) \, q(b) + r(b) = 0.q(b) + c = c \, ,$$

showing that the remainder is $p(b)$. The second part of the theorem follows immediately. ∎

9.4.11 *Corollary*

A polynomial over the field F of degree greater than 1 which has a root in F is reducible. Moreover, a polynomial of degree 2 or 3 over F is irreducible over F if, and only if, it has no root in F.

Proof. Suppose $p(x)$ is a polynomial of degree greater than 1 with a root b in F. By Theorem 9.4.10, $x - b$ is a factor of $p(x)$. Since $\partial(p(x)) > 1$, the other factor of $p(x)$ is of positive degree and so $p(x)$ is reducible.

Suppose now that $p(x)$ is of degree 2 or 3 and assume $p(x)$ is reducible, say

$$p(x) = s(x)t(x)$$

where $s(x)$ and $t(x)$) are of positive degree. Then

$$\partial(p(x)) = \partial(s(x)) + \partial(t(x))$$

and since the degree of $p(x)$ is 2 or 3, it follows that one of $\partial(s(x))$ and $\partial(t(x))$ is 1, say $s(x) = cx + d$ with $c \neq 0$. Hence $-c^{-1}d$ is a root of $p(x)$. ∎

Remark. (i) If the degree of a polynomial is greater than 3, then it is quite possible for it to be reducible and have no root in the field. For example, the real polynomial $x^4 + 2x^2 + 1$ has no real roots but is reducible over the field of real numbers since

$$x^4 + 2x^2 + 1 = (x^2 + 1)(x^2 + 1) \, .$$

(ii) If a polynomial

$$p(x) = (x-b)^m g(x)$$

and b is not a root of $g(x)$, we say that b is a *root of multiplicity m*.

9.4.12 Corollary

If $f(x)$ is a polynomial over the field F of degree $n \geq 0$, then, counting multiplicities, $f(x)$ has at most n roots in F.

Proof. If $\partial f(x) = 0$ or 1, then $f(x)$ has either 0 or 1 root, respectively, in F.

Assume inductively that the claim holds for polynomials of degree less than n for some $n \geq 2$, and let $\partial f(x) = n$. If $f(x)$ is irreducible, there is nothing to prove since $f(x)$ has no roots in F by the Factor Theorem. Assume therefore that

$$f(x) = g(x)h(x) \text{ where } r = \partial g(x) < n \text{ and } s = \partial h(x) < n.$$

By the induction hypothesis, $g(x)$ has at most r roots in F and $h(x)$ has at most s roots in F. Therefore $f(x)$ has at most $r + s = n$ roots in F and the result follows by induction. ∎

9.5 Polynomials Over More Familiar Fields

We now specialize to the more familiar fields of the rational numbers, the real numbers and the complex numbers and prove several theorems dealing with the question of irreducibility of polynomials over these fields.

The next few results lead to Gauss's Lemma which says that a polynomial over $\mathbf{Z}$ is irreducible over the rationals if and only if, it is irreducible over $\mathbf{Z}$.

9.5.1 Definition

A polynomial with coefficients in $\mathbf{Z}$ is said to be *primitive* if no prime divides all of its coefficients.

Remark. Every polynomial over **Z** is expressible as the product of an integer and a primitive polynomial by just factoring out the greatest common divisor of the coefficients of the polynomial.

9.5.2 *Example*

The polynomial $3x^2 + 6x + 4$ is primitive while $15x^4 - 5x^3 + 10x + 30$ is not but can be expressed as $5(3x^4 - x^3 + 2x + 6)$ where $3x^4 - x^3 + 2x + 6$ is primitive.

9.5.3 *Lemma*

Let S and T be commutative rings with unity and $f : S \to T$ a (ring) homomorphism. Then the map $\theta : S[x] \to T[x]$ defined by

$$\theta(\sum_{j=0}^{n} a_j x^j) = \sum_{j=0}^{n} f(a_j)x^j$$

is also a (ring) homomorphism which agrees with f on S.

Proof.
  (i) Using the fact that f is a homomorphism, we have

$$\theta(\sum a_j x^j + \sum b_j x^j) = \theta[\sum (a_j + b_j)x^j]$$

$$= \sum f(a_j + b_j)x^j = \sum f(a_j)x^j + \sum f(b_j)x^j$$

$$= \theta(\sum a_j x^j) + \theta(\sum b_j x^j).$$

  (ii) Similarly,

$$\theta[(\sum a_j x^j)(\sum b_j x^j)] = \theta(\sum c_j x^j) = \sum f(c_j)x^j$$

where

$$c_j = \sum_{k+l=j} a_k b_l \ \text{ and } \ f(c_j) = \sum_{k+l=j} f(a_k)f(b_l).$$

Also,

$$\theta(\sum a_j x^j)\theta(\sum b_j x^j) = (\sum f(a_j)x^j)(\sum f(b_j)x^j) = \sum d_j x^j$$

where $d_j = \sum_{k+l=j} f(a_k)f(b_l)$. Hence,

$$\theta[(\sum a_j x^j)(\sum b_j x^j)] = \theta(\sum a_j x^j)\theta.(\sum b_j x^j).$$

Finally, if a is a constant polynomial, $\theta(a) = f(a)$. ∎

9.5.4 Theorem

The product of two primitive polynomials is primitive.

Proof. Let $a(x)$ and $b(x)$ be primitive polynomials and suppose $c(x) = a(x)b(x)$ is not primitive. Then there exists a prime number p such that p divides every coefficient of $c(x)$.

Let $v: \mathbf{Z} \to \mathbf{Z}_p$ be the natural homomorphism and let $\theta: \mathbf{Z}[x] \to \mathbf{Z}_p[x]$ be as defined in the lemma. Then

$$\theta(c(x)) = \overline{0} = \theta(a(x))\theta(b(x)).$$

But $\mathbf{Z}_p[x]$ is an integral domain since $\mathbf{Z}_p$ is a field and so either $\theta(a(x)) = \overline{0}$ or $\theta(b(x)) = \overline{0}$. But this implies that either p divides every coefficient of $a(x)$ or p divides every coefficient of $b(x)$, a contradiction. Therefore $c(x)$ is primitive. ∎

9.5.5 Lemma

If $p(x)$ and $q(x)$ are primitive polynomials and $mp(x) = nq(x)$ where m and n integers, then $p(x) = \pm q(x)$.

Proof. The gcd of the coefficients of $mp(x)$ (respectively, $nq(x)$) is $|m|$ (respectively, $|n|$) and so $|m| = |n|$. ∎

The next result is the cornerstone of Eisenstein's Irreducibility Criterion, the subject of Theorem 9.5.7.

9.5.6 Theorem (Gauss's Lemma)

If $a(x)$ is a polynomial with integer coefficients and $a(x) = b(x)c(x)$ with $b(x)$ and $c(x)$ polynomials over the rationals $\mathbf{Q}$, then there exist polynomials $B(x)$ and $C(x)$ with integer coefficients such that $a(x) = B(x)C(x)$. Moreover, $B(x)$ and $C(x)$ are constant multiples of $b(x)$ and $c(x)$, respectively. (More succinctly, if a polynomial over $\mathbf{Z}$ can be factored over $\mathbf{Q}$, then it can be factored over $\mathbf{Z}$.)

Proof. We may assume without loss of generality that $a(x)$ is primitive. Let h (respectively, k) be the lcm of the denominators of the coefficients of $b(x)$ (respectively, $c(x)$). Then

$$hb(x) = \hat{b}(x) \in \mathbf{Z}[x] \text{ and } kc(x) = \hat{c}(x) \in \mathbf{Z}[x].$$

By the remark following 9.5.2, $\hat{b}(x) = mB(x)$ and $\hat{c}(x) = nC(x)$ where m and n are integers and $B(x)$ and $C(x)$ are primitive. Hence,

$$a(x) = \frac{mn}{kh} B(x)C(x)$$

and so

$$kha(x) = mnB(x)C(x)$$

where $B(x)C(x)$ is primitive by 9.5.4 and $a(x)$ is primitive by assumption.

By Lemma 9.5.5, $a(x) = \pm B(x)C(x)$, that is, either

$$a(x) = B(x)C(x) \text{ or } a(x) = -B(x)C(x).$$

In the latter case, setting $-B(x) = B'(x)$, we have $a(x) = B'(x)C(x)$ and the theorem is proved. ∎

9.5.7 *Theorem (Eisenstein's Irreducibility Criterion)*

Let $a(x) = \sum_{j=0}^{n} a_j x^j$ be a polynomial with integer coefficients having the following properties:

   (i) There exists a prime p such that $p \mid a_j, j = 0, 1, ..., n-1$ and p does *not* divide a_n;

   (ii) p^2 does *not* divide a_0 (and so $a_0 \neq 0$).

Then $a(x)$ is irreducible over $\mathbf{Q}$.

Proof. By Gauss's Lemma, it is sufficient to show that $a(x)$ cannot be factored over $\mathbf{Z}$. Assume, by way of contradiction, that

$$a(x) = b(x)c(x)$$

where

$$b(x) = \sum_{j=0}^{m} b_j x^j, \ c(x) = \sum_{j=0}^{k} c_j x^j \in \mathbf{Z}[x]$$

$m > 0$, $k > 0$ and p does not divide either b_m or c_k since $b_m c_k = a_n$.

Let $v : \mathbf{Z} \to \mathbf{Z}_p$ be the natural homomorphism and $\theta : \mathbf{Z}[x] \to \mathbf{Z}_p[x]$ the homomorphism defined in Lemma 9.5.3. Then, since $p \mid a_j, j = 0,1,\ldots,n-1$ and p does not divide a_n, we have $\theta(a(x)) = \bar{a}_n x^n \neq \bar{0}$ and so

$$\bar{0} \neq \bar{a}_n x^n = (\sum_{j=0}^{m} \bar{b}_j x^j)(\sum_{j=0}^{k} \bar{c}_j x^j).$$

Since $\mathbf{Z}_p$ is a field, each polynomial of positive degree over $\mathbf{Z}_p$ is uniquely expressible as a constant times a product of monic irreducible polynomials (Theorem 9.4.6). Therefore

$$\sum_{j=0}^{m} \bar{b}_j x^j = \bar{b}_m x^m, \ \sum_{j=0}^{k} \bar{c}_j x^j = \bar{c}_k x^k, \text{ where } m > 0 \text{ and } k > 0.$$

Hence $\bar{b}_0 = \bar{c}_0 = \bar{0}$, from which it follows that $p \mid b_0$ and $p \mid c_0$. On the other hand, $a_0 = b_0 c_0$ since $a(x) = b(x)c(x)$ and thus, $p^2 \mid a_0$, a contradiction. Therefore, $a(x)$ is irreducible over $\mathbf{Q}$. ∎

9.5.8 *Examples*

(i) $x^2 - 2$ is irreducible over $\mathbf{Q}$ and so $\sqrt{2} \notin \mathbf{Q}$.
(ii) $x^3 - 6x^2 + 15x - 30$ is irreducible over $\mathbf{Q}$. ∎

We now tackle the problem of irreducibility over the complex numbers $\mathbf{C}$ (see Appendix A). This turns out to be quite easy if we are willing to accept a theorem first proved by Gauss and which goes by the name of the Fundamental Theorem of Algebra. We shall not give a proof of this theorem since it is most easily proved using the theory of functions of a complex variable. There are proofs which are completely algebraic except for the assumption that any cubic polynomial over the reals has a real root. The assumption, however, depends on the notion of continuity.

The algebraic portion of the proof is quite intricate and is based on Galois Theory, a topic encountered either in graduate school or in the last year of an honors degree in mathematics.

9.5.9 *Theorem (Fundamental Theorem of Algebra)*

Every polynomial of positive degree over the complex numbers **C** has a root in **C**. ∎

9.5.10 *Corollary*

The only irreducible polynomials over **C** are the polynomials of degree 1 (sometimes called *linear polynomials*).

Proof. Let $p(x)$ be an irreducible polynomial over **C**. By the Fundamental Theorem of Algebra, $p(x)$ has a root b, say, in **C**. By Theorem 9.4.10, $p(x) = (x-b)q(x)$ for some polynomial $q(x)$ over **C**. Since $p(x)$ is irreducible, $q(x)$ is a constant, and so $p(x)$ is linear. ∎

9.5.11 *Corollary*

Every polynomial of positive degree n over **C** is uniquely expressible in the form
$$c(x - a_1)(x - a_2)...(x - a_n).$$

Proof. Apply Theorem 9.4.6 and the previous corollary. ∎

Note that this corollary is an existence theorem. It says nothing about the problem of actually finding the factorization.

Armed with these facts, we are now in a position to deal with the problem of irreducibility of polynomials over the real numbers. The key to tackling this problem is to consider a polynomial with real coefficients as a polynomial over **C**. This is allowable since a real number is a complex number whose imaginary part is zero. This means that $\mathbf{R}[x] \subset \mathbf{C}[x]$.

We repeat here some of the definitions given in Appendix A.

9.5.12 *Definition*

The *complex conjugate* of the complex number $z = a + ib$ is the complex number $\overline{z} = a - ib$. The *real part* of the complex number $z = a + ib$, denoted $\operatorname{Re} z$, is the real number a and its *imaginary part*, denoted $\operatorname{Im} z$ is b. If $b \neq 0$ the complex number is said to be *imaginary*. If $b = 0$, then it is *real*.

The following facts are easy to prove and are left as an exercise for the reader

9.5.13 *Theorem*

Let $z = a + ib$, and $w = c + id$ be complex numbers. Then:
 (i) z is real if, and only if, $z = \overline{z}$.
 (ii) $z.\overline{z} = \overline{z}.z = a^2 + b^2$.
 (iii) $z + \overline{z} = 2\operatorname{Re} z$.
 (iv) $z - \overline{z} = 2i\operatorname{Im} z$.
 (v) $\overline{z.w} = \overline{z}.\overline{w}$.
 (vi) $\overline{z + w} = \overline{z} + \overline{w}$.

Proof. Exercise. ■

9.5.14 *Lemma*

Let $a(x) = \displaystyle\sum_{j=0}^{n} a_j x^j$ be a polynomial with real coefficients. If b is an imaginary root of $a(x)$, so is $\overline{b}$, the complex conjugate of b.

Proof. Since b is a root, we have $\displaystyle\sum_{j=0}^{n} a_j b^j = 0$. Taking conjugates of both sides of this equation and bearing in mind that:

(i) $\bar{a}_j = a_j$, $j = 1, 2, ..., n$ (9.5.13 (i));

(ii) the conjugate of a sum is the sum of the conjugates (9.5.13 (vi));

(iii) the conjugate of a product is the product of the conjugates

(9.5.13 (v)), we get $\displaystyle\sum_{j=0}^{n} a_j \bar{b}^{\,j} = 0$. Therefore, $\bar{b}$ is a root of $\displaystyle\sum_{j=0}^{n} a_j x^{j}$. ∎

The above result can be more succinctly stated as: The imaginary roots of a real polynomial occur in conjugate pairs.

9.5.15 *Theorem*

If $a(x)$ is a polynomial over the reals **R** which is irreducible over **R**, then $a(x)$ is either linear or quadratic.

Proof. By 9.5.11, $a(x)$ factors as a product of linear factors over **C**. If $a(x)$ has a real root r, by 9.4.10, $a(x) = (x - r)q(x)$ for some real polynomial $q(x)$. But $a(x)$ is irreducible over **R** and so $q(x)$ is a constant. Hence $a(x)$ is linear.

Assume therefore that $a(x)$ has no real roots and let b be an imaginary root. By the lemma, $\bar{b}\,(\neq b)$ is also a root. By 9.5.11, $x - b$ and $x - \bar{b}$ are factors occurring in the factorization of $a(x)$ over **C**. Hence $(x - b)(x - \bar{b}) = x^2 - 2\operatorname{Re}b(x) + b\bar{b} = p(x)$ is a real quadratic factor of $a(x)$. Since $a(x)$ is irreducible, $a(x)$ must be a constant times this polynomial, proving that $a(x)$ is quadratic. ∎

To round things off, let us determine which real quadratic polynomials are irreducible over **R**.

9.5.16 *Theorem*

The real polynomial $ax^2 + bx + c$ is irreducible over **R** if, and only if, $b^2 - 4ac < 0$.

Proof. Assume $ax^2 + bx + c$ is reducible over **R**. By 9.4.11, $ax^2 + bx + c$ has a root in **R**. Solving using the usual quadratic formula, the roots are

$$(-b \pm \sqrt{b^2 - 4ac}) \Big/ 2a$$

and if either of these is to be real, then $b^2 - 4ac \geq 0$.

Conversely, suppose $b^2 - 4ac \geq 0$. Then the roots $(-b \pm \sqrt{b^2 - 4ac}) \Big/ 2a$ are both real and so, by 9.4.10, $ax^2 + bx + c$ is reducible. ∎

Combining Theorem 9.5.15 with Theorem 9.4.6, we get:

9.5.17 *Theorem*

Every real polynomial of positive degree is uniquely expressible as a nonzero real number times a product of monic linear or irreducible quadratic factors. ∎

Once again, these theorems are existence theorems and are of little help in actually computing the factorization. Nevertheless, they are very powerful results which are often used to answer some deep theoretical questions.

9.6 Factor Rings of the Form $F[x] \Big/ (g(x))$, F a Field

Earlier on, we proved that $\mathbf{Z} \Big/ (p)$, the quotient ring **Z** mod the principal ideal generated by p, is a field if, and only if, p is prime. More generally, we proved that an element $\bar{a} \in \mathbf{Z} \Big/ (n)$ is a unit if, and only if, $\gcd(a, n) = 1$. The next theorems establish analogous results for $F[x]$ where F is a field.

9.6.1 *Theorem*

An element $b(x)+(p(x))=\overline{b(x)} \in \left.F[x]\middle/_{(p(x))}\right.$ is a unit if, and only if, $\gcd(b(x), p(x))=1$.

Proof. Suppose $\gcd(b(x), p(x))=1$. Then there exist polynomials $q(x)$ and $a(x)$ such that
$$a(x)p(x)+b(x)q(x)=1.$$
Then
$$\overline{1} = \overline{a(x)p(x)+b(x)q(x)} = \overline{a(x)}.\overline{p(x)}+\overline{b(x)}.\overline{q(x)}$$
and so
$$\overline{1} = \overline{a(x)}.\overline{0}+\overline{b(x)}.\overline{q(x)}.$$
Thus $\overline{b(x)}.\overline{q(x)}=\overline{1}$ proving that $\overline{b(x)}$ is a unit with inverse $\overline{q(x)}$.

Conversely, suppose $\overline{b(x)}$ is a unit with $\overline{q(x)}$ as inverse. Then
$$\overline{b(x)}.\overline{q(x)} = \overline{1}$$
and so
$$b(x)q(x)-1\in (p(x)).$$
Hence there exists $a(x)$ such that
$$b(x)q(x)-1= a(x)p(x).$$
Therefore,
$$b(x)q(x)-a(x)p(x)=1$$
and so $\gcd(b(x), p(x))=1$. ∎

The following is the analogue of the theorem: $\left.\mathbf{Z}\middle/_{(p)}\right.$ is a field, if, and only if, p is prime.

9.6.2 *Theorem*

If F is a field and $p(x)$ a polynomial over F, then $\left.F[x]\middle/_{(p(x))}\right.$ is a field if, and only if, $p(x)$ is irreducible over F.

Proof. Suppose $p(x)$ is irreducible and let $\overline{q(x)} \in \dfrac{F[x]}{(p(x))}$ where $\overline{q(x)} \neq \overline{0}$. Then $p(x)$ is not a factor of $q(x)$ and so $\gcd(p(x), q(x)) = 1$. Therefore, by the previous theorem, $\overline{q(x)}$ is a unit and so $\dfrac{F[x]}{(p(x))}$ is a field.

Conversely, assume that $p(x)$ is reducible, say $p(x) = a(x)b(x)$ where $a(x)$ and $b(x)$ are polynomials of positive degree. Then $\overline{a(x)} \neq \overline{0} \neq \overline{b(x)}$ and $\overline{0} = \overline{p(x)} = \overline{a(x)}.\overline{b(x)}$, and so $\dfrac{F[x]}{(p(x))}$ has nonzero zero divisors. Therefore, $\dfrac{F[x]}{(p(x))}$ cannot be a field. ∎

In $\dfrac{Z}{(p)}$ every coset has a representative b where $0 \le b < p$. A similar statement can be proved for the cosets of $\dfrac{F[x]}{(p(x))}$.

9.6.3 *Theorem*

If $p(x)$ is a polynomial of degree n over the field F, then every coset of $\dfrac{F[x]}{(p(x))}$ has a unique representative of degree less than n.

Proof. Consider the coset $\overline{a(x)} = a(x) + (p(x))$ with representative $a(x)$. By the Division Algorithm,

$$a(x) = p(x)q(x) + r(x)$$

where $\partial(r(x)) < n$. Hence

$$\overline{a(x)} = \overline{p(x)q(x) + r(x)} = \overline{p(x).q(x)} + \overline{r(x)} = \overline{0}.\overline{q(x)} + \overline{r(x)} = \overline{r(x)}$$

and so $r(x)$ is a representative of the coset of degree less than n.

To prove uniqueness, suppose $\overline{r(x)} = \overline{s(x)}$ with $\partial(r(x)) < n$, $\partial(s(x)) < n$. Then

$$r(x) - s(x) \in (p(x))$$

and so

$$r(x) - s(x) = p(x)q(x)$$

for some $q(x) \in F[x]$. Therefore,

$$\partial(r(x) - s(x)) = \partial(p(x)) + \partial(q(x)).$$

But $\partial(r(x) - s(x)) < n$ and so $\partial(p(x)) + \partial(q(x)) < n$. Hence, since $\partial(p(x)) = n$, we conclude that $\partial(q(x)) < 0$, from which it follows that $q(x) = 0$. Thus $r(x) - s(x) = 0$ and so $r(x) = s(x)$. ∎

Remark. Theorem 9.6.3 implies that there are as many elements in $F[x] \big/ (p(x))$ as there are polynomials over F of degree less than the degree of $p(x)$. In case F is infinite, this does not say much other than that $F[x] \big/ (p(x))$ is infinite.

If F is finite, however, we can compute the exact size of $F[x] \big/ (p(x))$. Indeed, if F is a field with k elements and $\partial(p(x)) = n$, then $F[x] \big/ (p(x))$ has k^n elements since a polynomial of degree less than n has n coefficients and each of these can be chosen in k ways.

9.6.4 *Examples*

(i) Consider the polynomial $p(x) = x^2 + x + 1$ over $\mathbf{Z}_2 = \{0,1\}$. (Note that we have omitted the bars over the elements of $\mathbf{Z}_2$. From now in we shall often do this.) Since no element of $\mathbf{Z}_2$ is a root of $p(x)$, it follows by Corollary 9.4.11 that $p(x)$ is irreducible over $\mathbf{Z}_2$. By Theorem 9.6.2, $\mathbf{Z}_2[x] \big/ (p(x))$ is a field. By the remark above, it contains as many elements as there are polynomials over $\mathbf{Z}_2$ of degree less than 2. Once again, omitting bars to denote cosets, we have

$$\mathbf{Z}_2[x] \big/ (p(x)) = \{0, 1, x, x+1\}.$$

We compute the Cayley Tables for addition and multiplication of this field below.

Table 9.1 Addition for $\mathbf{Z}_2[x]\big/(x^2+x+1)$

+	0	1	x	$x+1$
0	0	1	x	$x+1$
1	1	0	$x+1$	x
x	x	$x+1$	0	1
$x+1$	$x+1$	x	1	0

Table 9.2 Multiplication in $\mathbf{Z}_2[x]\big/(x^2+x+1)$

$\cdot$	0	1	x	$x+1$
0	0	0	0	0
1	0	1	x	$1+x$
x	0	x	$x+1$	1
$x+1$	0	$x+1$	1	x

The reader may be puzzled over the Cayley Table for multiplication, so we explain how the various entries were computed. First, observe that since we are dealing with the field $\mathbf{Z}_2$ and $1 = -1$ in this field, there is no need to use the minus sign.

Next, in $\dfrac{\mathbf{Z}_2[x]}{(p(x))}$, $x^2+x+1=0$. (Strictly speaking we should write $x^2+x+1=\overline{0}$.) From this we deduce

$$(*)\; x^2 = x+1,\; x^2+1 = x \text{ and } x^2+x = 1.$$

Therefore, in computing $x(x+1) = x^2+x$, we replace x^2+x by 1. Indeed, whenever the exponent of x is greater than 1, we use (*) to reduce the exponent. Alternatively, we can use the Division Algorithm to obtain

$$a(x) = (x^2+x+1)q(x) + r(x)$$

where $\partial r(x) < 2$.

Note that $\left(\mathbf{Z}_2[x]\Big/(p(x))\,,+\right)$ is another instance of the Klein 4-group

while the multiplicative group of nonzero elements of $\mathbf{Z}_2[x]\Big/(p(x))$ is a

cyclic group of order 3 having x and $x+1$ as generators.

We shall see (9.6.5 and 9.6.9) that this is typical for finite fields.

(ii) The polynomial $p(x) = x^2 + 1$ is irreducible over the real

numbers $\mathbf{R}$ so that, according to Theorem 9.6.2, $\mathbf{R}[x]\Big/(x^2+1)$ is a field.

Again, omitting bars, the elements of $\mathbf{R}[x]\Big/(x^2+1)$ are of the form

$a + bx$. Addition and multiplication are performed as follows:

$$(a+bx)+(c+dx) = a+c+(b+d)x;$$
$$(a+bx)(c+dx) = ac+bdx^2+(ad+bc)x.$$

But in $\mathbf{R}[x]\Big/(x^2+1)$, $x^2+1 = 0$ and so $x^2 = -1$. Therefore,

$$(a+bx)(c+dx) = ac-bd+(ad+bc)x.$$

These computations should ring a bell with the reader. They should remind him/her of the complex numbers. In fact, the map

$f : \mathbf{C} \to \mathbf{R}[x]\Big/(x^2+1)$ defined by $f(a+ib) = a+bx$ is an isomorphism.

Therefore, the field $\mathbf{R}[x]\Big/(x^2+1)$ is, from an algebraic standpoint, the

field of complex numbers.

(iii) Let us compute the multiplicative inverse of $a(x) = x^2 + x + 1$ in

$\mathbf{Z}_5[x]\Big/(p(x))$ where

$$p(x) = x^3 + x^2 + x + 1 = (x^2+1)(x+1).$$

Since $p(x)$ is not irreducible, $\mathbf{Z}_5[x]\Big/(p(x))$ is not a field. Nevertheless, if

$a(x)$ and $p(x)$ are relatively prime, $a(x)$ will have an inverse.

An easy computation by repeated application of the Division Algorithm yields $(x^2+x+1)^{-1} = 4x$. We check this by finding the product:

$$(x^2+x+1)(4x) = 4x^3+4x^2+4x = 4(x^3+x^2+x+1)-4 = -4 = 1$$

since
$$\overline{x^3 + x^2 + x + 1} = \overline{0}.$$
Also, $\overline{-4} = \overline{1}$ in $\mathbf{Z}_5$. ∎

We end this section by proving two theorems which shed light on both the additive and multiplicative structures of a finite field.

9.6.5 *Theorem*

A finite field has p^n elements for some prime p.

Proof. We give two proofs of this.

(i) A finite field F is clearly of characteristic a prime p since otherwise it would contain a copy of $\mathbf{Q}$. Let $P(\cong \mathbf{Z}_p)$ be the prime subfield. Then we can view F as a finite-dimensional vector space over P. (See Appendix B.)

If the dimension is n and $\alpha_1, \alpha_2, ..., \alpha_n$ is a basis, then there are a total of p^n linear combinations of the α's and so $|F| = p^n$.

(ii) Since the additive order of every nonzero element of F is p, we can apply Theorem 6.5.6 to deduce that, as an additive group,

$$F = <x_1> \oplus <x_2> \oplus ... \oplus <x_n>$$

where each cyclic subgroup is of order p. Therefore $|F| = p^n$. ∎

The previous result gives us a clear picture of the additive structure of a finite field. What about the multiplicative structure of the group of nonzero elements of the field? We first prove a lemma and then show that it is cyclic.

9.6.6 *Definition*

Let G be a finite group. Then $m = \max\{|g| \mid g \in G\}$ is called the *exponent* of G.

9.6.7 *Lemma*

If G is a finite abelian group with exponent m, then $x^m = e$ for all $x \in G$.

Proof. Let g be an element of order m and suppose x is an element such that $x^m \neq e$. Let

$$m = \prod_{i=1}^{s} p_i^{k_i}, \; k_i \geq 0 \text{ for all } i \text{ and } |x| = \prod_{i=1}^{s} p_i^{l_i}, \; l_i \geq 0 \text{ for all } i.$$

Since $x^m \neq e$, it follows that $|x|$ does not divide m. Therefore, for some $j, l_j > k_j$. Now

$$|g^{p_j^{k_j}}| = p_1^{k_1} p_2^{k_2} ... p_{j-1}^{k_{j-1}} p_{j+1}^{k_{j+1}} ... p_s^{k_s} \text{ and } |x^{p_1^{l_1} p_2^{l_2} ... p_{j-1}^{l_{j-1}} p_{j+1}^{l_{j+1}} ... p_s^{l_s}}| = p_j^{l_j}.$$

Since these two orders are relatively prime, it follows that

$$|g^{p_j^{k_j}} x^{p_1^{l_1} p_2^{l_2} ... p_{j-1}^{l_{j-1}} p_{j+1}^{l_{j+1}} ... p_s^{l_s}}| = p_1^{k_1} p_2^{k_2} .. p_j^{l_j} ... p_s^{k_s},$$

(why?) a contradiction since $p_1^{k_1} p_2^{k_2} ... p_j^{l_j} ... p_s^{k_s} > m$. ∎

9.6.8 *Corollary*

If G is a finite abelian group, then G is cyclic iff $|G| = $ exponent of G.

Proof. Suppose the exponent m of G is equal to its order and let $|g| = m$. Then $|<g>| = m$ and so $G = <g>$ is cyclic.

The converse is obvious. ∎

9.6.9 *Theorem*

The group of nonzero elements of a finite field F is cyclic.

Proof. Let $F^* = F \setminus \{0\}$ be of order n and let the exponent of F^* be m. By the lemma, the polynomial $x^m - 1$ has n roots in F^*. By Theorem 9.4.12 $x^m - 1$ has at most m roots in F^*. Hence $n \leq m$ and since m is clearly at most n we have $m = n$. By the corollary above, F^* is cyclic. ∎

Unlike groups, where there are many pairwise nonisomorphic groups of a given order, there is only one field of order p^n. We quote the theorem below and postpone its proof until the next chapter.

9.6.10 *Theorem*

Any two finite fields of the same order are isomorphic. ∎

9.7 Exercises

1) Apply the Division Algorithm to the following pairs of polynomials over the given fields.

  (i) $x^4 - 3x^3 + 5x^2 + x + 2$, $x^2 + 4x - 3$ over (a) $\mathbf{R}$; (b) $\mathbf{Z}_7$;

  (ii) $x^5 + 4x^2 - 8$, $x^3 - 3x^2 + x + 10$ over (a) $\mathbf{R}$; (b) $\mathbf{Z}_{11}$.

2) Find the gcd of the following pairs of polynomials and express it as a linear combination of the given polynomials.

  (i) $x^3 + x^2 + x$, $x^4 + 2x^2 + 2x + 1$ over $\mathbf{Z}_3$ and over $\mathbf{Z}_5$.

  (ii) $2x^3 + x - 2$, $x^2 + 2x + 1$ over $\mathbf{Z}_5$ and over $\mathbf{Z}_7$.

  (iii) $x^3 - 6x^2 + x + 4$, $x^5 - 6x + 1$.

3) (Rational Root Theorem) Let $\sum_{j=0}^{n} a_j x^j$ be a polynomial with integral coefficients and suppose that the rational number p/q is a root of $\sum_{j=0}^{n} a_j x^j$ where p and q are relatively prime integers. Prove that $p \mid a_0$ and $q \mid a_n$.

4) Show by means of an example that there is a nonzero polynomial of degree 2 over $\mathbf{Z}_4$ which has more than 2 roots in $\mathbf{Z}_4$ (Of course, $\mathbf{Z}_4$ is not a field).

5) Find all irreducible quadratic and cubic polynomials over the field $\mathbf{Z}_2$. Also find at least two quartic (degree 4) irreducible polynomials over the same field.

6) Let I be the set of all polynomials in $\mathbf{Z}[x]$ with even constant term. Prove that I is an ideal of $\mathbf{Z}[x]$ which is not principal.

7) Use Eisenstein's Irreducibility Criterion to prove that $x^6 + x^5 + x^4 + x^3 + x^2 + x + 1$ is irreducible over the rationals. [Hint: As the polynomial stands, we cannot apply Eisenstein's Irreducibility Criterion. If, however, we transform it by the substitution $y + 1 = x$, we obtain a polynomial to which the criterion can be applied. More generally, it can be proved in a similar manner that the polynomial $\sum_{j=0}^{p-1} x^j$ is irreducible over the rationals if p is prime.]

8) Prove that $\mathbf{Q}[\sqrt{2}] = \{a + b\sqrt{2} \mid a, b \in \mathbf{Q}\}$ is a subfield of $\mathbf{Q}$ which is isomorphic to $\mathbf{Q}[x]\big/(x^2 - 2)$. (Hint: Define a map $f : \mathbf{Q}[x] \to \mathbf{Q}[\sqrt{2}]$ by $f(\sum_{j=0}^{n} a_j x^j) = \sum_{j=0}^{n} a_j (\sqrt{2})^j$ and show that it is a homomorphism onto $\mathbf{Q}[\sqrt{2}]$. Find the kernel of f and apply the First Isomorphism Theorem).

9) (i) Find a representative of the coset $\overline{x^7 + x^4 + x + 1}$ in $\mathbf{Z}_2[x]\big/(x^3 + x + 1)$ of degree less than 3.

   (ii) Find the multiplicative inverse of $\overline{x^2 + 1}$ in the field $\mathbf{Z}_2[x]\big/(x^3 + x + 1)$.

10) Find two irreducible polynomials $p(x)$ and $q(x)$ each of degree 2 over $\mathbf{Z}_3$ and construct the fields $\mathbf{Z}_3[x]\Big/(p(x))$ and $\mathbf{Z}_3[x]\Big/(q(x))$ each of order 9. By Theorem 9.6.10, these two fields should be isomorphic. Can you find an isomorphism between them?

11) Prove that the Binomial Theorem holds in any commutative ring.

12) Prove that if F is a finite field of characteristic p, then the map $f : F \to F$ defined by $f(a) = a^p$ is an automorphism of F which fixes each element of the prime field. It is called the Frobenius automorphism.

13) If p is a prime, prove that the polynomial $x^p - x$ factors as a product of p linear factors over Z_p.

14) Prove that if $p(x)$ and $q(x)$ are relatively prime polynomials each dividing $s(x)$, then $p(x)q(x)$ divides $s(x)$.

15) Construct a field of order 8 and compute its Cayley Tables.

16) If F is a field and $p(x)$ a polynomial over F, prove that every element of $F[x]\Big/(p(x))$ is either a unit or a divisor of zero.

17) Are the following polynomials irreducible over the given field? Give reasons.
    (i) $3x^8 - 4x^3 + 6x^2 - 2x + 10$ over $\mathbf{Q}$;
    (ii) $x^3 + 3x^2 + x + 1$ over $\mathbf{Z}_5$;
    (iii) $x^5 + 15$ over $\mathbf{Q}$;
    (iv) $x^4 - 6x^3 + 3x^2 + 9x + 12$ over $\mathbf{R}$;
    (v) $x^3 - 3x^2 + 4x - 2$ over $\mathbf{C}$.

18) Find all irreducible polynomials of degree 3 over $\mathbf{Z}_3$.

19) List all polynomials of degree 4 over $\mathbf{Z}_2$ and factor each into a product of irreducible polynomials

20) Use Gauss's Lemma and the factor theorem to prove that $\sqrt[m]{n} \in \mathbf{Q}$ iff n is the mth power of an integer.

21) Suppose $f(x) \in \mathbf{Z}[x]$ and let p be a prime. If $\overline{f(x)}$ is the image of $f(x)$ under the homomorphism $\theta : \mathbf{Z}[x] \to \mathbf{Z}_p[x]$ (see Lemma 9.5.3), $\partial f(x) = \partial \overline{f(x)} \geq 1$ and $\overline{f(x)}$ is irreducible over $\mathbf{Z}_p$, then $f(x)$ is irreducible over $\mathbf{Q}$. Show by means of an example that the condition $\partial f(x) = \partial \overline{f(x)}$ Is essential.

22) With the help of the previous problem, show that each of the following polynomials is irreducible over $\mathbf{Q}$
  (i) $3x^4 + 2x^3 - 6x^2 + 5x - 9$.
  (ii) $5x^4 - 15x^3 + 6x^2 - 4x + 25$.
  (iii) $\sum\limits_{i=0}^{4} a_i x^i$ where each a_i is an odd integer.

23) If $x^2 - 1$ is divided into $x^{100} + x^{37} + 2$ in $\mathbf{Z}[x]$, what is the remainder?

24) Let F be a field and set
$$I = \{a_0 + a_1 x + a_2 x^2 + \ldots + a_n x^n \mid a_0 + a_1 + a_2 + \ldots + a_n = 0\}.$$
Show that I is an ideal and find a generator for I.

25) Recall that a maximal ideal of a ring R is an ideal $M \neq R$ which is not properly contained in any ideal other than R. Determine the maximal ideals of each of the following:

  (i) $\mathbf{Q} \oplus \mathbf{Q}$.
  (ii) $\mathbf{Z}_6[x] \Big/ (x^2)$.
  (iii) $\mathbf{R}[x] \Big/ (x^2 - 3x + 2)$.

26) Prove that two polynomials $p(x)$ and $q(x)$ over $\mathbf{Z}[x]$ are relatively prime over $\mathbf{Q}[x]$ if, and only if, $\mathbf{Z}[x]p(x) + \mathbf{Z}[x]q(x)$, the ideal generated by $p(x)$ and $q(x)$, contains an integer.

27) Factor $x^5 + x + 1$ into irreducible factors over $\mathbf{Z}_2$.

28) Prove that the kernel of the homomorphism $f : \mathbf{Z}[x] \to \mathbf{R}$ defined by $f(p(x)) = p(2 + \sqrt{3})$ is a principal ideal and find a generator.

29) Let $p(x) = x^3 - 2$ and $q(x) = x^3 + 2$ be polnomials in $\mathbf{Z}_7[x]$. Show that both are irreducible over $\mathbf{Z}_7$ and exhibit an isomorphism from $\mathbf{Z}_7[x]\!\Big/\!(p(x))$ to $\mathbf{Z}_7[x]\!\Big/\!(q(x))$.

30) Let R be a commutative ring in which $a^2 = 0$ implies $a = 0$. Show that if $p(x) = \sum_{i=0}^{n} a_i x^i \in R[x]$ is a nonzero zero-divisor, then there is a nonzero element b in R such that $ba_i = 0$ for all $i, 0 \le i \le n$.

31) Let I be an ideal of the commutative ring R. Prove that
$$I[x] = \{\sum a_i x^i \mid a_i \in I\}$$
is an ideal of $R[x]$. Moreover, show that
$$\frac{R[x]}{I[x]} \cong \left(R\Big/I\right)[x].$$

32) Prove that a polynomial in $\mathbf{Z}_2[x]$ has a factor $x + 1$ iff it has an even number of nonzero terms.

33) Let $p_i(x), 1 \le i \le n$ be irreducible polynomials over the field F and let $q(x) = \prod_{i=1}^{n} p_i(x)$. Prove that
$$F[x]\!\Big/\!(q(x)) \cong \bigoplus_{i=1}^{n} F[x]\!\Big/\!(p_i(x)).$$

Abstract Algebra

34) If φ is an automorphism of $F[x]$ such that $\varphi(a) = a$ for all $a \in F$, prove that there exist $b \neq 0$, $c \in F$ such that
$$\varphi(f(x)) = f(bx + c).$$

35) Prove that $x^3 + 3x + 2$ is irreducible over **Q.**

36) For a given n, find an automorphism of **C**$[x]$ of order n.

Chapter 10

Field Extensions

10.1 Introduction

In number theory, algebraic geometry, combinatorial mathematics and other areas of mathematics, it often happens that the field we are operating in is too confining and we need to find a bigger field which will give us more scope. In this section we shall discuss how to extend fields and we shall show how our results can be applied to finite fields.

10.2 Definitions and Elementary Results

10.2.1 *Definition*

Let E and F be fields. We say E is *an extension of* F if there exists a monomorphism $\varphi : F \to E$. The map φ is called the *embedding monomorphism*.

Note. If $E \supseteq F$, then E is an extension of F with embedding monomorphism $\varphi : F \to E$ defined by $\varphi(a) = a$.

10.2.2 *Example*

If F is a field and $p(x)$ is an irreducible polynomial over F, then, by Theorem 9.6.2, $F[x] \big/ (p(x))$ is a field and the map $\varphi : F \to F[x] \big/ (p(x))$

defined by $\varphi(a) = \bar{a}$ is clearly a monomorphism and so $F[x] \big/ (p(x))$ is an extension of F. ∎

Students sometimes have difficulties in dealing with the notion of an extension, particularly since mathematicians tend to think of F as actually being contained in the extension. Strictly speaking, an isomorphic copy of F, and not F itself, is contained in the extension. However, the following theorem shows that we can construct a field which actually contains F itself and is isomorphic to the extension field.

10.2.3 *Theorem*

If E is an extension of F with embedding monomorphism φ, there exists a field $\hat{E}$ containing F and an isomorphism $\hat{\varphi}: \hat{E} \to E$ extending φ, i.e $\hat{\varphi}$ restricted to F is φ.

Proof. Let J be a set disjoint from F in 1:1 correspondence with the set $E \setminus \varphi(F)$ and θ a bijection from J to $E \setminus \varphi(F)$. Set $\hat{E} = J \cup F$ and Define the map $\hat{\varphi}: \hat{E} \to E$ by

$$\hat{\varphi}(a) = \begin{cases} \varphi(a) \text{ if } a \in F \\ \theta(a) \text{ if } a \in J. \end{cases}$$

Clearly $\hat{\varphi}$ is a bijection since both φ and θ are 1:1. Now define two operations $\oplus$ and $\bullet$ on $\hat{E}$ as follows:

$$a + b = \hat{\varphi}^{-1}[\hat{\varphi}(a) + \hat{\varphi}(b)]$$

$$a \bullet b = \hat{\varphi}^{-1}[\hat{\varphi}(a).\hat{\varphi}(b)].$$

We claim that $(\hat{E}, \oplus, \bullet)$ is a field such that $\oplus$ and $\bullet$ restrict respectively to $+$ and $.$ on F.

Indeed, suppose a and b are in F. Then $\hat{\varphi}(a) = \varphi(a)$ and $\hat{\varphi}(b) = \varphi(b)$. Furthermore, $a + b \in F$ and so

$$\hat{\varphi}(a + b) = \varphi(a + b) = \varphi(a) + \varphi(b)$$

since φ is a homomorphism
Therefore

$$a \oplus b = \hat{\varphi}^{-1}[\hat{\varphi}(a) + \hat{\varphi}(b)] = \hat{\varphi}^{-1}[\varphi(a + b)]$$

$$= \varphi^{-1}[\varphi(a + b)] = a + b$$

since $\hat{\varphi}^{-1} = \varphi^{-1}$ on $\varphi(F)$ because $\varphi(a+b) \in \varphi(F)$. A similar argument shows that $a \bullet b = a.b$ for all for all $a,b \in F$.

The proof that $(\hat{E}, \oplus, \bullet)$ is a field is left as an exercise for the reader. ■

An irreducible polynomial of degree greater than 1 over a field F has no root in F but may have a root in an extension field. A natural question which arises is: Given an irreducible polynomial f over some field F, does there exist an extension field of F containing a root of f? The following result, due to Kronecker, is of fundamental importance.

10.2.4 *Theorem* (*Kronecker*)

If $p(x)$ is an irreducible polynomial over the field F, there exists an extension field E of F containing a root of $p(x)$.

Proof. By Theorem 9.6.2 $F[x]\big/(p(x))$ is a field and $\varphi : F \to F[x]\big/(p(x))$ defined by

$$\varphi(a) = \overline{a} = a + (p(x))$$

is clearly an embedding map. If

$$p(x) = \sum_{i=0}^{n} a_i x^i, \text{ then } \overline{0} = \overline{p(x)} = \sum_{i=0}^{n} \overline{a}_i \overline{x}^i$$

and so $\overline{x}$ a root of the polynomial $\sum_{i=0}^{n} \overline{a}_i X^i$ in the indeterminate X over $\varphi(F)$. By the previous theorem, we can construct a field $\hat{E} \supseteq F$ and an isomorphism $\hat{\varphi} : \hat{E} \to F[x]\big/(p(x))$ which extends φ. Let $u = \hat{\varphi}^{-1}(\overline{x})$. Then

$$\hat{\varphi}(\sum_{i=0}^{n} a_i u^i) = \sum_{i=0}^{n} \overline{a}_i \overline{x}^i = \overline{0}$$

whence $\sum_{i=0}^{n} a_i u^i = 0$ since $\hat{\varphi}$ is an isomorphism. ■

Having constructed an extension field of F containing a root of the irreducible polynomial $f(x)$, we can repeat the procedure a number of times to obtain all the other roots of $f(x)$. We show precisely how this is done.

10.2.5 *Definition*

Let $f(x)$ be a polynomial over the field F and E an extension of F. We say $f(x)$ *splits over* E if $f(x)$ factors as a product of linear factors over E.

10.2.6 *Corollary*

If $f(x)$ is a polynomial over F, then there is an extension field E of F over which $f(x)$ splits.

Proof. If $\partial f(x) = 1$, there is nothing to prove. Assume inductively that if $\partial p(x) < n$, then there is a field over which $p(x)$ splits and suppose $\partial f(x) = n$. Let $g(x)$ be an irreducible factor of $f(x)$ and apply Kronecker's Theorem to obtain a field E_1 containing a root b of $g(x)$. By Theorem 9.4.10 $g(x) = (x-b)h(x)$ over E_1. Letting $s(x) = h(x)q(x)$ where $f(x) = g(x)q(x)$, we have

$$f(x) = (x-b)s(x)$$

over E_1. Since $\partial s(x) < n$, by the induction hypothesis, there is an extension field E of E_1 (and hence of F) over which $s(x)$ splits. Therefore there is a field E over which $f(x)$ splits. By induction, the result follows.∎

If F is a subfield of a field E, then E is a vector space over F. Indeed, $(E,+)$ is an abelian group and we define multiplication of a vector $\beta \in E$ by a scalar $a \in F$ as ordinary multiplication in the field E. Thus the vector β times the scalar a is $a\beta$. It is then a simple exercise to verify the axioms of a vector space (see Appendix B).

10.2.7 *Definition and notation*

(i) If E is an extension field of F, the dimension of the vector space E over F is denoted by $|E:F|$ and is called the *degree* of the extension. If $|E:F|$ is finite, we say that E is a *finite-dimensional* extension of F. We sometimes say more briefly that E/F is finite. (Read "E over F is finite").

(ii) If E is an extension of F and T is a subset of E, then $F[T]$ will denote the smallest subring of E containing both F and T and $F(T)$ the smallest subfield containing F and T.

As usual, when we speak of the smallest algebraic object, we must prove that it exists and, if possible, show what form the elements take.

10.2.8 *Theorem*

If E is an extension of F and T is a subset of E, then $F[T]$ and $F(T)$ exist.

Proof. If $\mathbb{S}$ (respectively $\mathbb{F}$) is the collection of all subrings (respectively, subfields) of E containing F and T, then

$$F[T] = \bigcap_{S \in \mathbb{S}} S \text{ and } F(T) = \bigcap_{K \in \mathbb{F}} K.$$

Note that $E \in \mathbb{S} \cap \mathbb{F}$ so that $\mathbb{S} \neq \phi \neq \mathbb{F}$. ∎

10.2.9 *Definition*

Let E and F be fields with $E \supseteq F$ and B a subset of E. The *span of B over F* is the subspace $\mathrm{sp}_F(B)$ of the vector space E over F generated by the vectors in B. The subspace $\mathrm{sp}_F(B)$ consists of all (finite) linear combinations of elements of B with coefficients from F (see Appendix B).

10.2.10 *Theorem*

Let E and F be fields with $E \supseteq F$ and T a subset of E. If B is the set of all finite products of elements of T, then

(i) $F[T] = \mathrm{sp}_F(B \cup 1)$;

(ii) $F(T) = \{ab^{-1} \mid a,b \in F[T], b \neq 0\}$.

Proof. (i) Clearly

$$\mathrm{sp}_F(B \cup 1) \subseteq F[T].$$

Conversely, since $1 \in B \cup 1$, it follows that F and T are contained in $\mathrm{sp}_F(B \cup 1)$. On the other hand, it is easy to see that $\mathrm{sp}_F(B \cup 1)$ is a subring of E and so, since $F[T]$ is the smallest subring containing F and T, it follows that $F[T] \subseteq \mathrm{sp}_F(B \cup 1)$. Thus $F[T] = \mathrm{sp}_F(B \cup 1)$.

(ii) It is obvious that $F[T] \subseteq F(T)$. Also, since $F(T)$ is a field, for each $b \in F[T]$, $b \neq 0$, we have $b^{-1} \in F(T)$. Hence

$$\{ab^{-1} \mid a,b \in F[T]\} \subseteq F(T).$$

On the other hand it is a routine exercise to show that $\{ab^{-1} \mid a,b \in F[T], b \neq 0\}$ is a field (it is, in fact, the field of quotients of the integral domain $F[T]$.) Therefore $F(T) \subseteq \{ab^{-1} \mid a,b \in F[T], b \neq 0\}$ since $F(T)$ is the smallest subfield containing F and T. Hence

$$F(T) = \{ab^{-1} \mid a,b \in F[T], b \neq 0\}. \blacksquare$$

Remark. If $T = \{u\}$ we write $F[u]$ and $F(u)$ instead of $F[\{u\}]$ and $F(\{u\})$.

Observe that $F[u]$ consists of all polynomials in u with coefficients in F while

$$F(u) = \left\{ \frac{p(u)}{q(u)} \mid p(x), q(x) \in F[x], q(u) \neq 0 \right\}.$$

It is always the case that $F[u] \subseteq F(u)$.

10.3 Algebraic and Transcendental Elements

10.3.1 *Definition*

If E is an extension of F and $u \in E$, there are two possibilities:

(i) There exists a polynomial $p(x)$ of positive degree such that u is a root of $p(x)$;

(ii) No such polynomial exists.

In the first case we say that u is *algebraic over* F and in the second that u is *transcendental over* F.

10.3.2 *Example*

$\mathbf{R}$ is an extension of $\mathbf{Q}$ and $\sqrt{2} \in \mathbf{R}$. Since $\sqrt{2}$ is a root of $x^2 - 2 \in \mathbf{Q}[x]$, $\sqrt{2}$ is algebraic over $\mathbf{Q}$.

However π and e (the base of the natural logarithm) are transcendental over $\mathbf{Q}$ (the proofs that they are transcendental are very difficult). ∎

If E is an extension of F and u is an element of E which is algebraic over F, then, among all polynomials of positive degree over F which have u as a root, there are some of least degree. Among these there is at least one which is monic.

10.3.3 *Theorem*

If E is an extension of F and u is an element of E which is algebraic over F, then a monic polynomial $p(x)$ over F of least degree having u as a root is irreducible and divides any polynomial which has u as a root.

Proof. Suppose $p(x) = a(x)b(x)$ where $a(x),\ b(x) \in F[x]$. Then

$$0 = p(u) = a(u)b(u).$$

But E is an integral domain, and so either $a(u) = 0$ or $b(u) = 0$. Since $\partial p(x)$ is minimal positive, it follows that $a(x)$ or $b(x)$ is constant, thus showing that $p(x)$ is irreducible.

Suppose $c(x) \in F[x]$ and $c(u) = 0$. By the Division Algorithm,

$$c(x) = q(x)p(x) + r(x), \ \partial r(x) < \partial p(x)$$

and so

$$0 = c(u) = q(u)p(u) + r(u).$$

But $p(u) = 0$, whence $r(u) = 0$. By the minimality of $\partial p(x)$, it follows that $r(x) = 0$. Hence $p(x)$ divides $c(x)$. ∎

10.3.4 *Corollary*

The polynomial $p(x)$ above is the unique monic polynomial of positive degree having u as root.

Proof. If $\hat{p}(x)$ is another monic polynomial of least positive degree such that $\hat{p}(u) = 0$, then by the argument above, each of $p(x)$ and $\hat{p}(x)$ divides the other. Since they are both monic and irreducible it follows that $p(x) = \hat{p}(x)$. ∎

10.3.5 *Definition*

The unique monic polynomial $p(x)$ over F of the preceding corollary is called *the irreducible polynomial of u over F* and is denoted by $\mathrm{Irr}(u, F)$.

10.3.6 *Examples*

(i) $\mathrm{Irr}(\sqrt{2}, \mathbf{Q}) = x^2 - 2$.
(ii) $\mathrm{Irr}(i, \mathbf{R}) = x^2 + 1$.
(iii) $\mathrm{Irr}(\frac{(-1 + i\sqrt{3})}{2}, \mathbf{R}) = x^2 + x + 1$.

10.3.7 *Theorem*

Let E be an extension of F and u an element of E algebraic over F. If $p(x) = \mathrm{Irr}(u, F)$ and $\partial p(x) = n$, then there is a ring isomorphism

$$\varphi : \left. F[x] \middle/ (p(x)) \right. \to F[u]$$

such that $\varphi[x+(p(x))]=u$ and $\varphi[a+(p(x))]=a$ for all $a\in F$. Moreover, $F[u]=F(u)$ and $|F[u]:F|=\partial p(x)$; that is, the degree of the extension is equal to the degree of the irreducible polynomial.

Proof. The substitution map (see 9.4.8) $\varphi_u:F[x]\to F[u]$ is a ring epimorphism and so, by the First Isomorphism Theorem (Theorem 8.3.1), $F[x]\big/\ker\varphi_u \cong F[u]$ under the isomorphism

$$\varphi[g+(p(x))]=\varphi_u(g)=g(u).$$

Now

$$\ker\varphi_u=\{g(x)\in F[x]\,|\,g(u)=0\}=(p(x))$$

by Theorem 9.7.13 and so

$$F[x]\big/(p(x)) \cong F[u].$$

Since $F[x]\big/(p(x))$ is a field by Theorem 9.6.2, it follows that $F(u)=F[u]$. Moreover, by Theorem 9.6.3, every coset $g(x)+(p(x))$ has a representative of degree less than $\partial p(x)=n$ and so

$$\mathrm{sp}_F\{\overline{1},\overline{x},\overline{x}^2,...,\overline{x}^{n-1}\}=F[x]\big/(p(x))$$

where, as usual, $\overline{g(x)}=g(x)+(p(x))$. Consequently,

$$\mathrm{sp}_F\{1,u,u^2...,u^{n-1}\}=F(u).$$

We show that $\{1,u,u^2,...,u^{n-1}\}$ is linearly independent over F. Suppose

$$\sum_{i=0}^{n-1}a_iu^i=0$$

for some $a_i\in F$, $i=1,2,...,n-1$. Then

$$\varphi[\sum_{i=0}^{n-1}a_ix^i+(p(x))]=0$$

and since φ is an isomorphism, we deduce

$$\sum_{i=0}^{n-1}a_ix^i+(p(x))=\overline{0}.$$

Therefore $p(x)$ divides $\sum_{i=0}^{n-1}a_ix^i$. But

$$\partial p(x) = n \text{ and } \partial[\sum_{i=0}^{n-1} a_i x^i] < n.$$

and so $a_i = 0$ for all i, $i = 1, 2, ..., n-1$, proving that $\{1, u, u^2, ..., u^{n-1}\}$ is a basis of $F(u)$ over F. Hence $| F(u) : F | = \partial(\text{Irr}(u, F))$. ∎

10.3.8 *Example*

If $i = \sqrt{-1} \in \mathbf{C}$, then $\text{Irr}(i, \mathbf{R}) = x^2 + 1$ and by Theorem 9 7.16

$$\mathbf{R}[i] \cong \mathbf{R}[x] \Big/ (x^2 + 1).$$

Moreover $\mathbf{R}[i] = \mathbf{C}$ and $|\mathbf{C} : \mathbf{R}| = 2 = \partial(x^2 + 1)$. ∎

We have dealt with the case where an element u of an extension field of F is algebraic over F. We now investigate the case where u is transcendental over F.

10.3.9 *Theorem*

If E is an extension of F and u is transcendental over F, then $F(u) \cong F(x)$, the field of quotients of the integral domain $F[x]$.

Proof. The substitution map $\varphi_u : F[x] \rightarrow F[u]$ is clearly a ring epimorphism. Since u is transcendental over F, $\ker \varphi_u = (0)$ and so φ is an isomorphism. It follows that $F(x) \cong F(u)$. ∎

Remark. The field $F(x)$ consists of the *rational functions* $f(x) \Big/ g(x)$ where $g(x) \neq 0$.

10.4 Algebraic Extensions

Extensions of a field F in which every element is algebraic over F play important roles in many branches of mathematics including algebraic geometry and number theory. In this section we shall derive their basic properties and apply the theory to finite fields.

10.4.1 *Definition*

An extension field E of F is said to be *algebraic over* F if every element of E is algebraic over F. In such a case we say that E / F is algebraic.

Clearly every field F is algebraic over itself since each element $a \in F$ is the root of $x - a$. Note also that if E is an extension of F and $u \in E$, then $u \in F$ if, and only if, $\partial \mathrm{Irr}(u, F) = 1$.

10.4.2 *Theorem*

Every finite-dimensional extension of a field F is algebraic over F.

Proof. Suppose $E \supseteq F$ and $|E : F| = n < \infty$. Let $u \in E \backslash F$ and consider the set $S = \{1, u, u^2, ..., u^n\}$. Since the vector space E is of dimension n over F, no set of elements of E with more than n elements can be linearly independent. Since $|S| = n + 1$ (counting repetitions), S is a linearly dependent set of vectors. Therefore there exist elements $a_i \in F$, $i = 0, 1, ..., n$, not all zero, such that $\sum_{i=0}^{n} a_i u^i = 0$ and so u is a root of the polynomial $p(x) = \sum_{i=0}^{n} a_i x^i$ of positive degree. Hence E / F is algebraic. ∎

The converse of the preceding theorem is not valid as the following example shows.

10.4.3 *Example*

The polynomial $x^{2^n} - 2$ is irreducible over the rationals $\mathbf{Q}$ by Eisenstein's Irreducibility Criterion (Theorem 9.5.7). Therefore $|\mathbf{Q}[\sqrt[2^n]{2}]:\mathbf{Q}| = 2^n$ by Theorem 9.7.16. It follows that

$$\mathbf{Q}[\sqrt[2^n]{2}] \subset \mathbf{Q}[\sqrt[2^{n+1}]{2}].$$

Let

$$K = \bigcup_{n=1}^{\infty} \mathbf{Q}[\sqrt[2^n]{2}].$$

Then K is an algebraic extension of $\mathbf{Q}$ which is clearly infinite dimensional over $\mathbf{Q}$. (The reader is urged to verify these statements). ∎

10.4.4 *Theorem*

If $F \subseteq E \subseteq K$, then $|K:F|$ is finite if, and only if, $|E:F|$ and $|K:E|$ are finite, in which case $|K:F| = |K:E|.|E:F|$.

Proof. If $|K:F|$ is finite, then every element of K is expressible as a finite F-linear combination of some finite basis B. Therefore, every element of K is an E-linear combination of elements of B and so $|K:E|$ is finite. Moreover, since E is clearly an F-subspace of K, it follows that $|E:F|$ is finite.

Conversely, suppose
$$|K:E| = m \text{ and } |E:F| = n$$
and let $\{\alpha_1, \alpha_2, ..., \alpha_m\}$ be a basis of K over E, $\{\beta_1, \beta_2, ..., \beta_n\}$ a basis of E over F. We show that $S = \{\alpha_i \beta_j \mid 1 \le i \le m, 1 \le j \le n\}$ is a basis of K over F. If $u \in K$, then

$$u = \sum_{i=1}^{m} e_i \alpha_i \text{ where } e_i \in E, 1 \le i \le m.$$

But

$$e_i = \sum_{j=1}^{n} f_{ij} \beta_j \text{ where } f_{ij} \in F, 1 \le i \le m, 1 \le j \le n.$$

Hence

$$u = \sum_{i=1}^{m} \left(\sum_{j=1}^{n} f_{ij} \beta_j \right) \alpha_i = \sum_{i,j} f_{ij} \alpha_i \beta_j.$$

Therefore

$$\mathrm{sp}_F \{ \alpha_i \beta_j \mid 1 \le i \le m, 1 \le j \le n \} = K.$$

To show linear independence, suppose

$$\sum_{i,j} d_{ij} \alpha_i \beta_j = 0, 1 \le i \le m, 1 \le j \le n.$$

Then

$$\sum_{i=1}^{m} \left(\sum_{j=1}^{n} d_{ij} \beta_j \right) \alpha_i = 0.$$

But $\sum_{j=1}^{n} d_{ij} \beta_j$ is in E for each $i, 1 \le i \le m$ and since the set of vectors

$\{ \alpha_1, \alpha_2, ..., \alpha_m \}$ is linearly independent over E, it follows that

$$\sum_{j=1}^{n} d_{ij} \beta_j = 0 \text{ for each } i, 1 \le i \le m.$$

Similarly, since d_{ij} is in F and the vectors $\beta_1, \beta_2, ..., \beta_n$ are linearly independent over F, we conclude that $d_{ij} = 0$ for all $i, j, 1 \le i \le m, 1 \le j \le n$. Therefore S is a basis of K over F and so
$$|K:F| = |K:E|.|E:F|. \blacksquare$$

Any finite-dimensional extension of the rationals, $\mathbf{Q}$, is known as an algebraic number field. The study of these fields is the central topic of algebraic number theory. All these fields are contained in the field of all algebraic numbers.

10.4.5 *Theorem*

If E is an extension of F and A is the set of all elements in E which are algebraic over F, then A is a subfield containing F.

Proof. Since each element of F is algebraic over F, $F \subseteq A$.

Suppose $a, b \in A$ and consider the field $F(a,b)$. Since a is algebraic over $F, |F(a):F| = \partial \mathrm{Irr}(a, F)$ by Theorem 9.7.16. Also,

b is algebraic over $F(a)$ since it is algebraic over F (why?). Therefore $|[F(a)](b):F(a)|=\partial\mathrm{Irr}(b,F(a))|$ by Theorem 9.7.16. Therefore, by the preceding result,

$$|F(a,b):F| = \partial\mathrm{Irr}(a,F).\partial\mathrm{Irr}(b,F(a))<\infty.$$

By Theorem 9.7.20, $F(a,b)$ is an algebraic extension of F. Therefore $F(a,b)\subseteq A$ and so A is a subfield of E. ∎

10.4.6 *Definition*

The subfield A of the previous theorem is called the *algebraic closure of F in E*.

We saw in Corollary 9.7.6 that, given a polynomial $f(x)\in F[x]$, there is an extension field E of F over which $f(x)$ splits. This is equivalent to saying that all the roots of $f(x)$ are in E. A smallest field containing F and all the roots of $f(x)$ is called a *splitting field of f(x) over F*. More precisely:

10.4.7 *Definition*

Let $f(x)\in F[x]$. An extension field E of F is a *splitting field of* $f(x)$ *over* F if:
  (i) $f(x)$ splits over F;
  (ii) if $u_1,u_2,...,u_n$ are the roots of $f(x)$, then $E=F(u_1,u_2,...,u_n)$.

We aim to prove that any two splitting fields of a polynomial $f(x)$ over F are isomorphic so that we shall be able to speak of *the* splitting field rather than *a* splitting field.

10.4.8 *Lemma (See Lemma 9.5.3)*

If $\varphi:F\to\hat{F}$ is an isomorphism of fields, then φ may be extended to an isomorphism $\hat{\varphi}$ from $F[x]$ to $\hat{F}[x]$. ∎

10.4.9 *Lemma*

Let $\varphi: F \to \hat{F}$ be an isomorphism of fields and $p(x) \in F[x]$ a polynomial irreducible over F. As in the lemma above, let $\hat{\varphi}$ denote the isomorphism from $F[x]$ to $\hat{F}[x]$ which extends φ and set $\hat{p}(x) = \hat{\varphi}(p(x))$.

If E and $\hat{E}$ are extensions of F and $\hat{F}$, respectively, with E containing a root u of $p(x)$ and $\hat{E}$ containing a root $\hat{u}$ of $\hat{p}(x)$, then φ can be extended to an isomorphism $\tilde{\varphi}$ from $F[u]$ to $\hat{F}[\hat{u}]$ such that $\tilde{\varphi}(u) = \hat{u}$.

Proof. It is clear that $\hat{p}(x)$ is irreducible over $\hat{F}$. Define

$$\theta : {F[x]}\Big/{(p(x))} \to {\hat{F}[x]}\Big/{(\hat{p}(x))}$$

by

$$\theta[f(x) + (p(x))] = \hat{\varphi}[f(x)] + (\hat{p}(x)).$$

We show the map is well defined. Indeed, suppose

$$f(x) + (p(x)) = g(x) + (p(x)).$$

Then

$$p(x) \mid [f(x) - g(x)] \text{ and so } \hat{\varphi}(p(x)) \mid \hat{\varphi}(f(x) - g(x)).$$

Therefore

$$\hat{p}(x) \mid [\hat{\varphi}(f(x)) - \hat{\varphi}(g(x))]$$

and so

$$\hat{\varphi}(f(x)) + (\hat{p}(x)) = \hat{\varphi}(g(x)) + (\hat{p}(x)).$$

(It is left as an exercise for the reader to show that θ is a ring homomorphism onto ${\hat{F}[x]}\Big/{(\hat{p}(x))}$ and that $\theta[x + (p(x))] = x + (\hat{p}(x)).$)

Since ${F[x]}\Big/{(p(x))}$ is a field and θ is not the zero map, it follows that θ is injective and so is an isomorphism from ${F[x]}\Big/{(p(x))}$ to ${\hat{F}[x]}\Big/{(\hat{p}(x))}.$

By Theorem 9.7.16, there are isomorphisms

$$\lambda : F[u] \to {F[x]}\Big/{(p(x))} \text{ and } \mu : {\hat{F}[x]}\Big/{(\hat{p}(x))} \to \hat{F}[\hat{u}]$$

such that

$$\lambda(u) = x + (p(x)) \text{ and } \mu[x + (\hat{p}(x))] = \hat{u}.$$

Thus $\mu\theta\lambda$ is an isomorphism from $F[u]$ onto $\hat{F}[\hat{u}]$ and

$$\mu\theta\lambda(u) = \mu\theta[x + (p(x))] = \mu[x + (\hat{p}(x))] = \hat{u}.$$

Moreover,

$$\mu\theta\lambda(a) = \mu\theta(a + (p(x))) = \mu[\varphi(a) + (\hat{p}(x))] = \varphi(a)$$

that is, $\mu\theta\lambda$ extends φ.

The various maps involved are shown below.

$$F[u] \xrightarrow{\ \lambda\ } F[x] \Big/ (p(x)) \xrightarrow{\ \theta\ } \hat{F}[x] \Big/ (\hat{p}(x)) \xrightarrow{\ \mu\ } \hat{F}[u] \ \blacksquare$$

We are now in a position to prove a general theorem which leads to the desired result that any two splitting fields of a polynomial are isomorphic.

10.4.10 *Theorem*

Let $\varphi : F \to \hat{F}$ be an isomorphism of fields, $f(x)$ a polynomial over F and $\hat{f}(x) = \hat{\varphi}[f(x)]$ where $\hat{\varphi}$ is the extension of φ to $F[x]$ (see Lemma 9.5.3.)

Let E be a splitting field of $f(x)$ over F and $\hat{E}$ a splitting field of $\hat{f}(x)$ over $\hat{F}$. Then φ can be extended to an isomorphism of E onto $\hat{E}$.

Proof. The reader will find the following diagram helpful in following the proof which is by induction on $\partial f(x)$.

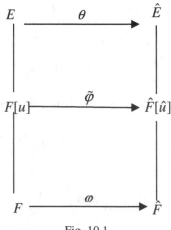

Fig. 10.1

If $\partial f(x) = 1$, $F = E$ and $\hat{F} = \hat{E}$ and there is nothing to prove. Assume the theorem is valid if the degree of the polynomial in question is less than $\partial f(x)$. Let $p(x)$ be an irreducible factor of $f(x)$ over F. Then $\hat{p}(x) = \hat{p}(x) = \hat{\varphi}[p(x)]$ is an irreducible factor of $\hat{f}(x)$ over $\hat{F}$.

If u is a root of $p(x)$ in E and $\hat{u}$ is a root of $\hat{p}(x)$ in $\hat{E}$, then, by the previous lemma, φ can be extended to an isomorphism

$$\tilde{\varphi}: F[u] \to \hat{F}[\hat{u}] \text{ with } \tilde{\varphi}(u) = \hat{u}.$$

Therefore, over $F[u]$, $p(x) = (x-u)q(x)$ and over $\hat{F}[\hat{u}]$, $\hat{p}(x) = (x-\hat{u})\hat{q}(x)$. Moreover, since $\tilde{\varphi}(u) = \hat{u}$, we have (in the notation of 9.7.27) $\hat{\tilde{\varphi}}[(x-u)q(x)] = (x-\hat{u})\hat{q}(x)$. Also, since $\tilde{\varphi}$ agrees with φ on F, $\hat{f}(x) = \hat{\tilde{\varphi}}(f(x))$. Thus, if

$$f(x) = (x-u)q(x)s(x),$$
then
$$\hat{f}(x) = \hat{\tilde{\varphi}}(f(x)) = (x-\hat{u})\hat{q}(x)\hat{s}(x)$$

Set $g(x) = q(x)s(x)$. Clearly E is a splitting field of $g(x)$ over $F[u]$ and $\hat{E}$ is a splitting field of $\hat{g}(x) = \hat{\tilde{\varphi}}(g(x))$ over $\hat{F}[\hat{u}]$. But $\partial g < \partial f$ and so, by the induction hypothesis, $\tilde{\varphi}$ can be extended to an isomorphism θ of E onto $\hat{E}$. Since $\tilde{\varphi}$ is an extension of φ, it follows that θ is an extension of φ and so, by induction, the result is proved. ∎

10.4.11 *Corollary*

If $f(x)$ is a polynomial over F and E and $\hat{E}$ are splitting fields of $f(x)$ over F, then there is an F-isomoprhism from E to $\hat{E}$ that is, an isomorphism $\varphi: E \to \hat{E}$ which restricts to the identity on F.

Proof. Let $F = \hat{F}$ and $\varphi =$ the identity in the above. ∎

Notation. We shall write $\mathrm{spl}(f(x), F)$ for the splitting field of f over F.

Given a field F, is there a "universal" splitting field of F? That is, is there a field E such that every polynomial over E splits over E? There is in fact such a field and it is unique up to isomorphism. It is known as the *algebraic closure of F* and is usually denoted by $\bar{F}$. For example, the algebraic closure of the real field is the field of complex numbers.

In order to construct the algebraic closure of a field we need transfinite induction which we do not have at our disposal at this stage. Suffice it to say that, roughly speaking, we construct the splitting field of the set of all polynomials.

10.5   Finite Fields

Recall that to construct a finite field of order p^n, p a prime, we form $\mathbf{Z}_p[x]\Big/_{(g(x))}$ where $g(x)$ is an irreducible polynomial over $\mathbf{Z}_p$ of degree n. We showed that any finite field is of order p^n but left open the question of the existence of such fields for any p and any n. This is tantamount to asking if there are irreducible polynomials over $\mathbf{Z}_p$ of any degree.

Before we can answer these questions we have to broach the subject of multiple roots. To do this, we introduce the notion of the formal derivative of a polynomial.

10.5.1   *Definition*

The *formal derivative* of a polynomial

$$f(x) = \sum_{i=0}^{n} a_i x^i$$

is defined to be

$$f'(x) = \sum_{i=1}^{n} i a_i x^{i-1}.$$

Although the derivative as defined above is independent of the notion of limit, it nevertheless has two important properties in common with the familiar derivative of the calculus.

10.5.2 *Theorem*

If $f(x)$ and $g(x)$ are polynomials over the field F, then
  (i) $(f(x) + g(x))' = f'(x) + g'(x)$;
  (ii) $(f(x)g(x))' = f'(x)g(x) + f(x)g'(x)$.
  (iii) $[(f(x))^n]' = n(f(x))^{n-1} f'(x)$.

Proof. Exercise. ∎

10.5.3 *Lemma*

If $f(x)$ and $g(x)$ are polynomials over the field F and E is an extension field of F, then the gcd of $f(x)$ and $g(x)$ over F is precisely the gcd of $f(x)$ and $g(x)$ over E; that is, gcd's don't change as the field gets bigger.

Proof. Let $h(x) = \gcd(f(x), g(x))$ over F. Then there exist polynomials $a(x)$ and $b(x)$ over F such that

$$h(x) = a(x)f(x) + b(x)g(x).$$

If $\hat{h}(x) = \gcd(f(x), g(x))$ over E, then, since $\hat{h}(x) \mid f(x)$ and $\hat{h}(x) \mid g(x)$, we have

$$f(x) = \hat{h}(x)r(x) \text{ and } g(x) = \hat{h}(x)s(x)$$

where $r(x)$ and $s(x)$ are polynomials over E. Hence, over E,

$$h(x) = a(x)r(x)\hat{h}(x) + b(x)s(x)\hat{h}(x) = (a(x)r(x) + b(x)s(x))\hat{h}(x)$$

and so $\hat{h}(x) \mid h(x)$ over E. But $h(x) \in F[x] \subseteq E[x]$, $h(x) \mid f(x)$ and $h(x) \mid g(x)$ over F and so also over E. Since

$$\hat{h}(x) = \gcd(f(x), g(x)) \text{ over } E,$$

it follows that $h(x) \mid \hat{h}(x)$. Therefore $h(x) \mid \hat{h}(x)$ and $\hat{h}(x) \mid h(x)$, proving that $\hat{h}(x) = h(x)$. ∎

The result above is counter-intuitive. When we go to a larger field, we expect some of the factors of $f(x)$ and $g(x)$, which are irreducible over F, to decompose over E and thus possibly to give rise to more common factors of $f(x)$ and $g(x)$.

The next result is quite surprising. It says that, for a polynomial $f(x) \in F[x]$, we can determine whether or not $f(x)$ has a repeated linear factor in some extension field of F without actually leaving F.

10.5.4 *Theorem*

Let $f(x)$ be a nonconstant polynomial over the field F. Then $f(x)$ has a repeated linear factor in some extension field if, and only if, $\gcd(f(x), f'(x)) \neq 1$.

Proof. Suppose $f(x)$ has a repeated linear factor in some extension field E. Then there exists $b \in E$ such that, over E, $f(x) = (x-b)^2 h(x)$. Hence

$$f'(x) = (x-b)^2 h'(x) + 2(x-b)h(x)$$

and so

$$(x-b) \mid f(x) \text{ and } (x-b) \mid f'(x).$$

It follows that $\gcd(f(x), f'(x)) \neq 1$ *over E*. By the previous lemma, greatest common divisors don't change as the field gets bigger and so $\gcd(f(x), f'(x)) \neq 1$ *over F*.

Conversely, suppose $\gcd(f(x), f'(x)) \neq 1$. Then there is a polynomial $q(x)$ of positive degree such that $q(x) \mid f(x)$ and $q(x) \mid f'(x)$. Let $E = \mathrm{spl}(q(x), F)$ and suppose b is a root of $q(x)$ in E. Then

$$(x-b) \mid q(x) \text{ and so } (x-b) \mid f(x) \text{ and } (x-b) \mid f'(x).$$

Therefore

$$f(x) = (x-b)r(x) \text{ and } f'(x) = (x-b)s(x).$$

But

$$f'(x) = [(x-b)r(x)]' = (x-b)r'(x) + r(x) \text{ and so } 0 = f'(b) = r(b).$$

Hence

$$(x-b) \mid r(x) \text{ and so } r(x) = (x-b)t(x).$$

Therefore

$$f(x) = (x-b)r(x) = (x-b)^2 t(x). \ \blacksquare$$

10.5.5 *Corollary*

If $f(x)$ is a nonconstant polynomial of degree n over a field F and $\gcd(f(x), f'(x)) = 1$, then $f(x)$ has n distinct roots in $\mathrm{spl}(f(x), F)$.

Proof. Over $E = \mathrm{spl}(f(x), F)$, $f(x) = \prod_{i=1}^{n}(x - b_i)$. By the previous result, $f(x)$ cannot have a repeated linear factor. Therefore all the b_i are distinct. ∎

We are just a lemma short of being able to prove the theorem quoted at the end of the previous chapter, namely, given any prime p and any positive integer n, there is one, and, up to isomorphism, only one field of order p^n.

10.5.6 *Lemma*

If a and b are elements of a field of characteristic p, then
$$(a+b)^{p^n} = a^{p^n} + b^{p^n}.$$

Proof. Using the Binomial Theorem which is valid in any commutative ring, we get
$$(a+b)^p = \sum_{k=0}^{p} \binom{p}{k} a^{p-k} b^k.$$

We show that $p \mid \binom{p}{k}$ for all k, $1 \le k \le p-1$. Indeed,
$$\binom{p}{k} = \frac{p!}{k!(p-k)!} = \frac{p(p-1)...(p-k+1)}{(p-k)!}$$
and since $1 \le k < p$, it follows that $1 \le p - k < p$. Hence $\gcd(p, (p-k)!) = 1$ and so
$$\frac{(p-1)(p-2)...(p-k+1)}{(p-k)!}$$
is an integer since $\binom{p}{k}$ is an integer. Therefore $p \mid \binom{p}{k}$ for all k, $1 \le k \le p-1$.

In a field of characteristic p, $lp.c = 0$ for all integers l and all elements c of the field. Thus
$$(a+b)^p = a^p + b^p.$$

Assume inductively that for some $s \geq 1$, $(a+b)^{p^s} = a^{p^s} + b^{p^s}$. Then

$$(a+b)^{p^{s+1}} = [(a+b)^{p^s}]^p = [(a^{p^s} + b^{p^s})]^p = a^{p^{s+1}} + b^{p^{s+1}}$$

where we have used the induction hypothesis at the second equality and the case proved above at the third. ∎

10.5.7 *Theorem*

If p is a prime and n a positive integer, there is one, and, up to isomorphism, only one field of order p^n.

Proof. Let $P = \mathbf{Z}_p$ and consider the polynomial $x^{p^n} - x \in P[x]$. Since $(x^{p^n} - x)' = -1$, we have $\gcd(x^{p^n} - x, (x^{p^n} - x)') = 1$ and so, by Corollary 9.7.34, $x^{p^n} - x$ has exactly p^n roots in $E = \mathrm{spl}(x^{p^n} - x, P)$.

Let S denote the set of roots of $x^{p^n} - x$ in E. Then $|S| = p^n$ and for any $a \in P$, $a^p = a$, $a^{p^2} = a^p = a$ and in general, $a^{p^n} = a$ by Fermat's Little Theorem, proving that $P \subseteq S$.

We now show that S itself forms a subfield of E. Let $a, b \in S$. Then $(ab)^{p^n} = a^{p^n} b^{p^n} = ab$ and so $ab \in S$. Also, if $b \neq 0$,

$$(b^{-1})^{p^n} = (b^{p^n})^{-1} = b^{-1} \text{ and } (-a)^{p^n} = -a^{p^n} = -a$$

if p is odd. If $p = 2$, $-a = a$ and the additive inverse is automatically in the field if a is.

Finally, $(a+b)^{p^n} = a + b$ by Lemma 9.7.35 and so $a + b \in S$. Hence S is a subfield of E containing P and all the roots of $x^{p^n} - x$ and so is the splitting field of $x^{p^n} - x$ over P, i.e., $S = E$. Hence S is a field of order p^n and is the splitting field of $x^{p^n} - x$ over $\mathbf{Z}_p$.

Suppose now that E and $\hat{E}$ are fields of order p^n and let P and $\hat{P}$ be their respective prime subfields. Then P and $\hat{P}$ are both isomorphic to $\mathbf{Z}_p$ and so isomorphic to each other via the isomorphism $\varphi : P \to \hat{P}$, say.

Since φ maps the unity of P to the unity of $\hat{P}$, it follows that $\hat{\varphi}(x^{p^n} - x) = x^{p^n} - x$ where $\hat{\varphi}$ is the extension of φ to $P[x]$ (see Lemma

9.5.3). But $E^* = E \setminus \{0\}$ and $\hat{E}^* = \hat{E} \setminus \{0\}$ are groups of order $p^n - 1$ and so if b is in either E^* or $\hat{E}^*$, $b^{p^n-1} = 1$. Therefore each element of E or $\hat{E}$ (including 0) is a root of $x^{p^n} - x$ whence $E = \mathrm{spl}(x^{p^n} - x, P)$ and $\hat{E} = \mathrm{spl}(x^{p^n} - x, \hat{P})$. By Theorem 9.7.28, φ can be extended to an isomorphism from E to $\hat{E}$. ∎

Since there is only one field of order p^n, we denote it by $GF(p^n)$. The "GF" stands for "Galois Field", in honor of Evariste Galois.

10.5.8 *Corollary*

Given any prime p and any positive integer n, there is an irreducible polynomial of degree n over $\mathbf{Z}_p$.

Proof. By the previous theorem there is a field E of order p^n. We know that E^* is cyclic (Theorem 9.6.9) with generator b, say. Then $\mathbf{Z}_p[b] = E$ and so $\partial\mathrm{Irr}(b, \mathbf{Z}_p) = |E : \mathbf{Z}_p| = n$. ∎

10.6 Exercises

1) (i) In the field $\mathbf{R}$ of real numbers, $\sqrt{2}$ and $\sqrt{3}$ are both algebraic over the rationals $\mathbf{Q}$ and so $\sqrt{2} + \sqrt{3}$ is also algebraic over $\mathbf{Q}$. Find a polynomial of degree 4 over $\mathbf{Q}$ which has $\sqrt{2} + \sqrt{3}$ as a root.

(ii) What is the degree of the irreducible polynomial of $\sqrt{2} + \sqrt{3}$ over $\mathbf{Q}$? Justify your answer.

(iii) What is the degree of the irreducible polynomial of $\sqrt{2} \cdot \sqrt{3}$ over $\mathbf{Q}$?

2) Show that $\sqrt{2} + \sqrt[3]{5}$ is algebraic over $\mathbf{Q}$ of degree 6.

3)   If E and F are fields with $E \supseteq F$ and a and b are algebraic over F of degrees m and n,   respectively,   and $\gcd(m,n) = 1$,   prove   that $|F(a,b): F| = mn$.

4) Prove that the splitting field E of a polynomial of degree n over F is of degree at most $n!$ over F.

5) Determine the degree of the splitting fields of the following polynomials over the rationals:
   (i) $x^4 + x^2 + 1$;
   (ii) $x^4 + 1$;
   (iii) $x^6 + 1$;
   (iv) $x^4 - 2$;
   (v) $x^5 - 1$;
   (vi) $x^9 + x^3 + 1$.

6) If p is prime, prove that $|\mathrm{spl}(x^p - 1, \mathbf{Q})| = p - 1$.

7) Prove that $\mathbf{Q}(\sqrt[3]{2})$ has no automorphisms other than the identity.

8) Prove that if m is an integer which is not a perfect square and $a + b\sqrt{m}$ (a and b rational) is a root of a polynomial $p(x)$ with rational coefficients, then $a - b\sqrt{m}$ is also a root of $p(x)$.

9) If
$$f(x) = x^3 + x + 1 \in \mathbf{Z}_2[x]$$
and b is a root of $f(x)$ in some extension field, how many   elements   are there in $\mathbf{Z}_2(b)$? Express each element of $\mathbf{Z}_2(b)$ in terms of b.

10) Let $\mathbf{Z}_2(b)$ be the field of the previous exercise.
   (i) Express b^5, b^{-2} and b^{100} in   the   form $c_0 + c_1 b + c_2 b^2$ where $c_i \in \mathbf{Z}_2$, $i = 0,1,2$.
   (iii) Show that $b+1$, $b^2 + 1$ and $b^2 + b + 1$ are all roots of $x^3 + x^2 + 1$.

11) Let F be a field and a and b elements of F with $a \neq 0$. If c is in some extension field of F, prove that $F(c) = F(ac + b)$.

12) If $f(x)$ is an irreducible polynomial of degree n over a field F of characteristic 0, prove that $f(x)$ has n distinct roots in its splitting field.

13) (Steinitz) Let F be a field of characteristic 0 and E an extension of F. If a and b are elements of E which are algebraic over F, prove that there is an element $c \in E$ such that $F(a,b) = F(c)$.
[Hint: Let $p(x) = \text{Irr}(a, F)$, $q(x) = \text{Irr}(b, F)$ with $\partial p(x) = m$, $\partial q(x) = n$ and suppose $a = a_1, a_2, ..., a_m$, and $b = b_1, b_2, ..., b_n$ are the roots of $p(x)$ and $q(x)$, respectively in $\text{spl}(p(x)q(x)) = K$ (note that the a_i and b_j are distinct by Problem 35).
Since $|F| = \infty$ we can choose $d \in F$ such that

$$d \notin \{(a - a_i)(b - b_j)^{-1} \mid 1 \le i \le m, 2 \le j \le n\}.$$

Then $a_i \neq a + d(b - b_j)$ for all $i, j, 1 \le i \le m, 2 < j < n$. Show that $F(a,b) = F(c)$ where $c = a + db$ as follows:

(i) Show that it is sufficient to prove that $b \in F(c)$.

(ii) Let $r(x) = p(c - dx)$ ($\in F(c)[x]$) and let $s(x) = \text{Irr}(b, F(c))$. Prove that $s(x)$ divides both $q(x)$ and $r(x)$.

(iii) Prove all the roots of $s(x)$ are in K.

(iv) Prove that b is the only element of K which is a common root of $r(x)$ and $q(x)$.

(v) Deduce that $s(x) = x - b$ and so $b \in F(c)$.]

14) Prove that if F is a field of characteristic 0 and $a_1, a_2, ..., a_n$ are elements of some extension field E of F, then there exists $c \in E$ such that $F(c) = F(a_1, a_2, ..., a_n)$. [Hint: Use the previous problem and induction].

15) Suppose $f(x) \neq g(x)$ are irreducible over F and b is a root of $f(x)$ in some extension field of F, prove that $g(x)$ is irreducible over $F(b)$.

16) Find the degree and a basis for $\mathbf{Q}(\sqrt{3}+\sqrt{5})$ over $\mathbf{Q}(\sqrt{15})$. Find the degree and a basis for $\mathbf{Q}(\sqrt{2}, \sqrt[3]{2}, \sqrt[4]{2})$ over $\mathbf{Q}$.

17) Find $\mathrm{Irr}(\sqrt{-3}+\sqrt{2}, \mathbf{Q})$ and $\mathrm{Irr}(\sqrt[3]{2}+\sqrt[3]{4}, \mathbf{Q})$.

18) If K is a finite extension of $\mathbf{R}$, prove that $K = \mathbf{C}$ or $K = \mathbf{R}$.

19) If $|E:\mathbf{Q}|=2$, prove that there is an integer d such that $E = \mathbf{Q}(\sqrt{d})$ where d is not divisible by the square of any prime.

20) Find $\mathrm{spl}(x^4 - x^2 - 2, \mathbf{Z}_3)$.

21) Prove that the maximum degree of any irreducible factor of $x^{p^n} - x$ over $\mathbf{Z}_p$ is n.

22) (i) Show that $x^3 + 2x + 1$ is irreducible over $\mathbf{Z}_3$.
    (ii) Prove that $\bar{x}$ is a generator of the cyclic group of nonzero elements of the field $\mathbf{Z}_3[x]\big/(x^3 + 2x + 1)$ under multiplication.
    (iii) Show that $x^3 + 2x + 2$ is irreducible over $\mathbf{Z}_3$ but that $\bar{x}$ is not a generator of the cyclic group of nonzero elements of $\mathbf{Z}_3[x]\big/(x^3 + 2x + 2)$ under multiplication. Find a generator.

23)
    (i) Prove that $x^5 + x^3 + 1$ is irreducible over $\mathbf{Z}_2$.
    (ii) Without calculating the multiplicative order of $|\bar{x}|$, explain why $\bar{x}$ is a generator of the multiplicative group of nonzero elements of the field $\mathbf{Z}_2[x]\big/(x^5 + x^3 + 1)$.

24) Denote the unique field of order p^n by $GF(p^n)$ (the Galois field of order p^n). Prove that $GF(p^n)$ contains an isomorphic copy of $GF(p^m)$ if, and only if, $m \mid n$.

25) If $g(x)$ is irreducible over $\mathbf{Z}_p = GF(p)$ and $g(x) \mid x^{p^n} - x$, prove that $\partial g(x) \mid n$.

26) Suppose $<b> = GF(p^{12})^* := GF(p^{12}) \setminus \{0\}$. For each subfield K of $GF(p^{12})$ find k such that $<b^k> = K^*$.

27) Suppose L and K are subfields of $GF(p^n)$ of orders p^s and p^t, respectively. How many elements are there in $L \cap K$?

28) Let $F = GF(q)$ where $q = p^n$, p a prime. If m divides $q - 1$ and $a \in F$, prove that the equation $x^m = a$ has either m distinct solutions in F or no solutions.

29) If E is an algebraic extension of F and R is a subring such that $F \subseteq R \subseteq E$, show that R must be a field.

30) Prove that the map $\varphi : GF(p^n) \to GF(p^n)$ defined by $\varphi(a) = a^p$ is an automorphism of $GF(p^n)$ of order n.

31) Let p be a prime. Prove that $x^{p^m} + x + 1$ divides $x^{p^n} + x + 1$ over $\mathbf{Z}_p$ iff $n = km$, m odd.

Chapter 11

Latin Squares and Magic Squares

11.1 Latin Squares

We begin with a practical problem. Suppose a researcher has three different fertilizers which he wishes to test on three different strains of wheat to determine which combination would give the maximum yield. How should he design the experiment?

Obviously he would want to test each strain with each type of fertilizer so he should divide up a field into a 3×3 matrix of nine plots, plant each strain of wheat in three different locations and fertilize each with a different fertilizer. To avoid skewing the results of the experiment because of the variability of the fertility of the soil in the nine different plots, he should:

(i) plant the strains of wheat in such a way that, in each row and each column of the matrix, each strain of wheat occurs once and once only;

(ii) make sure that he applies the fertilizers so that no two plots in the same row or column are fertilized with the same fertilizer.

We shall see that the solution of this problem leads quite naturally to *Mutually Orthogonal Latin Squares*.

Suppose we denote the three strains of wheat by $1,2,3$ and the three types of fertilizer by a,b,c and suppose we write $(1,c)$ to convey that strain 1 is being fertilized by fertilizer c. Then our task is to form a 3×3 matrix with entries

$$(1,a), (1,b), (1,c), (2,a), (2,b), (2,c), (3,a), (3,b), (3,c)$$

subject to the condition that every brand of fertilizer and every strain of wheat must occur once and once only in each row and in each column.

Let us first arrange the strains of wheat to satisfy the condition. Here is a possibility:

$$\begin{bmatrix} 1 & 2 & 3 \\ 2 & 3 & 1 \\ 3 & 1 & 2 \end{bmatrix}$$

Fig. 11.1

As for the fertilizers, we construct a 3×3 matrix with entries a,b,c so that when we superimpose the fertilizer matrix on the wheat matrix, we obtain all the pairs

$(1,a), (1,b), (1,c), (2,a), (2,b), (2, c), (3,a), (3,b), (3,c)$.

The following is such a matrix:

$$\begin{bmatrix} a & b & c \\ c & a & b \\ b & c & a \end{bmatrix}$$

Fig. 11.2

These considerations lead us quite naturally to the following definition.

11.1.1 *Definition*

(i) An $n\times n$ *Latin square over n symbols* taken from a set S of n elements is an $n\times n$ array of elements of S with the property that each of the symbols from S occurs once and once only in each row and in each column of the array. Just as with matrices, a Latin square whose entry in the ith row and j^{th} column is a_{ij} is denoted by $[a_{ij}]$. Often we take $S = \{1,2,3,...,n\}$.

(ii) Two $n\times n$ Latin squares $L_1 = [a_{ij}]$, $L_2 = [b_{ij}]$ over S are said to be *orthogonal* if the set $\{(a_{ij}, b_{ij}) \mid 1\le i \le n,\ 1\le j \le n\}$ consists of all the n^2 ordered pairs from $S \times S$.

(iii) A collection of $n\times n$ Latin squares is said to form a *set of mutually orthogonal Latin squares* (MOLS for short) if every pair of the collection is orthogonal.

In the above example, the first matrix is a Latin square over $\{1,2,3\}$ while the second is a Latin square over $\{a,b,c\}$. They are orthogonal since all 9 ordered pairs

$$(1,a),\ (1,b),\ (1,c),\ (2,a),\ (2,b),\ (2,\ c),\ (3,a),\ (3,b),\ (3,c)$$

occur when we superimpose the first matrix on the second.

11.1.2 *Theorem*

If $L_1 = [a_{ij}]$, $L_2 = [b_{ij}]$ are two orthogonal Latin squares over $\{1,2,3,...,n\}$ and π and τ are permutations of $1,2,3,...,n$ then the Latin squares $\pi(L_1) = [\pi(a_{ij})]$ and $\tau(L_2) = [\tau(b_{ij})]$ are also orthogonal.

Proof. It is sufficient to show that the pairs $(\pi(a_{ij}), \tau(b_{ij}))$, $1 \le i \le n, 1 \le j \le n$ are distinct; that is, that no two distinct pairs are the same. Suppose in fact that $(\pi(a_{ij}), \tau(b_{ij})) = (\pi(a_{rs}), \tau(b_{rs}))$. Then $\pi(a_{ij}) = \pi(a_{rs})$ and $\tau(b_{ij}) = \tau(b_{rs})$. Since π and τ are bijections, it follows that $a_{ij} = a_{rs}$ and $b_{ij} = b_{rs}$. Hence $(a_{ij}, b_{ij}) = (a_{rs}, b_{rs})$ and so, since L_1 and L_2 are an orthogonal pair, it follows that $(i,j) = (r,s)$. Therefore, $\pi(L_1) = [\pi(a_{ij})]$ and $\tau(L_2) = [\tau(b_{ij})]$ are orthogonal. ∎

11.1.3 *Theorem*

Any permutation of the rows (respectively, columns) of a Latin square results in a Latin square.

Proof. Exercise. (Hint: Since any permutation is a product of transpositions, it is sufficient to show that the "Latinness" is preserved if two rows (respectively, columns) are interchanged). ∎

11.1.4 *Theorem*

There are at most $n-1$ $n \times n$ MOLS.

Proof. It is clear that if we perform the same permutations of the rows and columns of every member of a collection of MOLS, then we obtain again a collection of MOLS. Thus we may assume that the first Latin square has its first row and column in ascending order, that is,

$$
L_1 = \begin{bmatrix} 1\,2\,3.....n \\ 2 \\ 3 \\ \\ n \end{bmatrix}
$$

Fig.11.3

Also, by applying appropriate permutations of 1, 2, 3,...,n to all the other Latin squares in the collection, we can arrange it so that the first row of all the Latin squares is $\begin{bmatrix} 1\,2\,3......n \end{bmatrix}$. All these operations we have performed transform the given collection of MOLS into another collection Ω of MOLS (by the previous two results).

Consider now another Latin square of Ω, say L_2. Since it is orthogonal to L_1, the entry in the 2nd row, 1st column of L_2 cannot be 1 or 2. (why?). Without loss of generality, suppose that entry is 3. Let L_3 be another member of Ω. Since L_3 is orthogonal to both L_1 and L_2, the entry in the 2nd row, 1st column of L_3 cannot be 1, 2 or 3 (why?). Again, without loss of generality, suppose it is 4.

Continuing in this manner we see that the entry in the 2nd row, 1st column of L_{n-1} must be n.. If there were an nth Latin square in Ω, then, by virtue of its being orthogonal to all the other Latin squares in Ω, its entry in the 2nd row, 1st column could not be 1, 2, 3,...., $n-1$ or n.. Clearly then, no such Latin square can exist and so there are at most $n-1$ MOLS in Ω. ∎

For what values of n can we find a collection of $n-1$ $n \times n$ MOLS? It is known that when n is the power of a prime there do exist $n-1$ $n \times n$ MOLS. When $n = 6$, it has been proved that we cannot even find *two* 6×6 orthogonal Latin squares, let alone five. For all other values of n bigger than 2, it is known that there are at least two orthogonal Latin squares but the existence of $n-1$ $n\times n$ MOLS for values other than powers of primes remains an open question.

We now show how to construct $n-1$ $n\times n$ MOLS when n is the power of a prime.

11.1.5 *Theorem*

If $n=p^m$ for some prime p, there exist $n-1$ $n\times n$ MOLS.

Proof. We know from Theorem 9.6.7 that there exists a field F of order n. Let $F=\{a_1\,(=1),a_2,a_3,...,a_n\,(=0)\}$ and set

$$L_k =[a_k a_i + a_j], k=1,2,...,n-1..$$

These are, of course, $n\times n$ arrays over the field F.

We show first that each L_k is a Latin square. Assume that $a_k a_i + a_j = a_k a_i + a_s$. Then, since we are in a field, we can cancel $a_k a_i$ to get $a_j = a_s$, which means that $j=s$. Therefore each element of F occurs once and once only in each row of L_k.

As for columns, suppose $a_k a_i + a_j = a_k a_t + a_j$. Again, because we are in a field and $a_k \neq 0$, we deduce that $a_i = a_t$ and so $i=t$. Hence each element of F occurs once and only once in each column of L_k and so, each L_k is a Latin square.

Suppose now that $k\neq s$. We show L_k is orthogonal to L_s. To do this it is sufficient to show that all the pairs $(a_k a_i + a_j, a_s a_i + a_j)$ are distinct. Let us therefore suppose that

$$(a_k a_i + a_j, a_s a_i + a_j) = (a_k a_u + a_v, a_s a_u + a_v)..$$

Then

$$a_k a_i + a_j = a_k a_u + a_v \text{ and } a_s a_i + a_j = a_s a_u + a_v$$

Subtracting the second equation form the first, we get $(a_k - a_s)a_i = (a_k - a_s)a_u$. Since $a_k \neq a_s$, we have $a_k - a_s \neq 0$ and so we can cancel $a_k - a_s$ to obtain $a_i = a_u$. Therefore, $i=u$ and cancelling $a_k a_i (= a_k a_u)$ from both sides of the first equation above, we deduce $a_j = a_v$ and so $j=v$. Therefore, L_k is orthogonal to L_s. ∎

11.1.6 *Example*

Let us apply this result to construct three 4×4 MOLS. We must first construct a field of 4 elements. To do so, we use the irreducible polynomial x^2+x+1 over $\mathbf{Z}_2$.

We know that the quotient ring $F = \dfrac{\mathbf{Z}_2[x]}{(x^2+x+1)} = \{0,1,x,x+1\}$

is a field. According to the theorem above, the first Latin square is just the body of the Cayley Table for addition. It is therefore

$$L_1 = \begin{bmatrix} 0 & 1 & x & x+1 \\ 1 & 0 & x+1 & x \\ x & x+1 & 0 & 1 \\ x+1 & x & 1 & 0 \end{bmatrix}.$$

Fig. 11.4

For the second Latin square, we compute $xa+b$ for all elements a and b in F. We obtain

$$L_2 = \begin{bmatrix} 0 & 1 & x & x+1 \\ x & x+1 & 0 & 1 \\ x+1 & x & 1 & 0 \\ 1 & 0 & x+1 & x \end{bmatrix}.$$

Fig. 11.5

For the third Latin square, we compute $(x+1)a+b$ for all elements a and b in F.

$$L_3 = \begin{bmatrix} 0 & 1 & x & x+1 \\ x+1 & x & 1 & 0 \\ 1 & 0 & x+1 & x \\ x & x+1 & 0 & 1 \end{bmatrix}.$$

Fig. 11.6

Simplify the notation by setting $1=1$, $2=x$, $3=x+1$ and $4=0$. Then the three MOLS over $1,2,3$ and 4 are:

$$L_1 = \begin{bmatrix} 4 & 1 & 2 & 3 \\ 1 & 4 & 3 & 2 \\ 2 & 3 & 4 & 1 \\ 3 & 2 & 1 & 4 \end{bmatrix} ; \quad L_2 = \begin{bmatrix} 4 & 1 & 2 & 3 \\ 2 & 3 & 4 & 1 \\ 3 & 2 & 1 & 4 \\ 1 & 4 & 3 & 2 \end{bmatrix} ; \quad L_3 = \begin{bmatrix} 4 & 1 & 2 & 3 \\ 3 & 2 & 1 & 4 \\ 1 & 4 & 3 & 2 \\ 2 & 3 & 4 & 1 \end{bmatrix} \blacksquare .$$

Fig. 11.7

11.2 Magic Squares

We end this chapter with a brief discussion of magic squares. The topic falls into the category of recreational mathematics and appears to have intrigued both amateur and professional mathematicians throughout the ages. The first known magic square was Chinese and dates back to about 2800 BCE. The most famous magic square is the 4×4 magic square which appears in a woodcut by Albrecht Durer entitled "Melancholia".

Before giving a general definition, we give an example

11.2.1 *Example*

This is an example of a 3×3 magic square.

$$\begin{bmatrix} 4 & 9 & 2 \\ 3 & 5 & 7 \\ 8 & 1 & 6 \end{bmatrix} .$$

Fig. 11.8

Observe the following:
(i) the entries are the positive integers 1 through $3^2 = 9$;
(ii) the sum of the numbers in each row is the same, namely, 15 (the square is *row-magic*);
(iii) the sum of the numbers in each column is the same, again 15 (the square is *column-magic*);
(iv) the sum of the numbers on each of the two diagonals is the same (the square is *diagonal-magic*).

11.2.2 *Definition*

A *magic square of order n* is an $n \times n$ array whose entries are the integers 1 through n^2 so arranged that the square is row-, column- and diagonal-magic.

Suppose we have two orthogonal $n \times n$ Latin squares over the set $\{0,1,2,...,n-1\}$. Superimpose one upon the other to obtain an $n \times n$ array A whose entries are all the pairs $(i,j), 0 \le i \le n-1, 0 \le j \le n-1$. By virtue of the properties of Latin squares, the sum of the first (respectively, second) components of the pairs occurring in any one row or column is

$$0+1+2+...+n-1 = \frac{n(n-1)}{2}.$$

Therefore, if we write the pairs omitting the brackets and the commas, the resulting pairs can be interpreted as integers given in base n. We thus obtain a square based on the integers 0 through $n^2 - 1$ which is row- and column-magic. At this point, let us take an example.

11.2.3 *Example*

Let us take the orthogonal Latin squares

$$L_2 = \begin{bmatrix} 4 & 1 & 2 & 3 \\ 2 & 3 & 4 & 1 \\ 3 & 2 & 1 & 4 \\ 1 & 4 & 3 & 2 \end{bmatrix}, \quad L_3 = \begin{bmatrix} 4 & 1 & 2 & 3 \\ 3 & 2 & 1 & 4 \\ 1 & 4 & 3 & 2 \\ 2 & 3 & 4 & 1 \end{bmatrix}$$

Fig. 11.9

of Example 11.1.6. Subtract 1 from each entry to get

$$\hat{L}_2 = \begin{bmatrix} 3 & 0 & 1 & 2 \\ 1 & 2 & 3 & 0 \\ 2 & 1 & 0 & 3 \\ 0 & 3 & 2 & 1 \end{bmatrix}, \quad \hat{L}_3 = \begin{bmatrix} 3 & 0 & 1 & 2 \\ 2 & 1 & 0 & 3 \\ 0 & 3 & 2 & 1 \\ 1 & 2 & 3 & 0 \end{bmatrix}.$$

Fig. 11.10

Superimposing $\hat{L}_3$ on $\hat{L}_2$, we obtain

$$\begin{bmatrix} 33 & 00 & 11 & 22 \\ 12 & 21 & 30 & 03 \\ 20 & 13 & 02 & 31 \\ 01 & 32 & 23 & 10 \end{bmatrix}.$$

Fig. 11.11

Interpreting this to the base 4, we get

$$\begin{bmatrix} 15 & 0 & 5 & 10 \\ 6 & 9 & 12 & 3 \\ 8 & 7 & 2 & 13 \\ 1 & 14 & 11 & 4 \end{bmatrix}$$

Fig. 11.12

We now add 1 to each entry to obtain the right range for the entries and obtain

$$\begin{bmatrix} 16 & 1 & 6 & 11 \\ 7 & 10 & 13 & 4 \\ 9 & 8 & 3 & 14 \\ 2 & 15 & 12 & 5 \end{bmatrix},$$

Fig. 11.13

a magic square of order 4. ■

In general, this procedure leads to a square which is row- and column-magic but may not be diagonal-magic. If we had chosen the Latin squares L_1 and L_2 of Example 10.1.6, we would not have obtained a diagonal-magic square.

11.3 Exercises

1) Construct four 8×8 MOLS and use two of them to construct a magic square of order 8.

2) Prove that in a magic square of order n, the entries in each row, column and diagonal must sum to $\dfrac{n(n^2 + 1)}{2}$.

3) Arrange the jacks, queens, kings and aces of a pack of cards in a 4×4 array so that each row and each column contains one card from each suit and one card from each rank.

4) Find a magic square of order 5.

5) Quality control would like to find the best type of music to play to its assembly line workers in order to reduce the number of faulty products. As an experiment, a different type of music is played on four days in a week, and on the fifth day no music at all is played. Design an experiment to last five weeks that will reduce the effect of the different days of the week.

6) Construct two MOLS of order 9.

7) Let $L = [l_{ij}]$ be a Latin square of order l over $\{1,2,3,...,l\}$ and $M = [m_{ij}]$ a Latin square of order m over $\{1,2,3,...,m\}$. Describe how to construct a Latin square of order lm over $\{(i, j) \mid 1 \le i \le l,\ 1 \le j \le m\}$.

8) Prove that a 3×3 magic square must have 5 at its centre.

9) A Latin square L is *self-orthogonal* if its transpose L^T is orthogonal to L.
    (i) Show that there is no 3×3 self-orthogonal Latin square.
    (ii) Give an example of a 4×4 self-orthogonal Latin square.

(iii) Prove that the diagonal entries of a self-orthogonal Latin square must be distinct.

10) Seven officers of different ranks are chosen from each of seven regiments. Arrange the forty-nine officers in a 7×7 array so that in each row and column of the array, every rank and every regiment is represented exactly once. (If we replace seven by six, there is no solution.)

11) A Latin square over $\{1, 2, ..., n\}$ is said to be in *standard form* if the entries in the first row and column are in ascending order.
  (i) Rewrite the Latin square

$$\begin{bmatrix} 1 & 3 & 4 & 2 \\ 3 & 1 & 2 & 4 \\ 2 & 4 & 3 & 1 \\ 4 & 2 & 1 & 3 \end{bmatrix}$$

in standard form by permuting rows and columns.
  (ii) Find a 4×4 Latin square in standard form orthogonal to the standardized one in (i).

12) This question refers to the three mutually orthogonal Latin squares appearing in Fig. 11.7.
  Apply the permutation (123) to the entries of the Latin square L_1, (12)(34) to the entries of L_2 and (13)(24) to the entries of L_3. Verify that we get another set of three mutually orthogonal Latin squares.

13) Complete the following Latin rectangle to a Latin square in at least two ways:

$$\begin{bmatrix} 1 & 2 & 3 & 4 \\ 2 & 1 & 4 & 3 \end{bmatrix}.$$

14) Prove that there is no Latin square orthogonal to

$$\begin{bmatrix} 1 & 2 & 3 & 4 \\ 2 & 3 & 4 & 1 \\ 3 & 4 & 1 & 2 \\ 4 & 1 & 2 & 3 \end{bmatrix}.$$

Chapter 12

Group Actions, the Class Equation
and the Sylow Theorems

12.1 Group Actions

Examples of group actions abound in group theory. Conjugation, left (or right) multiplication by elements of a group and multiplication of cosets by group elements are but a few of the specific examples of the general notion of a group action.

In this section we define the concept of a group action and obtain some important results which will subsequently be used to prove basic results in the theory of groups.

12.1.1 *Definition*

A *group action* (more precisely, a *left group action*) is a pair (G, X) where G is a group and X a set, together with a map $f : G \times X \to X$ satisfying:

(i) $f(gh, \alpha) = f(g, f(h, \alpha))$ for all $g, h \in G$, $\alpha \in X$;

(ii) $f(e, \alpha) = \alpha$ for all $\alpha \in X$ where e is the identity of G.

The image of (g, α) under f is denoted by $g\alpha$. In this simplified notation, the two conditions read:

(i)' $(gh)(\alpha) = g(h\alpha))$ for all $g, h \in G$, $\alpha \in X$;

(ii)' $e\alpha = \alpha$ for all $\alpha \in X$.

If (G, X) is an action, we say G *acts on* X.

The definition above should be reminiscent of real vector spaces wherein the real numbers "act" on the vectors, the "action" being referred to as scalar multiplication.

12.1.2 *Examples*

(i) The group S_n of permutations on $X = \{1, 2, 3, ..., n\}$ acts on X and (S_n, X) is an action.

(ii) Let $X = S(G)$ be the set of all subgroups of a group G. For $H \in S(G)$, $g \in G$, define $g.H$ to be gHg^{-1}. It is an easy exercise to prove that (G, X) is an action.

(iii) Same as the previous example except that we take $X = G$.

(iv) Let H be a subgroup of a group G and denote the set of left cosets of H in G by X. Define an action of G on X by: $(g, L) \rightarrow gL$; that is, the left coset L acted upon by g is L multiplied on the left by g.

■

We now prove a series of theorems which we shall interpret in the context of each of the examples listed above.

The reader will often come across the word "induced" in mathematics. It can be replaced by "gives rise to" or "brings about".

12.1.3 *Theorem*

Let (G, X) be an action. Then each element of G induces a permutation of X.

Proof. Given $g \in G$ and $\alpha \in X$, we have

$$g(g^{-1}\alpha) = (gg^{-1})\alpha = e\alpha = \alpha$$

where the first equality is a consequence of 11.1.1(i) while the second follows from 11.1.1(ii). Therefore the map induced by g is onto.

Next, suppose $g\alpha = g\beta$. Then

$$g^{-1}(g\alpha) = g^{-1}(g\beta)$$

and so,

$$(g^{-1}g)\alpha = (g^{-1}g)\beta$$

by 11.1.1 (i). Hence $e\alpha = e\beta$, which, by 11.1.1(ii), implies $\alpha = \beta$. Thus the map induced by g is injective. ∎

Remark. It is often the case that two or more different elements of G induce the same permutation. As an example, consider a group G acting on itself by conjugation. If G has a nontrivial center, then all the elements of the center of G induce the identity map.

12.1.4 *Definition*

(i) If different elements of G induce different permutations, then we say that the action is *faithful* or that G *acts faithfully* on X.

(ii) The *kernel* K of the action (G, X) is the set of all elements of G which induce the identity permutation on X.

We observe that the kernel of an action is trivial if, and only if, the action is faithful. (Proof?)

12.1.5 *Notation*

Let T be a subset of the group G and S a subset of X. Then $TS = \{t\sigma \mid t \in T, \ \sigma \in S\}$.

12.1.6 *Theorem*

The kernel K of the action (G, X) is a normal subgroup of G. Moreover, we may define an action of G/K on X, turning $\left(G/K, X\right)$ into a faithful action.

Proof. That K is a subgroup is left as an exercise for the reader. As for normality, let $\alpha \in X, g \in G$. Then

$$(gKg^{-1})\alpha = (g(K(g^{-1}\alpha))) = g(g^{-1}\alpha) = (gg^{-1})\alpha = e\alpha = \alpha$$

since $Kg^{-1}\alpha = g^{-1}\alpha$. Therefore $gKg^{-1} \subseteq K$ for all $g \in G$ and so K is normal in G.

Define an action of $G\!\big/\!_K$ on X by $(gK)\alpha = g\alpha$. This is a well defined action since any other representative of the coset gK is of the form gk for some $k \in K$ and so $(gk)\alpha = g(k\alpha) = g\alpha$. Moreover, the kernel of this action is clearly $\{K\}$ from which we deduce that the action is faithful. ■

12.1.7 *Definition*

If (G, X) is an action and $\alpha \in X$, the *orbit of α under G* is the set $G\alpha$, denoted by $\mathrm{Orb}_G(\alpha)$. The collection of all such subsets of X are the *orbits of G.*

12.1.8 *Theorem*

If (G, X) is an action, the orbits of G form a partition of X.

Proof. Since $\alpha \in G\alpha$, it follows that X is the union of the orbits. Suppose that $G\alpha \cap G\beta \neq \phi$. Then there exist elements g and h in G such that $g\alpha = h\beta$. Hence $\alpha = g^{-1}h\beta$ and so $G\alpha = (Gg^{-1}h)\beta$. Since $Gg^{-1}h = G$, it follows that. $G\alpha = G\beta$. ■

Every orbit of G is closed under the action of G; that is, $G(G\alpha) = G\alpha$. We may therefore consider $(G, G\alpha)$ as an action which we shall call a *subaction of* (G, X).

12.1.9 *Definition*

(G, X) is said to be *transitive* if there is only one orbit. In such a case, we say that G *acts transitively* on X.

Note that G acts transitively on each of its orbits.

So far we have placed no conditions on X. Henceforth we shall assume that X is finite and we shall obtain some important information about the sizes of the orbits of G.

12.1.10 *Definition*

Let (G, X) be an action. For $\alpha \in X$, the *stabilizer* of α, denoted G_α, consists of all elements of G which fix α; that is,

$$G_\alpha = \{ g \in G \mid g\alpha = \alpha \}.$$

12.1.11 *Theorem*

If (G, X) is an action, then $|G : G_\alpha| = |\mathrm{Orb}_G(\alpha)|$.

Proof. Define a map f from the set of left cosets of G_α in G to $\mathrm{Orb}_G(\alpha)$ by:

$$f(gG_\alpha) = g\alpha.$$

The map is well defined since any other representative of the coset gG_α is of the form gx where x is an element of G_α and so $gx\alpha = g\alpha$ given that x fixes α. Clearly the map is surjective and if $f(gG_\alpha) = f(hG_\alpha)$, we have $g\alpha = h\alpha$ from which we get $g^{-1}h\alpha = \alpha$. Hence $g^{-1}h \in G_\alpha$ proving that $gG_\alpha = hG_\alpha$. Therefore, f is bijective and the theorem is proved. ∎

12.1.12 *Corollary*

If G acts transitively on X, then $|G : G_\alpha| = |X|$ for all $\alpha \in X$.

Proof. This follows immediately from the previous theorem and the fact that the orbit of any element of X is X itself. ∎

We shall use the following result, due to Burnside, in Chapters 13 and 14 to great effect.

12.1.13 *Notation*

Let (G, X) be an action and $g \in G$. Then $\text{Fix}(g) = \{\alpha \in X \mid g\alpha = \alpha\}$.

12.1.14 *Theorem*

If (G, X) is an action with N orbits, then $|G| = \dfrac{1}{N} \sum_{g \in G} |\text{Fix}(g)|$.

Proof. Consider the set $S = \{(g, \alpha) \mid g \in G, \alpha \in X \text{ and } g\alpha = \alpha\}$. We count the number of elements in S in two ways. First, fix α. For each such α, the number of pairs in S of the form (g, α) is $|G_\alpha|$. Hence $|S| = \sum_{\alpha \in X} |G_\alpha|$. But if two elements of X are in the same orbit their stabilizers are of the same size by Theorem 12.1.10. Hence

$$\sum_{\xi \in \text{Orb}_G(\alpha)} |G_\xi| = |\text{Orb}_G(\alpha)| \, |G_\alpha| = |G : G_\alpha| \, |G_\alpha| = |G|.$$

Therefore, $|S| = N|G|$.

Next, fix g. For each such g, the number of pairs in S of the form (g, ξ) is $|\text{Fix } g|$. Thus $|S| = \sum_{g \in G} |\text{Fix}(g)|$ and so

$$|G| = \frac{1}{N} \sum_{g \in G} |\text{Fix}(g)|. \blacksquare$$

Observe that Burnside's result tells us that if G acts transitively on X, then, on average, each element of G fixes one symbol. Since the identity fixes all the elements of X, it follows that there must be at least one element of G which fixes no element of X.

12.2 The Class Equation of a Finite Group

The class equation of a group relates the order of the group to the orders and indices of certain subgroups, often affording us useful information about the group. As we shall see, it is a result which is often used in proofs by induction.

12.2.1 *Theorem (The Class Equation of a Group)*

If G is a finite group, $Z(G)$ its center and $C_G(g) = \{x \in G \mid gx = xg\}$, then

$$|G| = |Z(G)| + \sum_{g \notin Z(G)} |G : C_G(g)|$$

where the sum is taken over certain elements not in the center of G.

Proof. Let G act on G by conjugation. Then for a given $g \in G$,

$$G_g = \{x \mid x \in G \text{ and } xgx^{-1} = g\} = \{x \mid x \in G \text{ and } xg = gx\} = C_G(g).$$

Thus, by Theorem 12.1.10, the number of elements in the orbit of g is $|G : C_G(g)|$.

We gather together the singleton orbits. These consist of those elements g of G for which $|G : C_G(g)| = 1$, that is, those elements g such that $C_G(g) = G$. Therefore, an element g forms a singleton orbit if, and only if, g is in the center of G. Since G is the union of the orbits of the action (G, G), we obtain

$$|G| = |Z(G)| + \sum_{g \notin Z(G)} |G : C_G(g)|. \blacksquare$$

12.2.2 *Corollary*

A group of prime power order has a nontrivial center.

Proof. If $|G| = p^n$ for some prime p, by the class equation,

$$p^n = |Z(G)| + \sum_{g \notin Z(G)} |G : C_G(g)|.$$

If $g \notin Z(G)$, $C_G(g) \neq G$ and so each term in the sum $\sum_{g \notin Z(G)} |G : C_G(g)|$ is divisible by p. Therefore $|Z(G)|$ is divisible by p. $\blacksquare$

12.3 The Sylow Theorems

We use the class equation to prove an important result which is a first step in the proof of a partial converse of Lagrange's Theorem.

12.3.1 *Lemma*

If G is a finite abelian group whose order is divisible by a prime p, then G has an element of order p.

Proof. Let $|G| = pn$. We argue by induction on n. If $n = 1$, the order of G is p and any nonidentity element of G is of order p.

Assume the result is true for abelian groups of order less than pn for some $n > 1$ and let $|G| = pn$. Let g be any nonidentity element of G. If p divides the order of g, say $|g| = kp$, then $|g^k| = p$. Suppose therefore that $\gcd(|g|, p) = 1$. Then

$$\left| G/_{\langle g \rangle} \right| = pm, \ m < n$$

and so, by the induction hypothesis, $G/_{< g >}$ has an element, say $x < g >$, of order p. But $x < g >$ is the homomorphic image of x under the natural homomorphism and so $|x| = lp$. Hence $|x^l| = p$ and, by induction, the theorem follows. ∎

We now extend the lemma to arbitrary groups and to subgroups of order p^m rather than subgroups of order p.

12.3.2 *Theorem. (Sylow)*

If G is a finite group whose order is divisible by p^m, p a prime, $m \geq 0$, then G contains a subgroup of order p^m.

Proof. If $|G| = 1$, the theorem holds trivially with $m = 0$. We argue by induction on n, the order of G.

Assume the result is true for groups of order less than n for some $n > 1$ and let $|G| = n$. By the class equation, we have

$$n = |G| = |Z(G)| + \sum_{g \notin Z(G)} |G : C_G(g)|.$$

If n is relatively prime to p, there is nothing to prove. Thus assume that $n = p^m q$, $m \geq 1$ (p and q are not necessarily relatively prime).

If p divides $|Z(G)|$, by the lemma, $Z(G)$ has an element g of order p. Since $<g>$ is a normal subgroup of G, we may form $G/_{<g>}$, a group of order $p^{m-1}q$. By the induction hypothesis, $G/_{<g>}$ contains a subgroup $P/_{<g>}$ of order p^{m-1} and so G contains a subgroup, namely P, (by Theorem 6.4.13) of order p^m.

Assume therefore that p does not divide $|Z(G)|$. Then for some $g \in G \setminus Z(G)$, we have $\gcd(|G:C_G(g)|, p) = 1$. Therefore

$$|C_G(g)| < |G| \text{ and } |C_G(g)| = p^m s .$$

By the induction hypothesis, $C_G(g)$ (and hence G) contains a subgroup of order p^m. By induction the result follows. ∎

Recall that if q divides the order of a group, it is not necessarily the case that the group contains a subgroup of order q. The Sylow theorem we have just proved is the best partial converse of Lagrange's Theorem which can be proved for arbitrary groups. There are stronger theorems which hold for special classes of groups and we shall soon prove one such theorem for the class of finite abelian groups. In fact the full converse of Lagrange's Theorem holds for this class.

12.3.3 *Definition*

A subgroup P of a group G is a *p-subgroup* for some prime p if its order is a power of p. If $|G:P|$ is relatively prime to P, then P is a *Sylow p-subgroup*.

Remark. If $|G| = p^n q$ where $\gcd(p, q) = 1$, then every Sylow p-subgroup of G is of order p^n.

12.3.4 *Lemma*

If H and K are subgroups of a group G such that H normalizes K i.e., $hKh^{-1} = K$ for all $h \in H$, then HK is a subgroup of G and K is a normal subgroup of HK.

Proof. For all $h \in H$ we have $hK = Kh$ since $hKh^{-1} = K$. Therefore $HK = KH$ and so $HKHK = H^2 K^2 = HK$ proving closure. Moreover, $(hk)^{-1} = k^{-1} h^{-1} \in KH = HK$, and so HK is a subgroup of G. The normality of K in HK is easily proved and is left as an exercise. ∎

12.3.5 *Lemma*

If H and K are finite subgroups of G, then $|HK| = \dfrac{|H||K|}{|H \cap K|}$. (Note that we are not claiming that HK is a subgroup. In fact it usually is not.)

Proof. Let $T = \{(h,k) \mid h \in H, k \in K\}$. Define a map $f : T \to HK$ by $f(h,k) = hk$. Now

$$f^{-1}(h_1 k_1) = \{(h,k) \mid hk = h_1 k_1\} = \{(h,k) \mid h_1^{-1} h = k_1 k^{-1}\}.$$

But $h_1^{-1} h = k_1 k^{-1} \in H \cap K$ and so, $h_1^{-1} h = k_1 k^{-1}$ if, and only if, there exists $x \in H \cap K$ such that $h = h_1 x$ and $k = x^{-1} k_1$. Hence there are $|H \cap K|$ elements in each inverse image of the elements of HK. There are $|HK|$ elements in the image of f and so $|HK| = \dfrac{|T|}{|H \cap K|} = \dfrac{|H||K|}{|H \cap K|}$. ∎

The following corollary is obvious and we omit the proof.

12.3.6 *Corollary*

If H and K are p-subgroups for some prime p and HK is a subgroup, then HK is a p-subgroup.

Proof. Exercise. ∎

12.3.7 *Lemma*

If P is a p-subgroup of G and S is a Sylow p-subgroup with $P \subseteq N_G(S)$ or $S \subseteq N_G(P)$, then $P \subseteq S$. In particular, if S_1, S_2 are Sylow p-subgroups of G and $S_1 \subseteq N_G(S_2)$, then $S_1 = S_2$.

Proof. By Lemmas 12.3.4 and 12.3.5 and Corollary 12.3.6, PS is a p-subgroup. By Lagrange's theorem, no p-subgroup of G can properly contain a Sylow p-subgroup. Therefore, $PS = S$ and so $P \subseteq S$. The last statement follows immediately since $S_1 = S_1 S_2 = S_2$. ∎

12.3.8 *Theorem (Sylow)*

For a given prime p, the number of Sylow p-subgroups of a group G is congruent to 1 modulo p.

Proof. Let $\Omega = \{S_1, S_2, ..., S_q\}$ be the set of Sylow p-subgroups of G and let S_1 act on Ω by conjugation. Then clearly $\{S_1\}$ is a singleton orbit of this action.

Since the stabilizer of S_j, $j \neq 1$ is $N_G(S_j) \cap S_1$, the size of the orbit of S_j is $|S_1 : N_G(S_j) \cap S_1|$ by Theorem 12.1.10 and is either 1 or a nonzero multiple of p. If $|S_1 : N_G(S_j) \cap S_1| = 1$, then $S_1 \subseteq N_G(S_j)$ and so, by Lemma 12.3.7, $S_j = S_1$, a contradiction. Thus all orbits except $\{S_1\}$ have size a nonzero multiple of p and so the number of Sylow p-subgroups is congruent to 1 modulo p. ∎

12.3.9 *Theorem (Sylow)*

A p-subgroup of G is contained in a Sylow p-subgroup.

Proof. Let P be a p-subgroup of G and let $\Omega = \{S_1, S_2, ..., S_q\}$ be the set of Sylow p-subgroups. Let P act by conjugation on Ω and consider the orbit of some Sylow p-subgroup S under the action of P.

The stabilizer of S is $N_G(S) \cap P$ and so, by Theorem 12.1.11, the size of the orbit of S is $|P : N_G(S) \cap P|$. Since $|P|$ is a power of p, $|P : N_G(S) \cap P|$ is either 1 or a nonzero multiple of p. Not all orbits can be multiples of p since the number of Sylow p-subgroups is congruent to 1 modulo p by Theorem 12.3.8. Hence, for some S_j, $|P : N_G(S_j) \cap P| = 1$. Therefore $P \subseteq N_G(S_j)$ and so, by Lemma 12.3.7, $P \subseteq S_j$. ∎

The final result in this series of theorems is also due to Sylow and establishes the somewhat surprising fact that all the Sylow p-subgroups are conjugate in G.

12.3.10 *Theorem (Sylow)*

All Sylow p-subgroups are conjugate in G. Their number is congruent to 1 modulo p and is $|G : N_G(S)|$ where S is any Sylow p-subgroup of G.

Proof. Let G act on the Sylow p-subgroups of G by conjugation and suppose $\Omega = \{S_1, S_2, ..., S_q\}$ is an orbit of the action. We know $|\Omega| = |G : N_G(S_1)|$ since the stabilizer of S_1 is $N_G(S_1)$. Therefore, since $S_1 \subseteq N_G(S_1)$, it follows that $\gcd(|\Omega|, p) = 1$.

Let S be any Sylow p-subgroup and let it act on Ω by conjugation. The orbit containing S_j under this action is of size $|S : N_G(S_j) \cap S|$ and so is either a nonzero multiple of p or 1. Since $\gcd(|\Omega|, p) = 1$, there must be an orbit of size 1 which means that for some i $|S : N_G(S_i) \cap S| = 1$. Hence, $S \subseteq N_G(S_i)$ and so, by Lemma 12.3.7, $S = S_i$. Therefore, every Sylow p-subgroups is in Ω and consequently any two Sylow p-subgroups are conjugate in G. Moreover, $|\Omega| = |G : N_G(S_1)|$. ∎

12.4 Applications of the Sylow Theorems

In this section we obtain structure theorems for some of the simpler groups. The first theorem below on abelian groups establishes that the converse of Lagrange's Theorem holds for abelian groups and uses only the Sylow theorems. To tackle nonabelian groups we need additional results which we prove below.

12.4.1 *Theorem*

For every divisor of the order of an abelian group there is a subgroup of that order.

Proof. Let G be an abelian group of order $n = p_1^{n_1} p_2^{n_2} ... p_k^{n_k}$, p_i prime for all i and suppose m is a divisor of n. Then $m = p_1^{m_1} p_2^{m_2} ... p_k^{m_k}$ where $0 \le m_i \le n_i$ for all i, $1 \le i \le k$.. By Theorem 12.3.2, G contains subgroups of orders $p_1^{m_1}, p_2^{m_2}, ..., p_k^{m_k}$, say $P_1, P_2, ..., P_k$, respectively. Since every subgroup of an abelian group is normal, an appeal to Lemmas 12.3.4 and 12.3.5 and an easy induction shows that $P_1 P_2 ... P_k$ is a subgroup of G of order $p_1^{m_1} p_2^{m_2} ... p_k^{m_k}$. ∎

12.4.2 *Definition*

A group G is said to be *simple* if it is nonabelian and has no normal subgroups other than itself and $<e>$.

Simple groups are, in fact, not simple at all. If a group contains a nontrivial normal subgroup, then it can be factored out to yield a smaller group whose structure may be more transparent than the original group. It is then often possible to "lift" the information garnered from this simpler group to the original group. In a simple group, we cannot use this strategy and are left with the much harder problem of investigating the big group.

Group theorists have shown that every finite nonabelian simple group is either alternating or belongs to one of 16 families of simple groups of a certain type or one of 26 so-called sporadic groups.

We shall show that the first simple group is the alternating group on five symbols and, with the help of this result, we shall prove that all the alternating groups on five or more letters are simple.

The other families of simple groups are too difficult to investigate at this stage. We need the machinery of group representations, a powerful tool in the study of groups. Simple groups often occur in connection with some geometric structure and we are able to deduce properties of the groups from properties of the geometric structure.

12.4.3 *Theorem*

Let G act on the left cosets of the subgroup H as described in Example 12.1.2 (iv) above. Then the kernel K of the action is $\bigcap_{x \in G} x^{-1}Hx$. If $|G:H| = n$, then G/K is isomorphic to a subgroup of S_n.

Proof. An element g of G induces the identity map on the set of left cosets of H in G if, and only if,

$$gxH = xH \text{ for all } x \in G \text{ iff } x^{-1}gxH = H \text{ for all } x \in G$$
$$\text{iff } x^{-1}gx \in H \text{ for all } x \in G \text{ iff } g \in xHx^{-1} \text{ for all } x \in G.$$

Thus $K = \bigcap_{x \in G} xHx^{-1}$. By Theorem 12.1.5, G/K acts faithfully on the set of left cosets of H in G and so G/K is a group of permutations of the set of left cosets of H in G (of which there are n) and so, G/K is isomorphic to a subgroup of S_n. ■

12.4.4 *Corollary*

If H is a subgroup of G of index p, the smallest prime dividing the order of G, then H is a normal subgroup of G.

Proof. By 12.4.3, G/K is isomorphic to a subgroup of S_p where $K = \bigcap_{x \in G} x^{-1} H x$. Hence $|G/K|$ divides $p!$. Since p is the smallest prime dividing the order of G, it follows that $|G/K|$ divides p. Since $K \neq G$, we deduce $|G:K| = p = |G:H|$. Hence $K = \bigcap_{x \in G} x^{-1} H x = H$ and so H is normal since K is. ∎

Remark. The prime 2 is the smallest prime so that if the index of a subgroup is 2, that subgroup is normal, a result we proved earlier using elementary methods.

We now have the tools necessary to obtain some structure theorems for some restricted classes of groups.

12.4.5 *Theorem*

  (i) If $|G| = pq$ where p and q are primes, $p < q$ and p does not divide $q - 1$, then G is cyclic. If p divides $q - 1$, then G is generated by a and b, where $a^p = e = b^q$, $aba^{-1} = b^s$ and $s^p \equiv 1 \bmod q$.

  (ii) If $|G| = 30$, then G contains a cyclic subgroup of order 15 and both the Sylow 3-subgroup and the Sylow 5-subgroup are normal in G.

Proof. By Theorem 12.3.2, G has Sylow subgroups P and Q of orders p and q, respectively. Moreover, by Theorem 12.4.4, Q is normal in G and $PQ = G$ since $|PQ| = pq$ by Lemma 12.3.5. The number of Sylow p-subgroups is of the form $1 + kp$ for some nonnegative integer k and is a divisor of $|G| = pq$ by Theorem 12.3.8.

Since $\gcd(1 + kp, p) = 1$, it follows that $1 + kp$ divides q and since q is prime, we must have $1 + kp = 1$ or q.

If p does not divide $q-1$, then $k=0$, showing that P is also normal in G. In this case, if x and y are elements of P and Q, respectively, $xyx^{-1}y^{-1} \in P \cap Q = <e>$ and so, $xy = yx$.

Let a and b be generators of P and Q, respectively, and set $|ab| = n$. Then $e = (ab)^n = a^n b^n$ since a and b commute. Therefore $a^n = b^{-n}$ and this element is in $P \cap Q = <e>$. Thus $a^n = b^n = e$, proving that both p and q divide n. Therefore pq divides n since p and q are relatively prime.

On the other hand, $(ab)^{pq} = (a^p)^q (b^q)^p = e$ and so n divides pq. Therefore, $|ab| = pq$ and G is cyclic with generator ab.

Next, suppose p divides $q-1$. With the same notation as above, since $<b>$ is a normal subgroup of G, $aba^{-1} = b^s$. Conjugating again by a yields

$$a^2 b a^{-2} = ab^s a^{-1} = (aba^{-1})^s = b^{s^2},$$

and in general, $a^k b a^{-k} = b^{s^k}$. In particular, $b = a^p b a^{-p} = b^{s^p}$ and so $s^p \equiv 1 \bmod q$. If $s \equiv 1 \bmod q$, the group is cyclic as above.

By Theorem 9.6.9, the multiplicative group of $\mathbf{Z}_q^*$ is cyclic and so, since p divides $q-1$, we can find an element s, say, of order p. Therefore $s^p \equiv 1 \bmod q$ and the resulting group G generated by a and b and satisfying $aba^{-1} = b^s$ exists (see Problem 33 at the end of the chapter).

Moreover,

$$G = \{b^i a^j \mid 0 \le i < q-1, \ 0 \le j < p-1\}$$

and computations are performed as follows:

$$b^i a^j . b^m a^n = b^i a^j b^m a^{-j} a^{j+n}$$
$$= b^i (a^j b a^{-j})^m a^{j+n} = b^{i+s^j m} a^{j+n}.$$

It can be shown that if we choose a different element, s_1 of order p in $(\mathbf{Z}_q^*, .)$, the group we obtain is isomorphic to the one above.

(ii) There are 1 or 6 Sylow 5-subgroups and 1 or 10 Sylow 3-subgroups. If there are 6 Sylow 5-subgroups and 10 Sylow 3-subgroups, we would have 24 elements of order 5 and 20 elements of order 3 giving us 44 elements which is impossible since G has only 30 elements. Therefore either the Sylow 3-subgroup or the Sylow 5-subgroup is normal in G.

Let P_3 (resp. P_5) be a Sylow 3-subgroup (resp. a Sylow 5-subgroup) Since one at least is normal, $H = P_3 P_5$ is a subgroup of order 15 and so cyclic by (i). Moreover, since $|G:H| = 2$, H is normal in G by Corollary 12.4.4.

Since H is abelian, both P_3 and P_5 are normal in H and since conjugation by any element g in G induces an automorphism of H, it follows that $gP_3 g^{-1} = P_3$ and $gP_5 g^{-1} = P_5$ and so both P_3 and P_5 are normal in G. ∎

12.4.6 *Examples*

(i) Let $|G| = 6$. Since 2 divides $3 - 1$
$$G = < a,b \,|\, a^2 = e = b^3 ; aba^{-1} = b^s > \text{ where } s^2 \equiv 1 \bmod 3.$$
One solution of this congruence is obviously $s = 1$ which gives rise to a cyclic group. The other solution is $s = -1$.

With $s = -1$ we have the group
$$G = < a,b \,|\, a^2 = e = b^3 ; aba^{-1} = b^{-1} >.$$

The symmetric group
$$S_3 = < (12),(123) \,|\, (12)^2 = (1) = (123)^3 ; (12)(123)(12)^{-1} = (123)^{-1} >.$$

If we put $a = (12)$ and $b = (123)$, we see that the group G is isomorphic to S_3.

(ii) Let $|G| = 21$. Since 3 divides $7 - 1$
$$G = < a,b \,|\, a^3 = e = b^7 ; aba^{-1} = b^s \text{ where } s^3 \equiv 1 \bmod 7 >.$$

If $s \equiv 1 \bmod 7$, the group is cyclic. We want an element of order 3 in the multiplicative group $\mathbf{Z}_7^*$.

We first find an element in $\mathbf{Z}_7$ whose multiplicative order is 6. Computing the powers of $\overline{3}$ we get $\overline{1}, \overline{3}, \overline{3}^2 = \overline{2}, \overline{3}^3 = \overline{6}, \overline{3}^4 = \overline{4}, \overline{3}^5 = \overline{5}$. We are actually trying to solve $\overline{s}^3 = \overline{1}$ in $\mathbf{Z}_7$, but by knowing a generator of the multiplicative group of nonzero elements of $\mathbf{Z}_7$, we can find the elements of order 3. They are $\overline{2}$ and $\overline{4}$. These two cases give rise to
$$G_1 = < a,b \,|\, a^3 = e = b^7 ; aba^{-1} = b^2 \text{ and}$$
$$G_2 = < a,b \,|\, a^3 = e = b^7 ; aba^{-1} = b^4 >.$$

In G_2 if we set $c = b^2$, we get $G_2 = <a, c \mid a^3 = e = c^7; aca^{-1} = c^2 >$, showing that in fact these two groups are isomorphic. ∎

12.4.7 Theorem

If $G/Z(G)$ is cyclic, then G is abelian.

Proof. Letting $Z(G) = Z$, we have $G = \bigcup_{m \in \mathbf{Z}} a^m Z$ since G/Z is cyclic. If b and c are elements of G there exist z_1 and z_2 in Z and p and q in $\mathbf{Z}$ such that $b = a^p z_1$ and $c = a^q z_2$. Therefore

$$bc = a^p z_1 a^q z_2 = a^{p+q} z_1 z_2 = a^q a^p z_2 z_1 = a^q z_2 a^p z_1 = cb$$

since z_1 and z_2 are in the centre of G. ∎

12.4.8 Corollary

If G is of order p^2, then G is abelian.

Proof. By Corollary 12.2.2, the centre $Z(G)$ of G is not trivial. Hence $G/Z(G)$ is cyclic (of order 1 or p) and so G is abelian by the previous theorem. ∎

We prove one last theorem which will be useful in our quest for the first simple group.

12.4.9 Theorem

Let G be a finite group and H a subgroup of G such that $|G|$ does not divide $|G:H|!$. Then G is not simple.

Proof. By 12.4.3, $G/_K$, with $K = \bigcap_{x \in G} x^{-1}Hx \,(\subseteq H\,)$, is isomorphic to a subgroup of $S_{|G:H|}$. It follows by Lagrange that K cannot be trivial since $|G|$ does not divide $|G:H|!$. Therefore G is not simple. ∎

12.4.10 *Examples*

(i) If $|G| = 4p$ where p is a prime greater than 3 then G is not simple. For, if P is a Sylow p-subgroup, $|G:P| = 4$ and so $|G:P|! = 24$ and $4p$ does not divide 24 since $p > 3$.

(ii) Let $|G| = 36$. There are 1 or 4 Sylow 3-subgroups. If P_3 is one of the 4, then $|G:N_G(P_3)| = 4$ and since 36 does not divide $|G:N_G(P_3)|!$, G is not simple. ∎

12.4.11 *Theorem*

The first simple group is at least of order 60.

Proof. The reader can easily verify that, by applying one of the theorems proved above, all groups of order less than 60 are immediately seen to be not simple. We give a few examples of how to proceed by proving that groups of orders $12, 40, 56$ are not simple.

(i) If $|G| = 12$, the Sylow 2-subgroup P is of order 4 and so $|G:P| = 3$. But 12 does not divide $3!$ and so G cannot be simple by Theorem 12.4.9.

(ii) If $|G| = 40 = 8.5$, then G has only one Sylow 5-subgroup (the possibilities are $1, 6, 11, 16, 21$. But their number must divide 40, leaving 1 as the only choice). Thus the unique Sylow 5-subgroup must be normal.

(iii) Suppose $|G| = 56 = 8.7$. The number of Sylow 7-subgroups is 1 or 8.

If there are 8, then there are 48 elements of order 7. A Sylow 2-subgroup contains 8 elements and these, together with the 48 elements of order 7 account for all elements of G. Hence in this case there is only one Sylow 2-subgroup which is therefore normal.

If there is only 1 Sylow 7-subgroup, it is normal. Therefore in either case, G is not simple. ∎

We now set about proving that A_5, the alternating group on 5 letters, is simple. To that end, we first prove a pair of useful lemmas.

12.4.12 *Lemma*

If H is a subgroup of G of index 2 and K a conjugacy class of G entirely contained in H, then either K is a conjugacy class of H or K splits into two conjugacy classes of H of equal size. Moreover, if g is an element of K, then K is a conjugacy class of H if $C_G(g) \not\subset H$ while K splits into two if $C_G(g) \subseteq H$.

Proof. Suppose $C_G(g) \not\subset H$. Since H is normal in G, $G = HC_G(g)$ and so, by the Second Isomorphism Theorem,

$$G\big/H = HC_G(g)\big/H \cong C_G(g)\big/H \cap C_G(g).$$

Since $H \cap C_G(g) = C_H(g)$, it follows that $|C_G(g)| = 2|C_H(g)|$. Therefore,

$$|K| = |G:C_G(g)| = |G|\big/|C_G(g)| = |H|\big/|C_H(g)| = |H:C_H(g)|.$$

But $|H:C_H(g)|$ is the number of elements in the conjugacy class of g in H. Thus K is a conjugacy class in H.

If $C_G(g) \subseteq H$, then $C_G(g) = C_H(g)$ and so the number of elements in the conjugacy class of g in H is

$$|H:C_H(g)| = |H|\big/|C_H(g)| = |G|\big/2|C_G(g)| = \tfrac{1}{2}|G:C_G(g)| = \tfrac{1}{2}|K|.$$

Therefore, in this case, K splits into two conjugacy classes in H, each of the same size. ∎

12.4.13 *Lemma*

A_n is generated by the 3-cycles.

Proof. Any element of A_n is the product of an even number of transpositions. But $(ij)(jl) = (ijl)$, while if the transpositions are disjoint, say $(ij)(kl)$, we interpose $(jk)(jk) = (1)$ between (ij) and (kl) to get

$$(ij)(kl) = (ij)(jk)(jk)(kl) = (ijk)(jkl). \blacksquare$$

12.4.14 *Theorem*

The alternating group A_5 is simple.

Proof. Other than the identity, A_5 contains 24 5-cycles, 20 3-cycles and 15 permutations of the form $(ab)(cd)$. These are all conjugacy classes of S_5 contained in A_5. Since (45) commutes with (123), $C_{S_5}((123)) \not\subset A_5$ and so, by Lemma 12.4.11 it follows that the 20 3-cycles form a single conjugacy class of A_5. Moreover, all permutations of the form $(ab)(cd)$ belong to the same conjugacy class in A_5 since their number is odd.
  Also,
$$C_{S_5}((12345)) = < (12345 > \subseteq A_5$$
since $g(12345)g^{-1} = (g(1), g(2), g(3), g(4), g(5)) = (12345)$ if, and only if, $g = (12345)$. Therefore by Lemma 12.4 11, the 24 5-cycles split into two conjugacy classes in A_5, each with 12 elements.
  Suppose that $H \neq < (1) >$ is normal in A_5. We note first that the normality implies that if h is an element of H, then H contains all the conjugates of h in A_5. Thus if H contains a 3-cycle, it contains all 3-cycles which, according to Lemma 12.4.2, generate A_5. It therefore follows that if we can force a 3-cycle in H, then $H = A_5$ and the theorem is proved.
  Suppose first that H contains a permutation of the form $(ab)(cd)$. Then, since H is normal, it also contains $(ab)(de)$ where e is not a, b, c or d. Thus H contains

$$(ab)(cd)(ab)(de) = (cde)$$

and so $H = A_5$ in this case.

The other possibility is for H to contain a 5-cycle $(abcde)$. Then
$$(ab)(cd)(abcde)(ab)(cd) = (badce) \in H.$$

But then $(abcde)(badce) = (aec) \in H$ and so, again $H = A_5$. Therefore, A_5 is simple. ∎

We make use of this theorem to prove that A_n is simple if $n \geq 5$. On the other hand, A_4 has a normal subgroup, namely

$$\{(1), (12)(34), (13)(24), (14)(23)\}$$

while A_3 is cyclic and A_2 is trivial.

Surprisingly, it is the simplicity of A_n, $n \geq 5$ which is related to the insolvability by radicals of polynomial equations of degree greater than 4.

12.4.15   *Theorem*

A_n, $n \geq 5$ is simple.

Proof. We first show that A_6 is simple. Let $A_5^{(i)}$ denote the alternating group on $\{1, 2, ..., i-1, i+1, ..., 6\}$. Clearly $A_5^{(i)}$ is a simple subgroup of A_6.

Suppose H is a nontrivial normal subgroup of A_6 and let H contain a 3-cycle which we may assume to be (123), by re-numbering , if necessary. Since $(45) \in C_{S_6}((123)) \not\subset A_6$, by 12.4.12 all 3-cycles are in H, and so $H = A_6$.

Assume therefore that H contains no 3-cycles. Then H cannot contain any of the $A_5^{(i)}$.

Since $H \cap A_5^{(i)}$ is a normal subgroup of the simple group $A_5^{(i)}$, it follows that $H \cap A_5^{(i)} = <(1)>$ for all i, $i = 1, 2, ..., 6$. Therefore no element in $H \setminus <(1)>$ can fix any symbol. Moreover, in the decomposition of an element of H as a product of disjoint cycles, the cycle lengths must be the same.

For, if a permutation π of H has cycles of lengths s and t with $s > t$, then $\pi^t \neq (1)$ and π^t fixes at least t symbols, a contradiction. Hence the elements of H have one of the two forms

$$(12)(34)(56) \text{ or } (123)(456).$$

The first is an odd permutation and so all elements in H are of the form $(123)(456)$. By renumbering if necessary, let $(123)(456) \in H$. Since H is normal in A_6, $(124)(123)(456)(142) = (243)(156) \in H$ and so $(123)(456)$ $(243)(156) = (16254) \in H$, a contradiction. Hence $H = \langle (1) \rangle$, proving that A_6 is simple.

Assume inductively that A_n is simple for some $n \geq 6$ and let H be a normal subgroup of A_{n+1}. The very same arguments used in the case of $n = 6$ apply and so we conclude that the elements of $H \backslash \langle (1) \rangle$ fix no symbol and that the cycles in their decomposition as a product of disjoint cycles are of the same length.

If there is an element in H which is a product of 2-cycles, say

$$(12)(34)(56)c_4...c_s$$

then

$$(123)(12)(34)(56)c_4...c_s(132) = (23)(14)(56)c_4...c_s$$

is in H. The product,

$$(12)(34)(56)c_3...c_s(23)(14)(56)c_3...c_s$$

is not the identity and is also in H but fixes 5 and 6. Therefore, in this case $H = \langle (1) \rangle$.

We may therefore assume that each cycle of an element of H has length at least 3.

(i) Suppose first that H contains an element which is a product of 3-cycles. We may assume without loss of generality that $(123)(456)$ $c_3 c_4 ... c_k \in H$ where the c_i are all 3-cycles. Then

$$(124)(123)(456)c_3 c_4...c_k(142) = (243)(156)c_3 c_4...c_k \in H$$

and the product

$$(123)(456)(243)(156)\, c_3^2 c_4^2 ... c_k^2 \in H.$$

But this element is clearly not the identity but fixes 3, a contradiction.

(ii) We may assume that H contains an element of the form $(1234...)$ $c_2 c_3 ... c_k$. Conjugating by $(12)(34)$ we see that

$$(2143...)c_2 c_3 ... c_k \in H$$

and so the product

$$(1234...)(2143...)c_2^2 c_3^2 ... c_k^2 \in H.$$

But this product is not the identity and fixes 2, a contradiction.
Therefore A_{n+1} is simple and by induction A_n is simple for all n. ∎

We end this chapter by proving that there is no other simple group of order 60. In fact, we prove a stronger result, namely, that a group of order 60 which has more than one Sylow 5-subgroup is isomorphic to A_5.

   We offer this theorem not so much because of its importance, but rather as an example of how representing an abstract group as a group of permutations can help in unraveling the structure of the abstract group. In addition, this theorem will come in handy when we prove that the rotation group of the icosahedron is isomorphic to A_5. (See Theorem 13.3.2(iii).)

12.4.16 *Lemma*

If G is a group of order 60 with non-normal Sylow 5-subgroups, then no proper normal subgroup of G has order a multiple of 5.

Proof. Since the number of Sylow 5-subgroups is $1 + 5k$ and $k \neq 0$, we must have $k = 1$. Therefore there are 6 Sylow 5-subgroups.
   Let H be a proper normal subgroup of G such that 5 divides $|H|$. Then all 24 elements of order 5 are contained in H and so $|H| = 30$. By Theorem 12.4.5(ii), the Sylow 5-subgroup is normal in H and so normal in G, a contradiction. ∎

12.4.17 *Lemma*

If G is of order 60 with non-normal Sylow 5-subgroups, then G is embedded in A_6.

Proof. Let N be the normalizer of a Sylow 5-subgroup. Then $|G:N|=6$, and by Theorem 12.4.3, $G\Big/\bigcap_{x\in G} xNx^{-1}$ is embedded in S_6 with embedding map φ, say. Letting $K=\bigcap_{x\in G} xNx^{-1}$ we have $|K|=1, 2, 5$ or 10. By the previous lemma $|K|=1$ or 2.

If $|K|=2$, then $G\big/K$ is of order 30 and so, by Theorem 12.4.5(ii), $G\big/K$ has an element of order 15. But the image of $G\big/K$ under φ is in S_6 and S_6 has no element of order 15. Hence K is the trivial subgroup. Therefore G is embedded in S_6. Moreover G is generated by the 24 elements of order 5 and so $\varphi(G)$ is also generated by 24 elements of order 5 in S_6. But the elements of order 5 in S_6 are 5-cycles and these are even permutations. Hence G is embedded in A_6. ■

12.4.18 *Theorem*

If G is of order 60 with non-normal Sylow 5-subgroups, then G is simple.

Proof. Let H be a proper normal subgroup of G. By Lemma 12.4.16, $|H|=1, 2, 3, 4, 6$ or 12.

(i) If $|H|=2$ or 4, then $G\big/H$ (and so G) has an element of order 15 by Theorem 12.4.5. But A_6 has no such element and so H is not of order 2 or 4.

(ii) If $|H|=3$ or 6, then the Sylow 3-subgroup of G is normal. Let P be the Sylow 3-subgroup and Q a Sylow 5-subgroup. Then PQ is a subgroup of G of order 15, and so cyclic by Theorem 12.4.5, leading to the same contradiction as in (i).

(iii) Suppose $|H|=12$. The Sylow 2-subgroup cannot be normal in H. For, if it were, then it would be normal in G and we are back to the situation in (i).

Thus let P and Q be two Sylow 2-subgroups of H. Then $|PQ| = \frac{16}{|P \cap Q|}$. Since PQ is contained in H, it follows that $P \cap Q = <x>$ where $|x| = 2$. Also, $<P,Q> = H$ and so x is in the centre of H since both P and Q are abelian.

If y is an element of order 3, then $|xy| = 6$. But A_6 has no element of order 6. Therefore H is trivial and G is simple. ∎

12.4.19 *Theorem*

If G is of order 60 with non-normal Sylow 5-subgroups, then $G \cong A_5$.

Proof. If H is a proper subgroup of G, the simplicity of G implies that $\bigcap_{x \in G} xHx^{-1} = <e>$. Hence, if $|G:H| = n$, G is embedded in S_n by Theorem 12.4.3. Therefore $60 \leq n!$ and so, $n \geq 5$ and $|H| \leq 12$.

There are 3, 5 or 15 Sylow 2-subgroups of G. If P is one of them and $N = N_G(P)$, then $|G:N| = 3, 5$ or 15. Since $|G:N|$ must be at least 5, it follows that $|G:N| = 5$ or 15.

If there are 15 Sylow 2-subgroups, they cannot intersect pairwise trivially since we would then have 45 elements of orders 2 or 4 and these, together with the 24 elements of order 5 amount to too many elements.

Thus let P and Q be two Sylow 2-subgroups such that $P \cap Q = <x>$ where $|x| = 2$. Since $|PQ| = 8$, it follows that either $<P,Q> = G$ or $|< P,Q >| = 12$.

If $<P,Q> = G$, then $x \in Z(G)$, contradicting simplicity of G. If $|< P,Q >| = 12$, then $x \in Z(H)$ where $H = <P,Q>$. If y is an element of H of order 3, then xy is an element of order 6, a contradiction since by Lemma 12.4.17 G is embedded in A_6 and A_6 has no element of order 6.

Therefore $|G:N| = 5$ and so G is embedded in S_5. But, as in Lemma 12.4.17, since G is generated by its elements of order 5, it follows that the embedded image of G in S_5 is generated by the 5-cycles in S_5 and so G is embedded in A_5. Since $|G| = |A_5|$, $G \cong A_5$. ∎

12.5 Exercises

1) Let D_k be the group generated by the matrices

$$\begin{bmatrix} 0 & 1 \\ 1 & 0 \end{bmatrix} \quad \text{and} \quad \begin{bmatrix} \operatorname{cis}\dfrac{2\pi}{k} & 0 \\ 0 & \operatorname{cis}\dfrac{2\pi}{k} \end{bmatrix}$$

where $\operatorname{cis}\varphi = \cos\varphi + i\sin\varphi$. Show that D_k is of order $2k$ and list the conjugacy classes of D_k. (These groups are called the *dihedral groups* (see 5.4.2(v).)

2) Show that the group in Problem 1) is isomorphic to the group of symmetries of a regular k-gon.

3) Prove that if H and K are abelian subgroups of a group G, then $H \cap K$ is a normal subgroup of $< H, K >$.

4) Let G be a finite group of order n with a subgroup H of order m. If $H \cap xHx^{-1} = <e>$ for all x in $G \backslash H$ then there are exactly $n/m - 1$ elements in G which are not in any conjugate of H.

5) Prove that a finite group in which every proper subgroup is abelian is not simple.

6) Define subgroups $Z_k(G)$ of G recursively as follows: $Z_1(G) = Z(G)$ and, having defined $Z_1(G), Z_2(G), ..., Z_k(G)$, define $Z_{k+1}(G)$ by

$$Z_{k+1}(G) = v^{-1}\left(Z\left(G/Z_k(G) \right) \right)$$

where v is the natural homomorphism from G to $G/Z_k(G)$. Prove that if G is a finite p-group, for some positive integer n, $Z_n(G) = G$. (This series of normal subgroups of G is the *upper central series of* G.)

7) Prove that if H is a proper subgroup of a finite p-group G, then $N_G(H)$ properly contains H. Give an example of a group G with a subgroup H whose normalizer is H itself.

8) If G is a finite group such that, for each abelian subgroup A, $N_G(A) = C_G(A)$, then G is abelian.

9) Let G be a noncyclic group which is generated by two elements of order 2. Prove that G has a normal cyclic subgroup of index 2 in G. If G is finite and of order $2k$, then G is isomorphic to the dihedral group D_k of Problem 1.

10) If H and K are subgroups of a group G, then a double coset of the pair (H, K) is a subset of G of the form HxK. Prove that if both H and K are finite, then the number of elements in HxK is $|K||H : xKx^{-1} \cap H|$. [Hint: Let H act on the left cosets of K in G.]

11) If G is a subgroup of the symmetric group S_n which contains an odd permutation, then G has a subgroup of index 2.

12) If G is a group of order $2^k m$, m odd, and the Sylow 2-subgroups of G are cyclic, then G has a normal subgroup of order m. [Hint: Let G act on G by left multiplication and apply the preceding problem.]

13) If a group G of order 12 has no element of order 2 in its center, then G is isomorphic to A_4.

14) If (G, X) is an action with orbits $X_1, X_2, ..., X_n$, prove that $G\!/\!K$ is isomorphic to a subgroup of $G\!/\!K_1 \times G\!/\!K_2 \times ... \times G\!/\!K_n$ where K is the kernel of the action (G, X) and K_i the kernel of the subaction (G, X_i).

15) If (G, X) is an action and $\alpha \in X$, $g \in G$, then $gG_\alpha g^{-1} = G_{g\alpha}$.

16) If (G, X) is a transitive action, then all the stabilizers are conjugate to one another in G.

17) Let G be a subgroup of S_n of order p^k and suppose $n < p^2$. Prove that G is abelian and that every element of G is of order p.

18) If G is a finite p-group, prove that for each divisor m of the order of G there is a normal subgroup of order m.

19) Let P be a Sylow p-subgroup of a finite group G, $N_G(P)$ the normalizer of P in G and H a subgroup of G such that $N_G(P) \subseteq H$. Prove that the normalizer of H is H itself.

20) If G is a transitive subgroup of S_n, prove that G contains at least $n-1$ permutations which have no fixed points.

21) (i) If p is a prime and m and n positive integers with $gcd(m, p) = 1$, prove that $\binom{mp^n}{p^n}$ is relatively prime to p.

($\binom{k}{l}$ is the binomial coefficient).

(ii) Let G be a group of order $p^n m$ where p, m and n are as above. Let $\mathfrak{A}$ be the collection of all subsets of G of size p^n and let G act on G of size p^n and let G act on $\mathfrak{A}$ by:

$$g \bullet E = gE \text{ for all } g \in G, E \in \mathfrak{A}.$$

Prove that $(G, \mathfrak{A})$ is indeed an action and that there is an orbit Ω such that p does not divide $|\Omega|$.

(iii) Let E be an element of the orbit Ω in (ii). Prove that $G_E = \{g \in G \mid gE = E\}$ is of order p^n. (This gives another proof of the first Sylow theorem).

(iv) Let H be any p-subgroup of G and set $P = G_E$. By considering the action of H on Ω, prove that H is contained in some conjugate of P. (This establishes that any p-subgroup of G is contained in a sylow p-subgroup of G and that any two Sylow p-subgroups of G are conjugate.) [Hint: $gG_x g^{-1} = G_{gx}$.]

22) Prove that a noncyclic group of order 21 must have 14 elements of order 3.

23) Let P be a finite p-group and Q a nontrivial normal subgroup of P.

(i) By letting P act on Q by conjugation, prove that $|Q \cap Z(G)| \neq 1$.

(ii) Suppose $|P| = p^n$ and let $<e> = P_0 < P_1 < P_2 < ... < P_{s-1} < P_s = P$ be a maximal chain of normal subgroups; that is to say, we cannot lengthen the chain by introducing more normal subgroups. Prove that $s = n$ and $|P_t| = p^t$ for all t. (This establishes that a finite p-group has normal subgroups of all orders dividing the order of the group.)

24) Prove that a group of order 175 is abelian.

25) Prove that a group of order 105 has a subgroup of order 15.

26) Show that the centre of a group of order 60 cannot be of order 4.

27) Suppose $|G| = 60$ and G has a normal subgroup of order 2. Show that:

(i) G has normal subgroups of orders 6, 10 and 30.

(ii) G has subgroups of orders 12 and 20.

(iii) G has a cyclic subgroup of order 30.

28) If $|G| = 60$ and the Sylow 3-subgroup of G is normal, prove that the Sylow 5-subgroup is also normal.

29) Let G be a finite group, H a normal subgroup of G and P a Sylow p-subgroup of H. If Ω is the set of all Sylow p-subgroups of H, show that both H and G act transitively on Ω by conjugation and deduce that $G = N_G(P)H$.

30) What is the smallest possible odd order nonabelian group?

31) Find the class equation of the group of symmetries of a square.

32) Find the class equation of the symmetry group of a regular hexagon.

33) Referring to Theorem 12.4.5, show that if s_1 and s_2 are not congruent to 1 mod q, but $s_i^p \equiv 1 \bmod q$, $i = 1, 2$, then the groups

$$G_1 = <a, b \mid a^p = e, b^q = e \text{ and } aba^{-1} = b^{s_1}>$$

and

$$G_2 = <a, b \mid a^p = e, b^q = e \text{ and } aba^{-1} = b^{s_2}>$$

are isomorphic and each is isomorphic to the group G defined as follows.

$$G = \{(\overline{i}, \overline{\overline{j}}) \mid \overline{i} \in \mathbf{Z}_q \text{ and } \overline{\overline{j}} \in \mathbf{Z}_p\}$$

and

$$(\overline{i}, \overline{\overline{j}}).(\overline{k}, \overline{\overline{l}}) = (\overline{i + s^j k}, \overline{\overline{j + l}})$$

where s is not congruent to 1 mod q but $s^p \equiv 1 \bmod q$.

34) How many pairwise nonisomorphic groups of order 30 are there? Prove your assertion.

Chapter 13

Isometries

13.1 Isometries of $\mathbf{R}^n$

In this chapter we study briefly groups of isometries of $\mathbf{R}^n$ for arbitrary n and then focus on finite groups of isometries of $\mathbf{R}^2$ and finite groups of rotations in $\mathbf{R}^3$.

We begin with a quick review of the linear algebra involved. The reader who feels unsure of some of the more elementary ideas should consult Appendix B. The inner product will be the usual dot product which we shall denote by $<\ ,\ >$ or $.$ and the length of a vector will be denoted by $\|\ \|$. Whenever matrices appear, unless otherwise stated, they are assumed to be referred to the standard orthonormal basis of $\mathbf{R}^n$ consisting of the vectors ε_i, $i = 1, 2, ..., n$ where ε_i is the vector which has 1 in the ith coordinate position and 0's everywhere else.

13.1.1 *Definition*

A map $T : \mathbf{R}^n \to \mathbf{R}^n$ is an *isometry* of $\mathbf{R}^n$ if T preserves distances, i.e.,

$$\| \alpha - \beta \| = \| T(\alpha) - T(\beta) \|.$$

13.1.2 *Example*

For a fixed α the map $\xi \to \alpha + \xi$ for all $\xi \in \mathbf{R}^n$ is an isometry called a *translation.* ∎

At this point it is not entirely obvious that an isometry is bijective. That it is injective is clear since distinct elements, being at a positive distance apart, map to elements which cannot be at zero distance apart. Once it

341

has been established that an isometry is indeed bijective, we shall see that the set of all isometries under composition of maps forms a group called the *Euclidean Group of dimension* n and denoted by $E(n)$. The next result is most useful in getting a handle on this group.

13.1.3 *Theorem*

An isometry T which fixes the origin is a nonsingular linear transformation.

Proof. Since $\| \alpha - \beta \| = \| T(\alpha) - T(\beta) \|$, setting $\beta = 0$ we get

$$\| \alpha \| = \| T(\alpha) \| \text{ for all } \alpha \in R^n.$$

Now

$$< \alpha - \beta, \alpha - \beta > \; = \; \| \alpha - \beta \|^2 = \| T(\alpha) - T(\beta) \|^2 \text{ for all } \alpha, \beta \in \mathbf{R}.^n$$

Therefore

$$\| \alpha \|^2 - 2 < \alpha, \beta > + \| \beta \|^2 = \| T(\alpha) \|^2 - 2 < T(\alpha), T(\beta) > + \| T(\beta) \|^2$$

and since $\| T(\alpha) \|^2 = \| \alpha \|^2$ and $\| T(\beta) \|^2 = \| \beta \|^2$, it follows that

$$< \alpha, \beta > \; = < T(\alpha), T(\beta) > .$$

(Note that this says that if a map preserves distances, it also preserves angles.)

Consider

$$\| T(\alpha + \beta) - T(\alpha) - T(\beta) \|^2 .$$

Expanding, we get

$$\| T(\alpha + \beta) \|^2 + \| T(\alpha) \|^2 + \| T(\beta) \|^2 - 2 < T(\alpha + \beta), T(\alpha) >$$
$$-2 < T(\alpha + \beta), T(\beta) > +2 < T(\alpha), T(\beta) > .$$

But this is

$$< \alpha, \alpha > +2 < \alpha, \beta > + < \beta, \beta > + < \alpha, \alpha > + < \beta, \beta > -2 < \alpha, \alpha >$$
$$-4 < \alpha, \beta > -2 < \beta, \beta > +2 < \alpha, \beta > = 0.$$

Therefore, $\| T(\alpha + \beta) - T(\alpha) - T(\beta) \|^2 = 0$ and so

$$T(\alpha + \beta) = T(\alpha) + T(\beta).$$

Furthermore, if c is a scalar, we have

$$\| T(c\alpha) - cT(\alpha) \|^2 = <T(c\alpha) - cT(\alpha), T(c\alpha) - cT(\alpha)>.$$

Again, expanding, we get

$$<T(c\alpha), T(c\alpha)> -2c <T(c\alpha), T(\alpha)> +c^2 <T(\alpha), T(\alpha)>$$
$$= c^2 <\alpha,\alpha> -2c^2 <\alpha,\alpha> +c^2 <\alpha,\alpha> = 0.$$

Thus

$$\| T(c\alpha) - cT(\alpha) \|^2 = 0,$$

proving that $T(c\alpha) = cT(\alpha)$.

Therefore, T is a linear transformation. Furthermore T is bijective since an isometry is clearly injective and $\mathbf{R}^n$ is a finite-dimensional vector space. ∎

13.1.4 *Corollary*

Every isometry can be expressed as a linear transformation followed by a translation in a fixed direction.

Proof. If S is an isometry, then the map T defined by

$$T(\xi) = S(\xi) - S(0)$$

is an isometry fixing 0 and so is a linear transformation. Setting $S(0) = \alpha$, we have $S(\xi) = T(\xi) + \alpha$. ∎

From the foregoing, it follows that an isometry is a bijective map whose inverse is clearly an isometry. Moreover, the composition of two isometries is an isometry and so the set of isometries of $\mathbf{R}^n$ forms a group under composition of maps.

For a given isometry S it is easily seen that T and α in the corollary above are uniquely determined. Therefore, each isometry can be represented by an ordered pair, (T, α).

An isometry is a linear transformation if, and only if, $\alpha = 0$ and it is a translation if, and only if, $T = I$, the identity map.

If (S, α) and (T, β) are isometries, we see that

$$(S, \alpha)(T, \beta) = (ST, S(\beta) + \alpha)$$

and

$$(S, \alpha)^{-1} = (S^{-1}, -S^{-1}(\alpha)).$$

Also, the set of translations forms a subgroup of the group $E(n)$ of all isometries of $\mathbf{R}^n$. Moreover, since

$$(T, \beta)(I, \alpha)(T, \beta)^{-1} = (I, T(\alpha)),$$

it follows that the set of all translations of $\mathbf{R}^n$ forms a normal subgroup of $E(n)$ which we shall denote by $\mathbf{T}_n$.

Similarly, the set of all isometries fixing 0 is also a (not normal) subgroup of $E(n)$ which we shall denote by $O(n)$. This group is called the *orthogonal group of dimension n*. Observe that since $O(n).T = E(n)$ and $O(n) \cap T = <(I, 0)>$, we see that

$$E(n)\Big/_T \cong O(n)$$

by the second isomorphism theorem.

Recall that a linear transformation preserves distances if, and only if, its matrix representation relative to an orthonormal basis is an orthogonal matrix, i.e., a matrix A such that $AA^T = I$. (See Appendix B). The elements of $O(n)$ can be classified as follows: If $A \in O(n)$, then $\det A = \pm 1$ since

$$1 = \det I = \det AA^T = \det A \det A^T = (\det A)^2.$$

Therefore the map $\det : O(n) \to \{1, -1\}$ is an epimorphism (why is it onto?) with kernel the so-called *special linear group* $SO(n)$ of dimension n consisting of *rotations* (or *direct isometries*). An element of $O(n)$ which is not a rotation is called an *indirect* (or *opposite*) *isometry*. The index in $O(n)$ of $SO(n)$ is 2 since

$$O(n)\Big/_{SO(n)} \cong \{1, -1\},$$

where $\{1, -1\}$ is the multiplicative group of units of $\mathbf{Z}$.

13.1.5 *Theorem*

Let G be a finite subgroup of $E(n)$. Then G is isomorphic to a subgroup of $O(n)$.

Proof. Let $\alpha \in \mathbf{R}^n$ and consider the orbit $G\alpha$ of α under G. Since G is finite, this orbit is finite, say

$$G\alpha = \{\alpha = \alpha_1, \alpha_2, ..., \alpha_n\}.$$

Set

$$\beta = (\alpha_1 + \alpha_2 + ... + \alpha_n) \Big/ n$$

and let $(T, \gamma) \in G$. Then

$$(T, \gamma)(\beta) = T(\beta) + \gamma = \frac{1}{n}[T(\alpha_1 + \alpha_2 + ... + \alpha_n) + n\gamma]$$

$$= \frac{1}{n}[(T, \gamma)\alpha_1 + (T, \gamma)\alpha_2 + ... + (T, \gamma)\alpha_n].$$

Since

$$\{(T, \gamma)(\alpha_i) \mid i - 1, 2, ..., n\} = \{\alpha_1, \alpha_2, ..., \alpha_n\},$$

it follows that $(T, \gamma)(\beta) = \beta$. Taking a coordinate system with origin at β, we see that G is isomorphic to a subgroup of $O(n)$. ∎

13.2 Finite Subgroups of $E(2)$

We now show that the finite subgroups of $E(2)$ are either cyclic or dihedral. By what we have just proved, we need only show this for finite subgroups of $O(2)$.

13.2.1 *Lemma*

A matrix A is in $SO(2)$ if, and only if, it is of the form $\begin{pmatrix} \cos\theta & -\sin\theta \\ \sin\theta & \cos\theta \end{pmatrix}$.

Proof. Let

$$A = \begin{pmatrix} a & b \\ c & d \end{pmatrix} \in SO(2).$$

Then $a^2 + c^2 = b^2 + d^2 = 1$ and $ad - bc = 1$ since A is orthogonal and $\det A = 1$. Hence, there exist θ and φ such that

$$a = \cos\theta, \; c = \sin\theta, \; b = \sin\varphi \text{ and } d = \cos\varphi.$$

Also

$$\cos\theta\cos\varphi - \sin\theta\sin\varphi = 1.$$

Therefore

$$\cos(\theta + \varphi) = 1 \text{ and so } \theta + \varphi = 2k\pi,$$

and so

$$\cos\theta = \cos\varphi \text{ and } \sin\theta = -\sin\varphi$$

proving that A is of the stated form. ∎

13.2.2 Corollary

A is an indirect isometry if, and only if, A is of the form

$$\begin{pmatrix} \cos\theta & \sin\theta \\ \sin\theta & -\cos\theta \end{pmatrix}.$$

(Observe that an indirect isometry in $E(2)$ is of order 2 and is referred to as a *reflection*).

Proof. Since $\begin{pmatrix} 1 & 0 \\ 0 & -1 \end{pmatrix}$ is a reflection, it follows that any reflection is given by

$$\begin{pmatrix} \cos\theta & -\sin\theta \\ \sin\theta & \cos\theta \end{pmatrix} \begin{pmatrix} 1 & 0 \\ 0 & -1 \end{pmatrix} = \begin{pmatrix} \cos\theta & \sin\theta \\ \sin\theta & -\cos\theta \end{pmatrix}$$

since $|O(2):SO(2)| = 2$, and this is of the required form. ∎

Remark. $\begin{pmatrix} \cos\theta & \sin\theta \\ \sin\theta & -\cos\theta \end{pmatrix}$ is a reflection in a line through the origin making an angle of $\theta/2$ with the positive direction of the x-axis. (Proof?)

13.2.3 *Theorem*

Any finite subgroup of $SO(2)$ is cyclic.

Proof. Let G be a finite subgroup of $SO(2)$ and assume G is not trivial. Let $A = \begin{pmatrix} \cos\theta & -\sin\theta \\ \sin\theta & \cos\theta \end{pmatrix}$ where θ is the smallest positive value of all the angles occurring in matrices in G. For any $X \in G$, let $\alpha(X)$ denote the angle of rotation of X and let $B \in G$, $B \neq I$. with $\alpha(B) = \varphi$. Then $\varphi \geq \theta$ and so there exists a positive integer k such that $k\theta \leq \varphi < (k+1)\theta$. But $BA^{-k} \in G$ whence $\alpha(BA^{-k}) = \varphi - k\theta$ is one of the angles of rotation of an element of G. Since $0 \leq \varphi - k\theta < \theta$, it follows that $\varphi = k\theta$ and so $B = A^k$. Therefore $G = <A>$.∎

13.2.4 *Theorem*

If G is a subgroup of $O(2)$ of finite order containing a reflection, then G is isomorphic to a dihedral group.

Proof. Consider the homomorphism $\det : G \to \{1, -1\}$. The kernel, H is cyclic generated by B, say, by the previous result. Since G contains a reflection, we have $G = H \cup AH$ where A is a reflection in G.
Then

$$G = \{I, B, B^2, B^3, ..., B^{n-1}, A, AB, AB^2, ..., AB^{n-1}\}$$

where $n = |B|$.

Since AB^i, $i = 1, 2, ..., n-1$ are all reflections, they are each of order 2 and so $(AB)^{-1} = B^{-1}A^{-1} = AB$. Hence $ABA = B^{-1}$. In terms of generators and relations,

$$G = <A, B \,|\, A^2 = I, B^n = I, ABA = B^{-1}>.$$

Compare this presentation of the group with that given in Example 5.2.6(iv). The reader should show that B corresponds to $R_{2\pi/n}$ and A

corresponds to M_α where α is an appropriate angle (what is it?). Note that $M_\alpha R_{2\pi/n} M_\alpha = R_{2\pi/n}^{-1}$. ∎

13.3 The Platonic Solids

The analogue of a polygon in two dimensions is a *polyhedron* (many faces) in three dimensions. The cube is an example of a polyhedron. It consists of eight vertices (corners), twelve edges and six faces. We say a vertex is *incident* to an edge or a face if it lies on the edge or face. Similarly, an edge is incident to a face if it lies entirely in the face. In a cube, each vertex is incident to three edges and three faces. We call a polyhedron which has n faces an *n-hedron*. We say that it is a *regular n-hedron* if its faces are congruent regular polygons and each vertex is incident to the same number of faces.

13.3.1 *Examples*

(i) A regular hexahedron is just a fancy name for a cube;

(ii) A regular tetrahedron, shown from above in the diagram below, consists of four (tetra in Greek) faces, six edges and four vertices. The faces are equilateral triangles and each vertex is incident to three faces.

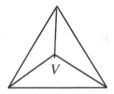

Fig. 13.1 A tetrahedron

(iii) A regular octahedron has eight faces each of which is an equilateral triangle. It has twelve edges and six vertices. Notice that its vital statistics are the same as those for a cube except that the number of faces of a cube is the number of vertices of tetrahedron and vice-versa.

There is a reason for this: if we take a point in the centre of each of the faces of a cube and join the six points judiciously with lines, these will form the edges of an octahedron. If we repeat the process on the octahedron, taking the centres of the triangles and joining them with care, we shall obtain a cube. Two polyhedra which are so related are said to be *dual*. It is easy to see that the dual of a tetrahedron is a tetrahedron. We say that a tetrahedron is *self-dual*.

Below is a diagram of an octahedron viewed from above. Below the plane of the page there is an exact replica of the pyramid with square base which lies above the page.

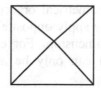

Fig. 13.2 An octahedron

Are there any other regular polyhedral? It turns out that there are only two more, namely the dodecahedron and the icosahedron, which are duals of each other. Below are diagrams of each.

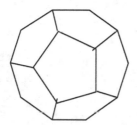 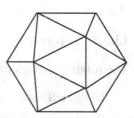

Fig. 13.3 A dodecahedron and an icosahedron

It is easy to see that these are the only regular polyhedra. For, at any vertex of the polyhedron, the sum of the angles between the edges lying in the faces incident to that vertex must be less than 360.

Suppose all the faces are equilateral triangles. An equilateral triangle has angles of 60 degrees and there must be at least three triangles incident at any vertex. Since there can be at most five triangles incident at any vertex, if a regular n-hedron has triangles for faces, there can be three, four or five triangles incident to any vertex.

If there are three triangles, the polyhedron is a tetrahedron; if four, we get an octahedron and 5 triangles yields an icosahedron.

If each face is a square, there can only be three faces incident to any vertex. We get a cube in this case.

A regular pentagon has angles of 108 degrees and so there can only be 3 faces incident at each vertex. This gives us a dodecahedron.

A regular n-gon has angles greater than or equal to 120 if n is greater than five, and so there are no regular polyhedra with regular polygons of more than five sides as faces.

In $\mathbf{R}^2$, given a rigid geometrical object, such as a square piece of cardboard, we can physically apply opposite isometries to it because we can move into the third dimension. For example, a rotation of 180 degrees about the diagonal can only be accomplished by lifting the square off the plane.

In three dimensions, to apply an opposite isometry on a rigid body while leaving it intact, would require a fourth dimension. Therefore when we speak of symmetries of a cube, say, we almost always mean those symmetries brought about by rotations.

We shall now compute the *rotation groups* − that is, the group of direct isometries − of the five regular solids which are known as the *Platonic Solids*.

13.3.2 *Theorem*

The rotation groups of the Platonic solids are:

(i) the tetrahedron: A_4;

(ii) the cube and the octahedron: S_4;

(iii) the dodecahedron and the icosahedron: A_5.

Proof. We use repeatedly Theorem 12.1.10, namely: If (G, X) is an action and $\alpha \in X$, then

$$|G| = |G : G_\alpha| |\mathrm{Orb}_G(\alpha)|.$$

(i) Let the rotation group G of the tetrahedron act on its vertices. In Fig 3.1 above $|G_V| = 3$ and the orbit of V has four elements. Therefore

$|G|=12$. Clearly G is a subgroup of S_4 which contains 8 3-cycles (two for each of the stabilizers of the 4 vertices). We know that the 3-cycles generate A_4 and so the rotation group of the tetrahedron is A_4.

Note that we need 7 axes to generate the 12 rotations, namely: one axis from each vertex to the centroid of the opposite face for a total of 4; one axis joining the midpoints of pairs of opposite sides for a total of 3.

(ii) Let G act on the 8 vertices of a cube. The stabilizer of a given vertex has 3 elements and the orbit of this element is of size 8. Therefore $|G|=24$.

On the other hand, we can think of G as acting on the 4 diagonals joining pairs of opposite vertices of the cube. If g is a nonidentity element of K, the kernel of this action, and the pairs of opposite vertices

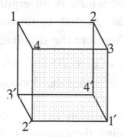

Fig. 13.4

are labeled as shown in the diagram, then $g\{i,i'\}=\{i,i'\}$, $i=1,2,3,4$. If, say, $g(2)=2$, then $g(2')=2'$ and so the axis of rotation is $22'$. But this rotation will take $4'$ to 3 or $4'$ to 1 and g takes $4'$ to $4'$or 4, a contradiction. Therefore $g(2)=2'$, $g(2')=2$ and similarly, for the other vertices. Hence g induces the permutation $(11')(22')(33')(44')$.

Rotations will preserve orientation and so will map a right-handed triad of vectors to another right-handed triad. But $2'1'$, $2'3'$ and $2'4$ form a right-handed triad while their images 21, 23, and $24'$ form a left-handed triad. Therefore such a g cannot exist. Hence G acts faithfully on the four diagonals and, since it is of order 24, $G \cong S_4$.

In this case we need 13 axes, namely: one axis for each of the joins of the centroids of pairs of opposite faces for a total of 3; one axis for each of the joins of pairs of opposite vertices for a total of 4; one axis for each of the joins of the midpoints of pairs of opposite edges for a total of 6.

Note that in each case the number of axes is given by $\dfrac{(|G|+2)}{2}$ (see the next section).

(iii) There are many ways of proving that the rotational symmetries of a dodecahedron or an icosahedron (they are the same since the two solids are duals of each other) form a group isomorphic to A_5, most of them of a geometric nature. However, since we have at our disposal Theorem 12.4.19, we need only show that the icosahedral (or dodecahedral) group is of order 60 and has non-normal Sylow 5-subgroups.

First, to determine its order, let G act on the faces of an icosahedron. There are 20 faces each of which is an equilateral triangle and so the rotation with axis the join of the centroids of two opposite faces, F_1 and F_2, is of order 3. Therefore the stabilizer of F_1, say, is of order 3. The orbit of F_1 is clearly of size 20 and so, by Theorem 12.2.10, the order of the rotation group of either the dodecahedron or the icosahedron is 60.

The faces of a dodecahedron are regular pentagons and so the rotation about an axis joining the centroids of two opposite faces is of order 5. Hence G has at least 6 Sylow 5-subgroups corresponding to the 6 pairs of opposite faces and so the Sylow 5-subgroups are not normal. By Theorem 12.4.18, the icosahedral group is isomorphic to A_5.

We compute the number of axes of rotation for the rotational symmetries of the icosahedron. There are 10 axes joining the centroids of pairs of opposite faces, 15 axes joining the midpoints of pairs of opposite edges and 6 joining pairs of opposite vertices for a total of 31. As remarked above, $31 = \dfrac{(|G|+2)}{2}$. ∎

13.4 Rotations in $\mathbf{R}^3$

By definition, a rotation in $\mathbf{R}^3$ is given by an orthogonal matrix A whose determinant is 1. We first show that the formal definition of rotation coincides with our intuitive notion.

13.4.1 *Theorem*

Let A be a rotation in $\mathbf{R}^3$. Then there is a nonzero vector fixed by A. and a basis $\{\beta_1, \beta_2, \beta_3\}$ of $\mathbf{R}^3$ relative to which the matrix of the linear transformation determined by A is of the form

$$\begin{pmatrix} 1 & 0 & 0 \\ 0 & \cos\theta & -\sin\theta \\ 0 & \sin\theta & \cos\theta \end{pmatrix}$$

which is clearly a rotation (about the fixed vector β_1) in the usual sense of the word.

Proof. We show that $\det(A - I) = 0$. Bearing in mind that $A^{-1} = A^T$, we have:

$$\det(A - I) = \det(A - I)^T = \det(A^T - I)$$
$$= \det(A^T - A^T A) = \det[A^T(I - A)]$$
$$= \det A^T \det(I - A) = \det(I - A).$$

In this line of equalities we have used that $\det B^T = \det B$ for all matrices B. Since we are in a space of odd dimension, $\det(I - A) = -\det(A - I)$. Therefore $\det(A - I) = 0$ and so the null space of the linear transformation $A - I$ contains a nonzero vector α. Thus $(A - I)\alpha^T = 0$ showing that $A\alpha^T = \alpha^T$.

We can extend α to an orthogonal basis $(\alpha_1(=\alpha), \alpha_2, \alpha_3)$ and by dividing these vectors by their respective lengths we obtain an orthonormal basis $B = (\beta_1, \beta_2, \beta_3)$. Relative to this basis, the matrix of the linear transformation determined by A is of the form

$$\begin{pmatrix} 1 & 0 & 0 \\ 0 & a & b \\ 0 & c & d \end{pmatrix} \text{ where } \det \begin{pmatrix} a & b \\ c & d \end{pmatrix} = 1 \text{ and } \begin{pmatrix} a & b \\ c & d \end{pmatrix} \text{ is orthogonal.}$$

By our work in the preceding section, $\begin{pmatrix} a & b \\ c & d \end{pmatrix} = \begin{pmatrix} \cos\theta & -\sin\theta \\ \sin\theta & \cos\theta \end{pmatrix}$ and so

A does indeed represent a rotation (in the usual sense of the word) about the fixed vector β_1. $\blacksquare$

Before we tackle the finite groups of rotations in $\mathbf{R}^3$, we cannot resist showing that an orthogonal matrix whose determinant is -1 can be described quite nicely in terms of rotations and reflections in planes. As pointed out above, these isometries cannot be achieved as motions in 3-space.

13.4.2 *Theorem*

If A is an orthogonal matrix with $\det A = -1$, then A represents a rotation followed by a reflection in a plane perpendicular to the axis of the rotation.

Proof. We show first that there is a nonzero vector α which is mapped to $-\alpha$. We have

$$\det(A + I) = \det(A + I)^T = \det(A^T + I)$$
$$= \det(A^T + A^T A) = \det[A^T (I + A)]$$
$$= \det A^T \det(I + A) = -\det(A + I).$$

It follows that $\det(A + I) = 0$ whence there exists $\alpha \neq 0$ such that $(A + I)\alpha^T = 0$ and so $A\alpha^T = -\alpha^T$. Extending this to an orthogonal basis and then dividing by the lengths of the respective vectors, we get an orthonormal basis $B=(\beta_1,\beta_2,\beta_3)$. Relative to this basis, the matrix of the linear transformation determined by A is of the form

$$\begin{pmatrix} -1 & 0 & 0 \\ 0 & a & b \\ 0 & c & d \end{pmatrix} \text{ where } \det\begin{pmatrix} a & b \\ c & d \end{pmatrix} = 1 \text{ and } \begin{pmatrix} a & b \\ c & d \end{pmatrix} \text{ is orthogonal.}$$

We argue in the same way we did in 13.3.1 and show that A is of the form

$$\begin{pmatrix} -1 & 0 & 0 \\ 0 & \cos\theta & -\sin\theta \\ 0 & \sin\theta & \cos\theta \end{pmatrix}.$$

This matrix is the product of two matrices:

$$\begin{pmatrix} -1 & 0 & 0 \\ 0 & 1 & 0 \\ 0 & 0 & 1 \end{pmatrix}\begin{pmatrix} 1 & 0 & 0 \\ 0 & \cos\theta & -\sin\theta \\ 0 & \sin\theta & \cos\theta \end{pmatrix}.$$

The first matrix represents a reflection in the plane determined by $\mathrm{sp}\{\beta_2,\beta_3\}$ while the second is a rotation about the axis determined by the vector β_1. ∎

Remarks. (i) The matrices above commute;

(ii) when $\theta = \pi$, we obtain the matrix $\begin{pmatrix} -1 & 0 & 0 \\ 0 & -1 & 0 \\ 0 & 0 & -1 \end{pmatrix}$ which is a reflection in the origin.

13.4.3 *Definition*

If G is a finite group of rotations of $\mathbf{R}^3$, associated with each element g of G is its axis of rotation. This axis intersects the unit sphere in two antipodal points which we call the *poles* of g.

13.4.4 *Lemma*

If G is a finite group of rotations of $\mathbf{R}^3$ of order n and Π is the set of poles of G, then G acts on Π.

Proof. We must show that if $\alpha \in \Pi$, then so is $g\alpha$ for all $g \in G$.

Since $\alpha \in \Pi$, there exists $x \in G$ such that $x\alpha = \alpha$. But

$$(gxg^{-1})(g\alpha) = gx(g^{-1}g\alpha) = gx\alpha = g\alpha.$$

Therefore $g\alpha \in \Pi$. ∎

13.4.5 *Theorem*

If G is a finite group of rotations of $\mathbf{R}^3$, then G is one of the following:
(i) cyclic;
(ii) dihedral;
(iii) A_4;
(iv) S_4;
(v) A_5.

Proof. Let the order of G be n and let Π be the set of poles of G. We count the number of elements in

$$T = \{(g,\alpha) \mid g \in G \setminus e, \ \alpha \in \Pi \text{ and } g\alpha = \alpha\}$$

two ways.

First, fix some α. The number of pairs in T with α as second coordinate is $|G_\alpha| - 1$ where $G_\alpha = \{g \in G \mid g\alpha = \alpha\}$ is the stabilizer of α. Therefore

$$|T| = \sum_{\alpha \in \Pi} (|G_\alpha| - 1).$$

On the other hand, fix some $g \neq e$. Then the number of pairs in T with g in the first coordinate position is 2. Hence $|T| = 2(n-1)$. Therefore

$$\sum_{\alpha \in \Pi} (|G_\alpha| - 1) = 2(n-1) \qquad (1).$$

We rewrite (1) as

$$\sum_{\alpha \in \Pi} |G_\alpha| - |\Pi| = 2(n-1) \quad (2).$$

Suppose there are s orbits under the action of G, say $\Omega_i, 1 \le i \le s$. We know that $|\Omega_i| = |G:G_{\alpha_i}|$ where $\alpha_i \in \Omega_i$ and so, if α and β are in the same orbit then we have $|G:G_\alpha| = |G:G_\beta|$ so that $|G_\alpha| = |G_\beta|$. Hence we can rewrite (2) in the form

$$\sum_{i=1}^{s} |G| - |\Pi| = 2(n-1)$$

since $|G:G_{\alpha_i}| \cdot |G_{\alpha_i}| = |G| = n$. We deduce from this that

$$|\Pi| = n(s-2) + 2 \quad (3).$$

Also (1) can be rewritten as

$$n\sum_{i=1}^{s} (1 - \frac{1}{|G_{\alpha_i}|}) = 2(n-1). \quad (4).$$

Setting $|G_{\alpha_i}| = p_i$, and dividing through by n, we obtain

$$2 - \frac{2}{n} = \sum_{i=1}^{s} (1 - \frac{1}{p_i}) \quad (5).$$

When $n=1$, G is trivial so we may assume that $n \ge 2$. Clearly

$$1 \le 2 - \frac{2}{n} < 2$$

and so

$$\sum_{i=1}^{s} (1 - \frac{1}{p_i}) < 2.$$

Since $p_i \ge 2$, it follows that

$$1 - \frac{1}{p_i} \ge \frac{1}{2}$$

whence

$$2 > \sum_{i=1}^{s} (1 - \frac{1}{p_i}) \ge \frac{s}{2}$$

and so $s = 2$ or 3.

(i) If $s = 2$, by (3), there are 2 poles and so the group must be cyclic.

Let $s = 3$, in which case, by Equation (3) above, $|\Pi| = |G|+2$ and so the number of axes required for the rotations is $\left(\dfrac{|G|+2}{2}\right)$. Then

$$1 - \frac{1}{p_1} + 1 - \frac{1}{p_2} + 1 - \frac{1}{p_3} = 2 - \frac{2}{n}$$

and so

$$\frac{1}{p_1} + \frac{1}{p_2} + \frac{1}{p_3} = 1 + \frac{2}{n}.$$

We may assume without loss of generality that $p_1 \le p_2 \le p_3$. If $p_1 \ge 3$,

$$\frac{1}{p_1} + \frac{1}{p_2} + \frac{1}{p_3} \le 1,$$

a contradiction since $\frac{2}{n} > 0$. Hence $p_1 = 2$ and so

$$\frac{1}{p_2} + \frac{1}{p_3} = \frac{1}{2} + \frac{2}{n}.$$

(ii) $s = 3$, $p_1 = 2$, $p_2 = 2$. Then $2p_3 = n$ and $|G_\alpha| = \frac{n}{2}$ for some pole α. It follows that G contains a cyclic group of order $\frac{n}{2}$ and it is easy to show that this results in the dihedral group of order n.

Suppose $p_2 \ge 4$. Then

$$\frac{1}{p_2} + \frac{1}{p_3} \le \frac{1}{4} + \frac{1}{4} = \frac{1}{2} < \frac{1}{2} + \frac{2}{n},$$

a contradiction. Therefore $p_2 = 3$.

(iii) $s = 3$, $p_1 = 2$, $p_2 = 3$, $p_3 = 3$ Then

$$\frac{1}{3} + \frac{1}{3} = \frac{1}{3} + \frac{2}{n}$$

and G is a group of order 12. There are 14 poles and so 7 axes of rotation. This yields the group of symmetries of the tetrahedron. (See 13.3.2(i).)

(iv) $s = 3$, $p_1 = 2$, $p_2 = 3$, $p_3 = 4$. Then

$$\frac{1}{3} + \frac{1}{4} = \frac{1}{2} + \frac{2}{n}$$

and so $\frac{2}{n} = \frac{1}{12}$. Hence G is of order 24 and there are 26 poles and 13 axes of rotation. Therefore G is the group of symmetries of a cube. (See 13.3.2(ii).)

(v) $s = 3$, $p_1 = 2$, $p_2 = 3$, $p_3 = 5$. Then

$$\tfrac{1}{3} + \tfrac{1}{5} = \tfrac{1}{2} + \tfrac{2}{n}$$

which, on solving for n yields $n=60$. This group has 62 poles and 31 axes of rotations and is easily seen to be the group of symmetries of a dodecahedron (or an icosahedron). Therefore G is isomorphic to the alternating group A_5 (see 13.3.2(iii)).

(vi) If $s = 3$, $p_1 = 2$, $p_2 = 3$, $p_3 \geq 6$, then

$$\tfrac{1}{3} + \tfrac{1}{p_3} \leq \tfrac{1}{3} + \tfrac{1}{6} = \tfrac{1}{2} < \tfrac{1}{2} + \tfrac{2}{n},$$

a contradiction. ∎

13.5 Exercises

1) Prove that the three orthonormal vectors u, β and γ taken in that order form a right-handed triad if, and only if, the scalar triple product $\alpha . \beta \times \gamma = 1$.

2) (i) Let T be a rotation with axis through the origin and α_1, α_2 and α_3 an orthonormal basis of $\mathbf{R}^3$.

Suppose $T(\alpha_i) = \alpha'_i$, $i = 1,2,3$. Prove that the three vectors $\alpha_1 - \alpha'_i$, $\alpha_1 - \alpha'_1$, $\alpha_2 - \alpha'_2$ and $\alpha_3 - \alpha'_3$ are coplanar and that the axis of rotation of T is a vector through the origin and orthogonal to the plane determined by $\alpha_1 - \alpha'_1$, $\alpha_2 - \alpha'_2$, $\alpha_3 - \alpha'_3$.

(ii) Using (i), find the axis of the rotation whose matrix relative to the standard basis is

$$
\begin{pmatrix}
\dfrac{1}{3} & -\dfrac{2}{3} & \dfrac{2}{3} \\[2mm]
\dfrac{2}{3} & \dfrac{2}{3} & \dfrac{1}{3} \\[2mm]
-\dfrac{2}{3} & \dfrac{1}{3} & \dfrac{2}{3}
\end{pmatrix}.
$$

(iii) Let S be an opposite isometry and α_1, α_2 and α_3 an orthonormal basis of $\mathbf{R}^3$. If $S(\alpha_i) = \alpha'_1,\ i = 1,2,3,$ prove that the vectors $\alpha_1 + \alpha'_1,$ $\alpha_2 + \alpha'_2,$ and $\alpha_3 + \alpha'_3,$ are coplanar and that the vector β through the origin and orthogonal to this plane maps under S to $-\beta$.

(iv) If the matrix B is obtained from the matrix above by replacing the third row by its negative, compute the vector β of (iii).

3) A cube has vertices at $(\pm 1, \pm 1, \pm 1)$. Find matrices of orders 2, 3 and 4 which represent rotations leaving the cube invariant. Find also the matrices which form the only normal subgroup of S_4.

4) Describe the group of isometries (direct and opposite) which leave:
   (i) the regular tetrahedron invariant;
   (ii) the cube invariant.

5) Find the matrices of all rotational symmetries of:
   (i) a box with vertices at $(\pm 1, \pm 2, \pm 1)$;
   (ii) a box with vertices at $(\pm 1, \pm 2, \pm 3)$.

6) If a polyhedron without "holes" has V vertices, E edges and F faces, then $V - E + F = 2$ (this is known as Euler's Formula for Polyhedra). If in a regular polyhedron, there are p faces incident with each vertex and q edges incident with each face, prove that $pV = qF = 2E$. Using this and Euler's Formula, show the number of edges and vertices of the Platonic solids are as stated.

7) If the rotation group G of the cube acts on the three axes joining the centroids of pairs of opposite faces, find the kernel K of the action. What familiar group is G/K isomorphic to?

8) Find two 2×2 matrices which, together generate the dihedral group of order $2n$.

9) Fill in the details of the proof given in Theorem 12.4.5.

10) The three pairs of opposite faces of a cube are painted three different colors. What is the rotation group of this cube?

11) The vertices of a tetrahedron are at $(1,1,1)$, $(1,-1,-1)$, $(-1,1,-1)$ and $(-1,-1,1)$. Determine the seven axes of the rotations which leave the tetrahedron invariant. Compute the matrices of the group of rotations of this tetrahedron.

12) Prove that the reflections form a single conjugacy class in the dihedral group D_n of order $2n$ when n is odd, but form two classes when n is even.

13) Let V denote the inner product space R^n endowed with the usual dot product. A subspace H of V of dimension $n-1$ is called a *hyperplane*.

(i) Prove that if H is a hyperplane, then there exists a nonzero vector α which is orthogonal to every vector in H. In such a case, we say that α is orthogonal to H.

(ii) Prove that if β is another vector orthogonal to H, then $\beta = c\alpha$ for some scalar c.

(iii) For a given vector α define a map $T_\alpha : V \to V$ by

$$T_\alpha(\xi) = \xi - \frac{2(\xi.\alpha)}{\alpha.\alpha}.$$

Prove that T_α is an orthogonal linear transformation which fixes every vector in some hyperplane H and maps α to $-\alpha$. (We say that T_α is a *reflection* in the hyperplane H.) Find a simple matrix of T_α referred to an appropriate basis.

(iv) A finite group is a *reflection group* if it is generated by reflections. Prove that the dihedral groups are reflection groups.

(v) Show that the symmetric group S_n is a reflection group.

[Hint: The symmetric group S_n can be thought of as a subgroup of the n-dimensional orthogonal group in the following way. If $\pi \in S_n$ and $\{\varepsilon_i \mid 1 \le i \le n\}$ is the standard basis, define a linear transformation $L_\pi : \mathbf{R}^n \to \mathbf{R}^n$ by

$$L_\pi(\varepsilon_i) = \varepsilon_{\pi(i)}, \, 1 \le i \le n.$$

Prove that each transposition gives rise to reflection in some hyperplane].

(vi) Find the hyperplane and a vector α orthogonal to it associated with the transposition (ij).

(vii) Regarding S_n as a subgroup of the orthogonal group, prove:

(a) The transpositions are the only reflections belonging to S_n.

(b) S_n fixes pointwise the line spanned by $\sum_{i=1}^{n} \varepsilon_i$ and maps the $(n-1)$-dimensional subspace W of vectors orthogonal to $\sum_{i=1}^{n} \varepsilon_i$ to itself.

(c) By (b) we can consider S_n as a reflection group acting on W. Prove that S_n fixes no nonzero vector of W.

Chapter 14

Polya-Burnside Enumeration

14.1 Introduction

We begin this chapter with a simple example which shows one type of problem this particular method of enumeration deals with. The method is based on Theorem 12.1.13, namely: If (G, X) is an action with N orbits, then

$$|G| = \frac{1}{N} \sum_{g \in G} |\text{Fix } g|.$$

14.1.1 *Example*

Given an unlimited supply of red medallions, blue medallions and yellow medallions which are to be strung on a silver chain to make a 4-medallion necklace, how many distinct 4-medallion necklaces can we make? Assume that medallions of the same color are indistinguishable.

As the problem stands, it is rather vague. Is one face of a medallion the same as the other, that is, are the medallions "reversible"? Does the chain have a clasp past which we cannot thread the medallions?

We start by assuming that one face of each medallion is indistinguishable from the other and that the necklace has a clasp past which the medallions cannot be threaded. We show some possibilities.

363

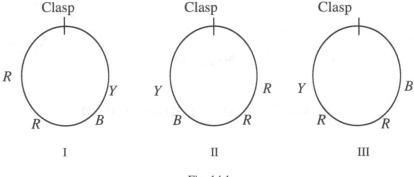

Fig. 14.1

We note that I and II are the same necklace since we get II from I by turning I through 180 degrees about an axis joining the clasp and B. This is tantamount to reflecting in the line determined by the clasp and B. We can do this since the medallions are the same on both faces.

Necklaces I and III are different. However, should we be able to thread the medallions past the clasp, I and III would be the same. Let us number the positions to be taken by the medallions by 1, 2, 3, 4 as shown in the diagram below.

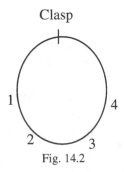

Fig. 14.2

A particular necklace can be thought of as a map f from $\{1,2,3,4\}$ to $\{R,B,Y\}$. Thus, for example, the map $f = \begin{pmatrix} 1 & 2 & 3 & 4 \\ R & R & B & Y \end{pmatrix}$ represents the necklace I above. The number of such maps is $3^4 = 81$ but distinct maps don't necessarily give rise to different necklaces. For example, the map

$g = \begin{pmatrix} 1 & 2 & 3 & 4 \\ Y & B & R & R \end{pmatrix}$ represents necklace II above, which, as we have

observed, is the same as necklace I.

How can we recognize that they are, in fact, the same by just looking at the maps? We note that $f \circ (14)(23) = g$, that is, we obtain g by composing f with the reflection in the line joining the clasp to a point between position 2 and position 3.

We can define a (right) group action of $G = \{(1), (12)(34)\}$ on the set of all functions from $\{1, 2, 3, 4\}$ to $\{R, B, Y\}$ by $f\pi = f \circ \pi$. It is then apparent that two necklaces will be the same if, and only if, they lie in the same orbit of the action. Therefore the number of different necklaces we can make is the number of orbits of this action. According to Burnside's result,

$$N = \frac{1}{|G|} \sum_{\pi \in G} |\text{Fix}\,\pi|.$$

In this example, $|\text{Fix}(1)| = 81$; $|\text{Fix}(14)(23)| = 9$ since for a map f to be in $\text{Fix}(14)(23)$ the values of f must be constant on the orbits of $< (14)(23) >$, i.e., $f(1) = f(4)$ and $f(2) = f(3)$. Therefore, the number of orbits is $\frac{1}{2}(81 + 9) = 45$ and so there are 45 distinguishable necklaces.

As another example, suppose there is no clasp and that there is a "right" side to each medallion so that we are not allowed to flip the necklace over. Then necklaces I and III are the same. This time the group in question is $\{ (1), (1234), (13)(24), (1432)\}$. In this case: $|\text{Fix}(1)| = 81$; $|\text{Fix}(1234)| = 3$; $|\text{Fix}(13)(24)| = 9$ and $|\text{Fix}(1432)| = 3$. Note that, for example,

$$\text{Fix}(1234) = \left\{ \begin{pmatrix} 1 & 2 & 3 & 4 \\ Y & Y & Y & Y \end{pmatrix}, \begin{pmatrix} 1 & 2 & 3 & 4 \\ B & B & B & B \end{pmatrix}, \begin{pmatrix} 1 & 2 & 3 & 4 \\ R & R & R & R \end{pmatrix} \right.$$

since the colors have to be constant in the orbit of (1234). Hence the number of orbits is $\frac{1}{4}(81 + 6 + 9) = 24$ and so the number of different necklaces is 24.

Finally, if there is no clasp and the medallions are "reversible", the number of different necklaces is the number of orbits of the of the (right) action (F, D_4) where F is the set of maps from $\{1, 2, 3, 4\}$ to $\{R, B, Y\}$

and D_4 is the dihedral group of order 8. We naturally consider D_4 as acting on $\{1,2,3,4\}$. Hence

$$D_4 = \{(1),(24),(13),(12)(34),(14)(23),(1234),(13)(24),(1432)\}.$$

Observe that the number of elements fixed by a given permutation is only dependent on the cycle structure of the permutation. Computing the various fixes, we obtain:

$$|\text{Fix}(1)| = 81; \ |\text{Fix}(24)| = 27; \ |\text{Fix}(12)(34)| = 9; \ |\text{Fix}(1234)| = 3.$$

Therefore, the number of orbits is $\frac{1}{8}(81 + 54 + 27 + 6) = 21$.

14.2   A Theorem of Polya

We generalize the situation described above and work towards a result due to G. Polya which will enable us to tackle a large variety of problems with relative ease.

14.2.1 *Definition*

Let G be a subgroup of S_n, the symmetric group on $X = \{1,2,...,n\}$ and $Y = \{C_1, C_2,..., C_m\}$, a set of m "colors". Then a *coloring* of the set X is a map from X to Y.

We define an action of G on Y^X by $f\pi = f \circ \pi$ thus yielding a right action (Y^X, G) .(That this is indeed an action is left as an exercise for the reader.)

The problem now is to determine the number of orbits of (Y^X, G).

14.2.2 *Definition*

Let $\pi \in S_n$ and suppose the decomposition of π as a product of disjoint cycles has s_i i-cycles, $i = 1,2,...,n$ where some of the s_i can be zero and $\sum_{i=1}^{n} is_i = n.$ Then the vector $v(\pi) = (s_1, s_2,....., s_n)$ is called the *cycle structure* of π. The *cycle index* of π, denoted $Z(\pi; x_1, x_2, ..., x_n)$ is the monomial $x_1^{s_1} x_2^{s_2} ... x_n^{s_n}$. The *cycle index of a subgroup* G of S_n, denoted $Z(G; x_1, x_2, ..., x_n)$, is the multinomial

$$\frac{1}{|G|} \sum_{\pi \in G} Z(\pi; x_1, x_2, ..., x_n).$$

We give a few examples to make these somewhat indigestible definitions a little more palatable.

14.2.3 *Examples*

(i) Suppose $\pi = (123)(456)(78910)$ is an element of S_{10}. Then
$v(\pi) = (0,0,2,1,0,0,0,0,0,0)$, $Z(\pi; x_1, x_2, ..., x_n) = x_3^2 x_4$.

(ii) Let $G = \{(1), (123), (132), (12), (13), (14)\}$. Then

$Z((1); x_1, x_2, x_3) = x_1^3$; $Z((123); x_1, x_2, x_3) = x_3$; $Z((12); x_1, x_2, x_3) = x_1 x_2$

and

$$(G; x_1, x_2, x_3) = \frac{1}{6}(x_1^3 + 2x_3 + 3x_1 x_2).$$

(iii) Let
$G = \{(1), (13), (24), (12)(34), (14)(23), (1234), (13)(24), (1432)\}$,

the dihedral group of order 8. Then

$$v((1)) = (4,0,0,0); \; v((12)) = (2,1,0,0);$$
$$v((12)(34)) = (0,2,0,0); \; v((1234)) = (0,0,0,1).$$
$$Z((1); x_1, x_2, x_3, x_4) = x_1^4; \; Z((12); x_1, x_2, x_3, x_4) = x_1^2 x_2;$$
$$Z((12)(34)); x_1, x_2, x_3, x_4) = x_2^2; \; Z((1234); x_1, x_2, x_3, x_4) = x_4.$$
$$Z(G; x_1, x_2, x_3, x_4) = \frac{1}{8}(x_1^4 + 2x_1^2 x_2 + 3x_2^2 + 2x_4). \; \blacksquare$$

In anticipation of Polya's result, observe that $Z(G; 3,3,3,3) = 21$ (compare with the last example above).

14.2.4 *Theorem (Polya)*

With the notation introduced above, the number of orbits of the action (Y^X, G) is $Z(G; m, m, ..., m)$.

Proof. We compute $|\text{Fix}\pi|$ where π has cycle structure $(s_1, s_2, ..., s_n)$. Now

$$\text{Fix}\pi = \{f \in Y^X \mid f\pi = f\} = \{f \in Y^X \mid f(\pi(i)) = f(i), i = 1, 2, ..., n\}.$$

If $(t_1 t_2, ..., t_k)$ is a cycle occurring in the factorization of π, then $f(\pi(t_i)) = f(t_i)$, i.e., $f(t_{i+1}) = f(t_i)$, $i = 1, 2, ..., k-1$, showing that for f to be in $\text{Fix}\pi$, it is necessary and sufficient for f to be constant on each orbit of π. Hence

$$|\text{Fix}\pi| = m^{s_1 + s_2 + ... + s_n} = Z(\pi; m, m, ..., m).$$

Therefore, using Burnside's result, the number of orbits is

$$\frac{1}{|G|}\sum_{\pi \in G} |\text{Fix}\pi| = \frac{1}{|G|}\sum_{\pi \in G} Z(\pi; m, m, ...m) = Z(G; m, m, ..., m). \qquad \blacksquare$$

14.2.5 *Example*

With an unlimited supply of red, blue and yellow beads, how many 5-bead necklaces without clasps can be made?

In this case the relevant action is (F, D_5) where F is the set of maps from $\{1, 2, 3, 4, 5\}$ to $\{R, B, Y\}$ and D_5 is the dihedral group of order 10.

We list the various cycle structures of the elements of D_5.

Table 14.1

Cycle Structure	Number
(5, 0, 0, 0, 0)	1
(1, 2, 0, 0, 0)	5
(0, 0, 0, 0, 1)	4

The cycle index of D_5 is $\frac{1}{10}(x_1^5 + 5x_1 x_2^2 + 4x_5)$ and so the number of necklaces is given by

$$\frac{1}{10}(3^5 + 5.3^3 + 4.3) = \frac{1}{10}(243 + 135 + 12) = 39.$$

Another slightly different type of problem which can be tackled using Burnside's result is exemplified in the following:

14.2.6 *Example*

How many distinguishable 5-bead necklaces without clasps consisting of 3 red beads and 2 blue beads can be made?

In this case, each necklace can be designated (not uniquely) by a bijective map from $\{1, 2, 3, 4, 5\}$ to $\{R, R, R, B, B\}$.

There are $\frac{5!}{3!2!} = 10$ such maps and we must find out how many different necklaces these give rise to.

Let F denote the set of these bijective maps and consider the (right) action of D_5 on F defined as above. Then, as before, the number of different necklaces is the number of orbits of the action (F, D_5). This time, however, it is harder to determine the "fixes" of the elements of D_5.

Using the cycle structures of the elements of D_5 listed in Table 14.1above, we compute the "fixes".

Clearly IFix(1)I=10. To calculate the other "fixes", we take (12)(34) as representative of the cycle structure $(1, 2, 0, 0, 0)$ and (12345) as representative of the cycle structure $(0, 0, 0, 0, 1)$.

Then

$$\text{IFix}(12)(34)(5)\text{I} = 2 \text{ and } \text{IFix}(12345)\text{I} = 0.$$

Therefore, the number of distinguishable necklaces is $\frac{1}{10}(10 + 5 \times 2) = 2$.

We observe that a map is fixed by π if, and only if, the colors are constant on each orbit of π. For example, in calculating IFix(12)(34)I, $f \in \text{Fix}(12)(34)$ if, and only if, $f(1) = f(2)$ and $f(3) = f(4)$. Since this time we do not have an unlimited supply of beads, the only options are:

$$\begin{pmatrix} 1 & 2 & 3 & 4 & 5 \\ R & R & B & B & R \end{pmatrix} \text{ and } \begin{pmatrix} 1 & 2 & 3 & 4 & 5 \\ B & B & R & R & R \end{pmatrix}.$$

We compute a further example which is more challenging.

14.2.7 *Example*

In how many ways can we paint the faces of a cube so that 2 of the faces are red, 2 blue, 1 yellow and 1 green?

We start by listing the cycle structures of the rotation group of a cube acting on the faces of the cube. The relevant group in this case is a subgroup of S_6.

Table 14.2

Cycle Structure	Number
(6, 0, 0, 0, 0, 0)	1
(2, 0, 0, 1, 0, 0)	6
(2, 2, 0, 0, 0, 0)	3
(0, 0, 2, 0, 0, 0)	8
(0, 3, 0, 0, 0, 0)	6

We now compute the sizes of the fixes of the elements of the group.

$$\text{Fix}(1) = \frac{6!}{2!2!} = 180; \ |\text{Fix}(1234)(5)(6)| = 0; \ |\text{Fix}(12)(34)(5)(6)| = 4;$$

$$|\text{Fix}(123)(456)| = 0; \ |\text{Fix}(12)(34)(56)| = 0.$$

Hence the number of ways the cube can be painted is $\frac{1}{24}\{180 + 12\} = 8$.

Suppose now that the faces of the cube are to be painted with the four colors above but with no stipulation as to the number of faces to be painted any given color. In this case, the cycle index of the group of rotations of a cube acting on the faces of the cube is

$$\frac{1}{24}(x_1^6 + 6x_1^2 x_4 + 3x_1^2 x_2^2 + 8x_3^2 + 6x_2^3).$$

Therefore, the number of ways of painting the cube is

$$\frac{1}{24}(4^6 + 6.4^3 + 3.4^4 + 8.4^2 + 6.4^3) = 240.$$

Incidentally, as a by-product we see that $x^6 + 3x^4 + 12x^3 + 8x^2$ is divisible by 24 for all positive integral values of x. ∎

As a final example, we show how Burnside's result can be used to compute the number of different so-called *switching functions*. To motivate the definition of a switching function, consider a "black box" with, say, three inputs and one output. Each input is either 0 or 1 and the output is also 0 or 1.

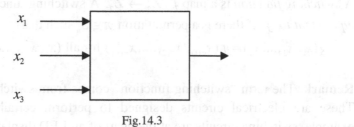

Fig.14.3

The "black box" may be considered as a map (or function) $f : \mathbf{Z}_2^3 \to \mathbf{Z}_2$. If g is another such map and there exists $\pi \in S_3$ such that $g(x_1, x_2, x_3) = f(x_{\pi(1)}, x_{\pi(2)}, x_{\pi(3)})$ for all $(x_1, x_2, x_3) \in \mathbf{Z}_2^3$, then the two associated boxes are essentially the same since the permutation merely changes the positions of the inputs. We say that two maps so related are equivalent. Our task is to determine how many pair-wise nonequivalent maps (or "black boxes") there are.

14.2.8 *Example*

We show first how a permutation acts on

$$\mathbf{Z}_2^3 = \{000,100,010,001,110,101,011,111\}.$$

Suppose $\pi = (123)$. Then

$$\pi(x_1, x_2, x_3) = (x_{\pi(1)}, x_{\pi(2)}, x_{\pi(3)}) = (x_2, x_3, x_1)$$

and so $\pi(000) = 000; \pi(100) = 001; \pi(001) = 010$; etc.

Suppose

$$f = \begin{pmatrix} 000 & 100 & 010 & 001 & 110 & 101 & 011 & 111 \\ 1 & 1 & 0 & 0 & 1 & 1 & 0 & 1 \end{pmatrix}.$$

We compute πf :

$$\pi f = \begin{pmatrix} 000 & 100 & 010 & 001 & 110 & 101 & 011 & 111 \\ 1 & 0 & 1 & 0 & 1 & 0 & 1 & 1 \end{pmatrix}.$$

14.2.9 Definition

A *switching function* is a map $f : \mathbf{Z}_2^n \to \mathbf{Z}_2$. A switching function g is *equivalent to* f if there is a permutation $\pi \in S_n$ such that

$$g(x_1, x_2,...,x_n) = f(x_{\pi(1)}, x_{\pi(2)},...,x_{\pi(n)}) \text{ for all } (x_1, x_2,...,x_n) \in \mathbf{Z}_2^n.$$

Remark. The term "switching function" comes from switching circuits. These are electrical circuits designed to perform certain tasks. For example, switching circuits are at the heart of an LED display on a clock. No current flowing through a particular section of the circuit is denoted by 0 while 1 denotes the presence of a current.

14.2.10 Theorem

The number of pair-wise nonequivalent switching functions $f : \mathbf{Z}_2^n \to \mathbf{Z}_2$ is

$$\frac{1}{n!}\left(\sum_{\pi \in S_n} 2^{Z(<\pi>;2,\,2,...,2)} \right).$$

Proof. We introduce two actions, namely:

(i) $(S_n, \mathbf{Z}_2^n)$ defined by $\pi(x_1,x_2,...,x_n) = (x_{\pi(1)},x_{\pi(2)},...,x_{\pi(n)})$;

(ii) $(S_n, \mathbf{Z}_2^{\mathbf{Z}_2^n})$ defined by $\pi f(x_1,x_2,...,x_n) = f(x_{\pi(1)},x_{\pi(2)},...,x_{\pi(n)})$.

We want the number of orbits of (ii) and this is

$$\frac{1}{n!}\sum_{\pi \in S_n} |\text{Fix } \pi|.$$

Clearly $f \in$ Fix π if, and only if, f is constant on the orbits of π in the first action. But the number of orbits of π in the first action is the number of colorings with two colors (1 and 0) of $\{1,2,...,n\}$ under the group $<\pi>$, that is, $Z(<\pi>; 2,2,...,2)$. Again, we have to compute the number of colorings by two colors (1 and 0) under the group S_n. But $|\text{Fix}\,\pi|$ in the second action is $2^{Z(<\pi>; 2, 2,..., 2)}$ and the result follows. ∎

14.2.11 *Example*

How many pair-wise nonequivalent switching functions of three variables are there?

In this case, the group is $S_3 = \{(1), (12), (13), (23), (123), (132)\}$. We compute $Z(<\pi>; 2, 2, 2)$ for each $\pi \in S_3$:

$$Z(<(1)>; 2,2,2) = 8;\quad Z(<(12); 2,2,2) = \frac{1}{2}(8+4) = 6;$$

$$Z(<(123)>;\ 2,2,2) = \frac{1}{3}(8+4) = 4.$$

Hence the number of pair-wise nonequivalent switching functions is

$$\frac{1}{6}\left(2^8 + 3.2^6 + 2.2^4\right) = 80 . \blacksquare$$

14.3 Exercises

1) Given n colors, how many colorings of the vertices of a cube are there?

2) Given an unlimited supply of red and blue beads, how many different 10-bead necklaces without clasps can be made? If the necklaces are to have a clasp past which the beads cannot be threaded, how many can then be made?

3) How many 7-bead necklaces without clasps can be made with 4 red beads and 3 blue beads?

4) A regular tetrahedron is constructed using 6 rods of equal length for the edges. If there are rods of n different colors, how many distinguishable tetrahedrons can be constructed?

5) In how many ways can we paint the faces of a regular tetrahedron if there are 4 colors available?

6) How many ways can the faces of a cube be colored with 6 colors, if no two faces are to be colored with the same color?

7) How many ways are there of coloring the faces of a cube so that 2 of the faces are blue, 1 white and 3 yellow?

8) Let (G, S) be an action where G is a cyclic group of order n generated by π. Prove that the number of orbits of this action is

$$\frac{1}{n} \sum_{d|n} |\operatorname{Fix}\pi^{n/d}| \; \varphi(d)$$

where $\varphi(d)$ is the number of positive integers less than or equal to d and relatively prime to d (φ is the so-called Euler φ-function).

9) How many pair-wise nonequivalent switching functions of 4 variables are there?

10) How many 4 by 5 quilts can be made from 20 square patches, 6 red and 14 yellow, assuming that the quilts cannot be turned over? If the quilts can be turned over, how many can then be made?

11) Prove that for any positive integers k and n, n divides $\sum_{d|n} k^{n/d} \varphi(d)$ where φ is the Euler φ-function. (Hint: Apply Problem 8 above). Hence deduce that $\sum_{d|n} \varphi(d) = n$.

12) In how many ways can 3 indistinguishable white and 2 indistinguishable black balls be placed in 3 boxes, two of which are indistinguishable?

13) The nine squares of a 3×3 chessboard are to be colored so that 2 of the squares are colored red and the rest blue. Under the action of the

dihedral group of order 8, how many distinguishable colorings are there?

14) How many ways are there of coloring the vertices of a regular pentagon with 3 colors if we are allowed to flip the pentagon over?

15) (This is a generalization of Theorem 12.1.13.) Let (G, X) be an action with s orbits $X_1, X_2, ..., X_s$ and suppose A is an additive abelian group. Define a map $\omega : X \to A$ which is constant on each of the s orbits of G and use ω to define a map $W : G \to A$ by $W(g) = \sum_{\alpha \in \text{Fix}g} \omega(g)$. Choose $\alpha_i \in X_i, i = 1, 2, ..., s$ and prove that

$$\sum_{g \in G} W(g) = |G| . \sum_{i=1}^{s} \omega(\alpha_i).$$

Observe that if the abelian group is the integers and $\omega(\alpha) = 1$, the equality above reduces to Theorem 12.1.13.

16) (i) Let $G = <(1, 2, 3, ..., n)> \subseteq S_n$. Prove that $(1, 2, 3, ..., n)^k$ is the product of $d = \gcd(k, n)$ cycles, each of length n/d.

   (ii) Using this, show that $Z(G; x_1, x_2, ..., x_n) = \frac{1}{n} \sum_{d|n} \varphi(d) x_d^{n/d}$.

17) Find the cycle index of the group of rotations of a cube acting on the vertices of the cube.

18) Find the cycle index of the group of rotations of a cube acting on the faces of the cube.

19) (i) If X and Y are disjoint sets with n and m elements, respectively, and (G, X) and (G, Y) are permutation groups, define an action of the direct product $G \times H$ on $X \cup Y$ by:

$$(g, h)\alpha = \begin{cases} g\alpha \text{ if } \alpha \in X \\ h\alpha \text{ if } \alpha \in Y \end{cases}.$$

Prove that this is indeed an action and determine the cycle index of $G \times H$ in terms of the cycle indices of G and H.

(ii) Using (i), determine the cycle index of $S_3 \times D_4$ where S_3 acts on $\{1, 2, 3\}$ and D_4 acts on the vertices of a square labelled a, b, c and d *clockwisewise* and consecutively.

20) (i) If G is a finite group, we can define a cycle index for G by considering the (left) action (G,G) where G acts on G by left multiplication. (For example, if G is the cyclic group generated by g, then the permutation induced by left multiplication by g is:

$$\begin{pmatrix} e & g & g^2 \\ g & g^2 & e \end{pmatrix}.$$

(ii) Let

$$G = \left\{ \begin{pmatrix} 1 & a & b \\ 0 & 1 & c \\ 0 & 0 & 1 \end{pmatrix} \mid a,b,c \text{ are in the field } \mathbf{Z}_3 \right\}.$$

Show that G is a group under ordinary matrix multiplication and compute the cycle index of G.

Chapter 15

Group Codes

15.1 Introduction

Messages consisting of sequences of binary digits (called *words*) transmitted over noisy channels are subject to error. Although we cannot entirely eliminate the noisy channel from causing such errors, we can minimize the probability of them being misinterpreted at the receiving end by *encoding* the words.

For example, suppose we want to send a message consisting of a sequence of two binary digits. The possible message words, each of *length* two, would be $00, 10, 01$ and 11. If the probability of receiving a 1 as a 0 or a 0 as a 1 is $.01$, then the probability of receiving the wrong word is $1 - (.99)^2 \doteq .02$. It is clear that the probability of receiving the wrong word increases with the length of the word and the length of the message.

The basic idea is to adjoin to the words which carry the information (in this case, $00, 10, 01$ and 11) *check digits* resulting in an *encoded word*. These digits are redundant in the sense that they do not carry any information but serve to detect and even correct errors. We think of the encoded words as consisting of two parts: the message at the tail-end and the check digits at the beginning of the word as we read from left to right. Any collection of such words is called a *code*.

In the example above let us adjoin a 0 or a 1 to each of the four words $00, 01, 10$ and 11 to get the four words $000, 101, 110$ and 011 with the property that each has an even number of 1's Note that our original four words constituted the collection of all words of length two on the *alphabet* {0,1}. On the other hand, the set of encoded words is a proper subset of the set of all words of length three. This is analogous to the situation in English where not all 26^4 "words" with four letters are words of the English language.

Suppose now that in the transmission of a message written in the code above, a single error occurs in one of the words. Then the received word will have an odd number of 1's and the receiver will detect that the word is not a word of the "language" and will ask for a re-transmission. This code is therefore called a *single-error-detecting* code. In fact the code will detect any odd number of errors but not an even number of errors.

Let us now compute the probability that a given transmitted word will be incorrectly interpreted at the receiving end. This will happen only when exactly two errors occur and the probability of this event is $\binom{3}{2}(0.1)^2(.99) = .000297$, a considerable improvement over .02, the probability of error if the uncoded words are transmitted. As is always the case, we don't get something for nothing: for the sake of accuracy we have increased transmission time. A moment's thought will convince the reader that this will always be the case: as accuracy increases, efficiency decreases.

It is often the case that a message can only be sent once. For example, it would hardly be practical to stop a modern high-speed computer each time an error is detected. In such situations error-detecting codes are therefore of little value. Instead error-correcting codes are used. These are codes which detect errors and then correct them according to some prescribed procedure. Once again we illustrate with a simple example.

Suppose we want our language to consist of the two words 0 and 1 where the probability of error is again .01. We encode the information words as 000 and 111 and stipulate the following *decoding scheme*: the received word is decoded according to majority rule, i.e., if the received word has more x's than y's, it will be read as xxx. What is the probability that a decoding error will occur?

Suppose xxx is transmitted. It will be incorrectly decoded if there is an error in at least two of the coordinate positions. This will happen with probability $\binom{3}{2}(.01)^2(.99)+(.01)^3 \doteq .000298$, an improvement over .01,

the probability of error if the uncoded words are transmitted. Once again, we have paid for accuracy by tripling transmission time.

The diagram below illustrates schematically the situation we have just described.

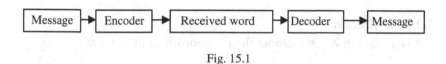

Fig. 15.1

The English language has considerable error correcting capacity. For example, if we come across the word "depcribe" we would have no difficulty in recognizing that there is a typographical error in the third letter and would correct it to read "describe". The reason we are able to do this is that there are no words different from but "close enough" to "describe". If on the other hand the word "cup" mistakenly appears as "cap", then, unless the context tells us, we do not even detect the error. In this case "cup" is "close to" "cap".

We take our cue from this discussion and conjecture: the "farther apart" the words of a code are, the more likely it will be that we shall be able to recognize and even correct errors. We shall prove a precise version of this conjecture after we have introduced the concept of Hamming distance between words.

15.2 Definitions and Notation

15.2.1 *Definition*

A *code* is a subset of $\mathbf{Z}_2^n$. The elements of a code are called *words* and the number of coordinate positions (in this case n) is the *length* of a word.

Abstract Algebra

Since $\mathbf{Z}_2^n$ is a ring, we make use of its algebraic structure so that, given two words α and β, we obtain a word $\alpha + \beta$.

15.2.2 Example

$\{0100, 1100, 1110, 1111\}$ is a code each of whose words is of length 4 and $01101 + 11011 = 10110$. ∎

15.2.3 Definition
If α is a word in $\mathbf{Z}_2^n$, we denote the ith coordinate of α by α_i.

  (i) The *support* $S(\alpha)$ of α is the set $\{i \mid \alpha_i = 1\} \subseteq \{1, 2, ..., n\}$;
  (ii) The *weight* $w(\alpha)$ of α is $|S(\alpha)|$;
  (iii) If α and β are in $\mathbf{Z}_2^n$, the *Hamming distance* $h(\alpha, \beta)$ is the number of coordinate positions where α and β differ.

15.2.4 Example

If $\alpha = 0110$, $\beta = 1101$, then

$$S(\alpha) = \{2, 3\}, \ S(\beta) = \{1, 2, 4\}$$
$$S(\alpha + \beta) = S(1011) = \{1, 3, 4\} \text{ and } h(\alpha, \beta) = 3. \ \blacksquare$$

Observe that in the example above,
$$S(\alpha + \beta) = S(\alpha) \cup S(\beta) \setminus S(\alpha) \cap S(\beta).$$
This is always the case as we see from the following theorem.

15.2.5 Theorem

$$S(\alpha + \beta) = S(\alpha) \cup S(\beta) \setminus S(\alpha) \cap S(\beta) \text{ for all } \alpha \text{ and } \beta \text{ in } Z_2^n.$$

Proof.
$$S(\alpha + \beta) = \{i \mid (\alpha + \beta)_i = \alpha_i + \beta_i = 1\}$$
$$= \{i \mid \alpha_i = 1 \text{ and } \beta_i = 0\} \cup \{i \mid \alpha_i = 0 \text{ and } \beta_i = 1\}$$
$$= [S(\alpha) \cap S(\beta)'] \cup [S(\alpha)' \cap S(\beta)]$$
$$= S(\alpha) \cup S(\beta) \setminus S(\alpha) \cap S(\beta). \ \blacksquare$$

Remark. $\alpha_i + \beta_i = 1$ iff the ith coordinates of α and β differ. Hence $|S(\alpha + \beta)|$ is the number of coordinate positions where α and β differ and so, $h(\alpha, \beta) = w(\alpha + \beta)$.

The following corollary is an easy consequence of the previous theorem and the proof is left to the reader (see Problem 1 at the end of this chapter).

15.2.6 *Corollary*

$$h(\alpha, \beta) = w(\alpha) + w(\beta) - 2\,|\,S(\alpha) \cap S(\beta)\,|.\ \blacksquare$$

The Hamming distance has the same properties as geometric distance.

15.2.7 *Theorem*

The Hamming distance h has the following properties:
  (i) $h(\alpha, \beta) = 0$ iff $\alpha = \beta$;
  (ii) $h(\alpha, \beta) = h(\beta, \alpha)$ for all $\alpha, \beta \in \mathbf{Z}_2^n$;
  (iii) $h(\alpha, \beta) + h(\beta, \gamma) \geq h(\alpha, \gamma)$ for all $\alpha, \beta \in \mathbf{Z}_2^n$ (the triangle inequality).

Proof. We prove only (iii) leaving (i) and (ii) as exercises. We have

$$h(\alpha, \gamma) = w(\alpha + \gamma) = w(\alpha + \beta + \beta + \gamma)$$
$$= w(\alpha + \beta) + w(\beta + \gamma) - 2\,|\,S(\alpha + \beta) \cap S(\beta + \gamma)\,|$$
$$\leq w(\alpha + \beta) + w(\beta + \gamma) = h(\alpha, \beta) + h(\beta, \gamma).\ \blacksquare$$

Let us assume that we can transmit messages only once. We must therefore devise some *complete decoding scheme*; that is, we must supply the receiver with an algorithm which will enable him/her to decode any received word, even if there are errors. This is in sharp contrast which obtains when we discussed the single-error-detecting code above where decoding does not take place at all if an error is detected.

How then are we to devise such a decoding scheme? We shall be considering only channels where the probability of error in any one symbol is independent of what happens to any other symbol. In such channels, it is less likely for $k+1$ errors to occur than for k to occur (see Problem 2 for a precise analysis).

We therefore adopt the following scheme: if β is received, find the nearest word α in the adopted code and decode β as α. If there is more than one codeword closest to β, make an arbitrary choice from among the code words with minimum distance from β. Of course, if β is itself a code word, it will be decoded as β. This method of decoding is called *nearest neighbor decoding*.

Suppose that we have chosen a code C and that the word α of this code is transmitted and β is received. The *error word* γ is defined by $\gamma = \alpha + \beta$. It is clear that γ has ones in precisely those coordinate positions where errors have occurred. Using this terminology we may describe nearest neighbor decoding as follows: Find all error words and choose the code word with associated error word of least weight.

15.2.8 *Definition*

(i) A code C is said to be *d-error detecting* if it recognizes all possible patterns of d or fewer errors; it is said to be *d-error-correcting* if, using the nearest neighbor decoding scheme, it corrects all possible patterns of d or fewer errors.

(ii) The *minimum Hamming distance* $\mu(C)$ *of a code* C is defined by

$$\mu(C) = \min\{h(\alpha,\beta)\,|\,\alpha,\beta \in C,\ \alpha \neq \beta\}.$$

15.2.9 *Example*

If $C = \{0000,0110,1001,1111\}$, then $\mu(C) = 2$. ∎

We shall now show precisely how the minimum Hamming distance of a code is related to its error-detecting and error-correcting capacities.

15.2.10 *Theorem*

A code C is
(i) d-error-detecting iff $\mu(C) \geq d+1$;
(ii) d-error-correcting iff $\mu(C) \geq 2d+1$.

Proof. (i) Suppose $\mu(C) \leq d$ and let α and β be different words in C with $h(\alpha, \beta) = k \leq d$. Then k errors in the appropriate coordinate positions in α will change it to β. Therefore C fails to detect these k errors.

Conversely, assume $\mu(C) \geq d+1$. If α is a code word and fewer than $d+1$ errors occur in transmission, the received word β is at a distance less than $d+1$ from α and so cannot be a member of the code.

(ii) Suppose now that C is d-error-correcting and that α and β are different code words with $h(\alpha, \beta) \leq 2d$. Let $h(\alpha, \beta) = k \leq 2d$.

If k is even, say $k = 2s$, $s \leq d$, then a certain pattern of s errors in the transmitted word α will produce a word γ with $h(\alpha, \gamma) = h(\beta, \gamma) = s$. Hence the received word γ could be erroneously decoded as β.

If k is odd, say $k = 2s+1$, $s < d$, then a certain pattern of $s+1$ errors in α will produce a word γ with $h(\beta, \gamma) = s$ and $h(\alpha, \gamma) = s+1 (\leq d)$. In this case the received word γ would not be decoded as α.

Conversely, suppose $\mu(C) \geq 2d+1$ and a pattern of d or fewer errors occurs in the transmitted word α. If γ is the resulting received word and β is any other codeword, by the triangle inequality,
$$h(\beta, \gamma) + h(\gamma, \alpha) \geq h(\beta, \alpha) \geq 2d+1.$$
Therefore
$$h(\beta, \gamma) \geq 2d+1-d = d+1 \text{ while } h(\alpha, \gamma) \leq d.$$
Hence nearest neighbor decoding will yield the correct word α. ∎

15.2.11 *Example*

Let
$$C = \{00000, 101100, 011010, 110001, 110110, 011101, 101011, 000111\}.$$
This code is a subgroup of $\mathbf{Z}_2^5$ and so it is closed under addition. Consequently, $\mu(C) = \min \{w(\alpha) \mid \alpha \in C, \ \alpha \neq 0\}$ (Why?). Therefore

$\mu(C) = 3$ and so, by the previous theorem, C is a single-error-correcting code. ∎

Two problems of practical importance which we should address are those of designing codes with high error-correcting capacity but which are efficient in the sense that the decoding scheme is easy to implement and transmission time is not seriously compromised. If we encode our message words at random, we must have a list of codewords at hand against which we have to compare each received word to determine whether or not it belongs to the code. This is obviously an expensive and time-consuming procedure. If, however, we choose a code with some mathematical structure or whose words have a certain pattern, then the encoding and recognition algorithms can be quite simple and efficient.

15.3 Group Codes

In this section we shall apply group theory to the construction of a certain class of codes. The notions of homomorphism and subgroup will play prominent roles in the theoretical development while it will be seen that cosets can be effectively used in the practical process of decoding.

15.3.1 *Definition*

A *group code* is any subgroup of $(\mathbf{Z}_2^n, +)$.

By Lagrange's theorem, a group code will consist of 2^k code words for some $k \leq n$. The ratio k/n is the *information rate* of the code and is a measure of its efficiency. Example 14.2.11 above is an example of a group code.

15.3.2 *Definition*

(See 8.4.6) Let $i_1, i_2, ..., i_k$ be a subset of $\{1, 2, ..., n\}$ where $i_u < i_v$ if $u < v$. Define the function $\Pi_S : \mathbf{Z}_2^n \to \mathbf{Z}_2^k$ as follows: if $\alpha = a_1 a_2 ... a_n$,

$$\Pi_S(\alpha) = a_{i_1} a_{i_2} \dots a_{i_k}.$$

15.3.3 *Example*

Let $S = \{1,3\} \subset \{1,2,3,4\}$. Then $\Pi_S(1011) = 11$; $\Pi_S(1101) = 10$. ∎

15.3.4 *Theorem*

If $\alpha = a_1 a_2 \dots a_n$, $\beta = b_1 b_2 \dots b_n$ and $S = \{i_1, i_2, \dots, i_k\}$ is a subset of $\{1, 2, \dots, n\}$, then Π_S is a group homomorphism and $\mathbf{Z}_2^n \big/ \ker \Pi_S \cong \mathbf{Z}_2^k$.

Proof. If $\alpha = a_1 a_2 \dots a_n$ and $\beta = b_1 b_2 \dots b_n$ then

$$\Pi_S(\alpha) = a_{i_1} a_{i_2} \dots a_{i_k}, \ \Pi_S(\beta) = b_{i_1} b_{i_2} \dots b_{i_k}$$

and

$$(\alpha + \beta)_i - \alpha_i + \beta_i.$$

Therefore

$$\Pi_S(\alpha + \beta) = (a_{i_1} + b_{i_1}, a_{i_2} + b_{i_2}, \dots, a_{i_k} + b_{i_k}) = \Pi_S(\alpha) + \Pi_S(\beta).$$

It is clear that Π_S is an epimorphism and so, by the First Isomorphism Theorem, we obtain the result. ∎

The next theorem shows that if C is a group code with 2^k code words contained in $\mathbf{Z}_2^n$, then there is a permutation of the coordinate positions of $\mathbf{Z}_2^n$ such that C is transformed into a code C' with the property that, given any word α of length k, there is exactly one word $\beta\alpha$ of C' of length n. It is clear that $|C| = |C'|$ and $\mu(C) = \mu(C')$. We shall call two codes, each of which is obtained from the other by a permutation of the coordinate positions *equivalent codes*. Since equivalent codes have identical properties with respect to error-correction and error-detection, whenever we speak of a group code, we shall tacitly assume that the above permutation has been effected.

15.3.5 *Example*

Consider the group code $C = \{000,001,100,101\}$. Permuting the first two coordinate positions we get $C' = \{0\!:\!00, 0\!:\!01, 0\!:\!10, 0\!:\!11\}$ where each element of $\mathbf{Z}_2^2$ appears to the right of the dotted line. In this case, if we take $S = \{1,3\}$, $\Pi_S(C) = \mathbf{Z}_2^2$.

15.3.6 *Theorem*

Let C be a group code of size 2^k contained in $\mathbf{Z}_2^n$. Then there exists a subset U of $\{1,2,...,n\}$ with k elements such that $\Pi_U(C) = \mathbf{Z}_2^k$.

Proof. We want to prove the following statement: For any $n \geq 1$, if C is a group code of size 2^k, $1 \leq k \leq n$, contained in $\mathbf{Z}_2^n$, there exists a subset S of $\{1,2,...,n\}$ with k elements such that $\Pi_S(C) = \mathbf{Z}_2^k$.

We proceed by induction. The statement is obviously true for $n = 1$. Suppose it is true for some n and for all $k, 1 \leq k \leq n$. Let C be a group code with 2^k elements contained in $\mathbf{Z}_2^{n+1}$.

If $k = n+1$ or $k = 1$ the theorem clearly holds and we may therefore assume that $2 \leq k \leq n$. Let $S = \{n+1\}$ and consider the homomorphism

$$\Pi_S : \mathbf{Z}_2^n \to \mathbf{Z}_2.$$

There are two cases to consider:

Case 1. $\Pi_S(C) = \{0\}$. This means that every word in C ends in 0. Delete the last coordinate of every word in C to obtain a group code C' consisting of words of length n. Moreover, $|C'| = 2^k$, $2 \leq k \leq n$ and therefore, by the induction hypothesis, there exists a subset T of $\{1,2,...,n\}$ such that

$$\Pi_T(C') = \mathbf{Z}_2^k.$$

Considering T as a subset of $\{1,2,...,n+1\}$, we have $\Pi_T(C) = \mathbf{Z}_2^k$. The result follows by induction.

Case 2. $\Pi_S(C) = \mathbf{Z}_2$. By the First Isomorphism Theorem,

$$\mathbf{C}\!\big/\!_{C \cap \ker \Pi_S} \cong \mathbf{Z}_2$$

and so exactly half the words in C end in 0 and half in 1. Those which end in 0 are the words of $\ker \Pi_S$ and those which end in 1 constitute the coset $(\ker \Pi_S \cap C) + \gamma$ where $\gamma = \beta 1$ is any word of C ending in 1. Therefore

$$C = \{\alpha_i 0 \mid i = 1, 2, ..., 2^{k-1}\} \cup \{\alpha_i 0 + \beta 1 \mid i = 1, 2, ..., 2^{k-1}\}.$$

Let $C^* = \{\alpha_i \mid i = 1, 2, ..., 2^{k-1} \subseteq \mathbf{Z}_2^n$. By the induction hypothesis, there exists a subset T of size $k-1$ of $\{1, 2, ..., n\}$ such that $\Pi_T(C^*) = Z_2^{k-1}$. Note that Π_T maps $\mathbf{Z}_2^n$ onto $\mathbf{Z}_2^{k-1}$ and is injective on C^*.

Now let $T' = T \cup \{n+1\}$ and consider $\Pi_{T'} : \mathbf{Z}_2^{n+1} \to \mathbf{Z}_2^k$ defined by

$$\Pi_{T'}(\alpha x) = \Pi_T(\alpha)\Pi_S(x).$$

Then

$$\Pi_{T'}(C) = \{\Pi_T(\alpha_i)0 \mid i = 1, 2, ..., 2^{k-1}\} \cup \{\Pi_T(\alpha_i)0 + \Pi_T(\beta)1 \mid i = 1, 2, ..., 2^{k-1}\}.$$

Since Π_T restricted to C^* is injective, $|\Pi_{T'}(C)| = 2^k$ and so

$$\Pi_{T'}(C) = \mathbf{Z}_2^k.$$

The result follows by induction. ∎

15.3.7 *Definition*

An (n, k) group code C is a group code contained in $\mathbf{Z}_2^n$ such that

$$\Pi_T(C) = \mathbf{Z}_2^k \text{ where } T = \{n - k + 1, n - k + 2, ..., n\}.$$

In other words, every word of $\mathbf{Z}_2^k$ occurs in the last k coordinate positions of some word of C.

Note. What we have proved above is that every group code is equivalent to an (n, k) code. The words at the tail-end of the code words are thought of as carrying the information while the words of length $n-k$ at the beginning are the check digits. To construct a code, we take the set of 2^k words of length k and add the check digits to the beginning.

15.3.8 *Example*

$\{000000, 101100, 011010, 110001, 110110, 011101, 101011, 000111\}$

is a $(6,3)$ group code. In this case, we start with all the eight words of length three and adjoin three check digits. ∎

15.4 Construction of Group Codes

It is apparent that we cannot construct these group codes by trial and error. In this section we show how to construct all (n,k) group codes using matrices over the field $\mathbf{Z}_2$.

15.4.1 *Definition*

An $n\times k$ matrix M over $\mathbf{Z}_2$ of the form $\begin{pmatrix} A \\ \cdots \\ I_k \end{pmatrix}$ where I_k is the identity matrix, is called an (n,k) *generator matrix*. The matrix M represents a linear transformation F_M from $\mathbf{Z}_2^k$ to $\mathbf{Z}_2^n$ and so its image is a subspace of $\mathbf{Z}_2^n$. In particular, it is a subgroup which is in fact an (n,k) code called the *code associated with* M. The matrix M is the *encoding matrix* since it encodes all words of length k as words of length n, each of which has $n-k$ check digits.

Remark. Subspaces and subgroups of vector spaces over $\mathbf{Z}_2$ are identical since the only scalars are 0 and 1. Similarly, linear transformations are the same as homomorphisms.

15.4.2 *Example*

$$M = \begin{pmatrix} 1 & 1 & 0 \\ 1 & 0 & 1 \\ \hline 1 & 0 & 0 \\ 0 & 1 & 0 \\ 0 & 0 & 1 \end{pmatrix}$$ is a $(5,3)$ generator matrix. This matrix represents a

linear transformation from $\mathbf{Z}_2^3$ to $\mathbf{Z}_2^5$. Its image is a $(5,3)$ group code which we now compute.

$$M\begin{pmatrix}1\\0\\0\end{pmatrix}=\begin{pmatrix}1\\1\\1\\0\\0\end{pmatrix};M\begin{pmatrix}1\\1\\0\end{pmatrix}=\begin{pmatrix}1\\0\\0\\1\\0\end{pmatrix};M\begin{pmatrix}0\\0\\1\end{pmatrix}=\begin{pmatrix}0\\1\\0\\0\\1\end{pmatrix};M\begin{pmatrix}1\\1\\0\end{pmatrix}=\begin{pmatrix}0\\1\\1\\1\\0\end{pmatrix};$$

$$M\begin{pmatrix}1\\0\\1\end{pmatrix}=\begin{pmatrix}1\\0\\1\\0\\1\end{pmatrix};M\begin{pmatrix}0\\1\\1\end{pmatrix}=\begin{pmatrix}1\\1\\0\\1\\1\end{pmatrix};M\begin{pmatrix}1\\1\\1\end{pmatrix}=\begin{pmatrix}0\\0\\1\\1\\1\end{pmatrix}.$$

This gives us the (5,3) group code
  {00:000, 11:100, 10:010, 01:001, 01:110, 10:101, 11:011, 00:111}.
The words of length 2 to the left of the dotted line are the check digits and those to the right the message digits. Its minimum Hamming distance is 2 which means that it is a single-error-detecting code. ∎

Remarks. (i) We observed earlier that the minimum Hamming distance of a group code C is $\min\{w(\alpha)\,|\,\alpha\in C,\alpha\neq 0\}$ (Why?).

(ii) Every column of the generator matrix is in the associated code (Just take $M\varepsilon_i^T$, $i=1,2,...,k$ where $\varepsilon_i=00...10...0$ is the word with 1 in the ith coordinate position and zeroes elsewhere)

We now ask whether every (n,k) group code arises as the associated code of some $n\times k$ generator matrix. The next theorem shows that indeed it does.

15.4.3 *Theorem*

If C is an (n,k) group code, there exists an (n,k) generator matrix M associated with this code.

Proof. Let ε_i be the word with 1 in the ith coordinate position and zeroes elsewhere. Since each word of length k occurs at the tail-end of some word of C, there exists β_i of length $n-k$ such that $\beta_i\varepsilon_i\in C$. Form the matrix

$$M = \begin{pmatrix} \beta_1^T & \beta_2^T & \cdots & \beta_k^T \\ \varepsilon_1^T & \varepsilon_2^T & \cdots & \varepsilon_k^T \end{pmatrix}.$$

The homomorphism F_M is injective since $M\alpha^T = \begin{pmatrix} D^T \\ \alpha^T \end{pmatrix}$ where

D is a word of length $n-k$. Therefore $|\operatorname{Im} F_M| = 2^k$.
But

$$\operatorname{Im} F_M = \; < \beta_i \varepsilon_i \mid i = 1, 2, \ldots, k > \; \subseteq C$$

since C is a subgroup and each $\beta_i \varepsilon_i \in C$. Therefore $\operatorname{Im} F_M = C$. ∎

From a theoretical standpoint it is unnecessary to think of the words of an (n,k) group code as consisting of a message word of length k to which have been adjoined the check digits. One merely considers C as a subgroup of size 2^k contained in $\mathbf{Z}_2^n$.

In practice, however, it is useful to think of the set of all words of length k as the vocabulary in terms of which the messages are written and the generator matrix as the device which adjoins the check digits to produce the code C.

The generator matrix affords an efficient way of producing the code without having to store the code words. We merely store the matrix in some electronic device and, to transmit a message, we input the appropriate words of length k which the device encodes as words of length n. The diagram below illustrates the procedure.

| Message Word | | Encoded Word | | Received Word | | Decoded Word |

Fig. 15.2

15.5   At the Receiving End

We have just seen that the generator matrix is a useful tool at the transmitting end. At the receiving end the so-called *parity-check* matrix obtained from the generator matrix is used to determine whether or not the received word is a code word.

15.5.1 *Definition*

Let $M = \begin{pmatrix} A \\ I_k \end{pmatrix}$ be an (n,k) generator matrix. The *parity-check* matrix P associated with M is defined by

$$P = (I_{n-k} \vdots A).$$

15.5.2 *Example*

If $M = \begin{pmatrix} 1 & 1 & 0 \\ 1 & 0 & 1 \\ 1 & 0 & 0 \\ 0 & 1 & 0 \\ 0 & 0 & 1 \end{pmatrix}$, then $P = \begin{pmatrix} 1 & 0 \vdots 1 & 1 & 0 \\ 0 & 1 \vdots 1 & 0 & 1 \end{pmatrix}$.

Observe that PM is the zero matrix.

15.5.3 *Theorem*

Let C be an (n,k) group code associated with the generator matrix $M = \begin{pmatrix} A \\ I_k \end{pmatrix}$ and let $P = (I_{n-k} \vdots A)$ be the corresponding parity-check matrix. Then F_P is an epimorphism from $\mathbf{Z}_2^n$ to $\mathbf{Z}_2^{n-k}$ whose kernel is C.

Proof. The diagram below should assist in following the proof.

$$\mathbf{Z}_2^k \xrightarrow{F_M} \mathbf{Z}_2^n \xrightarrow{F_P} \mathbf{Z}_2^{n-k}.$$

We show first that F_P is onto. If $\alpha = a_1 a_2 ... a_{n-k} \in \mathbf{Z}_2^{n-k}$, adjoin k zeroes to the tail-end of α to form the word $\beta = a_1 a_2 ... a_{n-k} 00...0$ of length n. Then $P\beta^T = \alpha^T$. By the First Isomorphism Theorem,

$$\mathbf{Z}_2^n \Big/ \ker F_P \cong \mathbf{Z}_2^{n-k} \text{ and so } |\ker F_P| = 2^k.$$

Since $C = \text{Im} F_M$, every word of C is of the form $\begin{pmatrix} A \\ I_k \end{pmatrix} \gamma^T = \begin{pmatrix} A\gamma^T \\ \gamma^T \end{pmatrix}$ and

$(I_{n-k} \vdots A)\begin{pmatrix} A\gamma^T \\ \gamma^T \end{pmatrix} = A\gamma^T + A\gamma^T = \underline{0}^T$ where $\underline{0}$ is the zero word of length

$n-k$. Hence $C \subseteq \ker F_P$. But $|C| = 2^k = |\ker F_P|$ and so $C = \ker F_P$. ∎

Summary. Given any (n, k) group code C there are matrices

$$M = \begin{pmatrix} A \\ \cdots \\ I_k \end{pmatrix} \text{ and } P = \left(I_{n-k} \vdots A \right)$$

such that

$$\alpha \in C \text{ iff } \alpha \in \operatorname{Im} F_M \text{ iff } \alpha \in \ker F_P.$$

Thus a group code can be viewed as the image or the kernel of a homomorphism. The matrix M is useful for transmission while P helps to determine whether or not an error has occurred in the received word.

15.6 Nearest Neighbor Decoding for Group Codes

So far the discussion has centred on the detection of errors. We now show how cosets help in the correction of received words.

An (n,k) group code C is a subgroup of $\mathbf{Z}_2^n$ of order 2^k and as such, we have a decomposition of $\mathbf{Z}_2^n$ into the left cosets of C in $\mathbf{Z}_2^n$. Let $\alpha_1, \alpha_2 ..., \alpha_m$, $m = 2^{n-k}$ be a complete set of coset representatives so that

$$\mathbf{Z}_2^n = (\alpha_1 + C) \cup (\alpha_2 + C) \cup ... \cup (\alpha_m + C).$$

Suppose now that a code word α is transmitted and received as $\beta = \alpha + \varepsilon$. If we can find ε, the *error word*, then the received word β can be correctly decoded as $\beta + \varepsilon$. We first ask: What are the possible values of the error word ε?

The received word β lies in some coset, say $\beta \in \alpha_j + C$. Then $\beta = \alpha_j + \gamma$ for some $\gamma \in C$. Therefore

$$\beta = \alpha + \varepsilon = \alpha_j + \gamma \text{ and so } \varepsilon = \alpha_j + (\alpha + \gamma).$$

Since α and γ are in C and C is a subgroup, $\alpha + \gamma \in C$. Therefore we have proved that *the error word is in the same coset as the received word.* Thus nearest neighbor decoding translates as follows for group codes:

(i) Find the coset to which the received word β belongs (this is of course $\beta + C$).

(ii) Find the word δ in $\beta + C$ of minimum weight and decode β as $\beta + \delta$.

15.6.1 *Definition*

The word of minimum weight in a coset is called the *coset leader*. If there is more than one word of minimum weight, choose one at random and call it the coset leader.

15.6.2 *Example*

Consider the $(4, 2)$ group code C with generator matrix

$$\begin{pmatrix} 1 & 1 \\ 0 & 1 \\ \hline 1 & 0 \\ 0 & 1 \end{pmatrix}.$$

Then $C = \{0000, 1010, 1101, 0111\}$ and the cosets of C in $\mathbf{Z}_2^4$ are:

$$\{0000, 1010, 1101, 0111\}, \{1000, 0010, 0101, 1111\},$$
$$\{0100, 1110, 1001, 0011\}, \{0001, 1011, 1100, 0110\}.$$

We draw up a table of the elements of $\mathbf{Z}_2^4$, listing the cosets as rows with the left-most element of each row the coset leader of that coset. We also list the elements of C in the first row and call such a table a *standard array* for the code. Observe that we have a choice for the coset leader of the second coset listed. We have chosen 1000 but could have chosen 0010.

Table 15.1   A standard array

	Coset leader				Syndrome
Coset	0000	1010	1101	0111	00
Coset	1000	0010	0101	1111	10
Coset	0100	1110	1001	0011	01
Coset	0001	1011	1100	0110	11

IGNORE THE LAST COLUMN FOR NOW

How a decoder uses a standard array. Suppose 1111 is received. The coset leader of the coset containing 1111 is 1000. Thus the word is decoded as $1111 + 1000 = 0111$ and the message word (the last two coordinate positions) is accepted as 11.

The reader will appreciate how much simpler this procedure is than having to compute the distance from the received word to each word of the code and picking the closest one.

What pattern of errors does this method correct? The answer is surprisingly simple.

15.6.3 *Theorem*

The error words the above decoding scheme corrects are precisely the coset leaders.

Proof. Let C be the group code in question and suppose α is transmitted and the error word ε occurs. The received word is then $\beta = \alpha + \varepsilon$ and the coset to which β belongs is $\beta + C = \varepsilon + \alpha + C$. Since $\alpha \in C$, $\alpha + C = C$ and so $\beta + C = \varepsilon + C$.

If ε is the coset leader of $\beta + C$, β is correctly decoded as $\beta + \varepsilon = \alpha + \varepsilon + \varepsilon = \alpha$. On the other hand, if $\varepsilon' \neq \varepsilon$ is the coset leader, β will be incorrectly decoded as $\beta + \varepsilon'$. ∎

15.6.4 *Example*

Referring to the standard array above, suppose 1010 is transmitted and 1000 is received. The error word, 0010 is not a coset leader and, by the theorem above, a decoding error will occur. Indeed 1000 is in the second coset whose coset leader is 1000 and so the received word will be decoded as $1000 + 1000 = 0000$.

Theorem 15.6.3 and the example indicate that if we want to be sure of correcting all patterns of one error, we should ensure that there is only one word of minimum weight in each coset. The code above fails to detect the error pattern 0010 because we had two choices for the coset leader of the coset $0010 + C$.

The standard array can be considerably simplified by introducing the concept of the syndrome of a word.

15.6.5 *Definition*

If C is an (n,k) group code with parity-check matrix P, for each $\alpha \in \mathbf{Z}_2^n$ define the *syndrome* of α to be the word σ (of length $n - k$) where $\sigma^T = P\alpha^T$.

15.6.6 *Theorem*

There is a one-one correspondence between the cosets of C in $\mathbf{Z}_2^n$ and the set of all syndromes.

Proof. By Theorem 15.5.3, $C = \ker F_P$ where F_P is an ipimorphism from $\mathbf{Z}_2^n$ to $\mathbf{Z}_2^{n-k}$. In the proof of the First Isomorphism Theorem (6.4.8), we saw that the map $\alpha + C \to F_P(\alpha)$ is the isomorphism which establishes that $\mathbf{Z}_2^n \Big/ C$ and $\mathbf{Z}_2^{n-k}$ are isomorphic. There is thus a one-one correspondence between the cosets of C in $\mathbf{Z}_2^n$ and the words of $\mathbf{Z}_2^{n-k}$. The latter are the syndromes of the words of $\mathbf{Z}_2^n$. ∎

15.6.7 *Example*

We compute the syndromes for the code whose standard array is displayed above. The generator matrix of C is

$$\begin{pmatrix} 1 & 1 \\ 0 & 1 \\ \hline 1 & 0 \\ 0 & 1 \end{pmatrix}$$

and its parity-check matrix is

$$\begin{pmatrix} 1 & 0 & 1 & 1 \\ 0 & 1 & 0 & 1 \end{pmatrix}.$$

The syndromes are:

$$P\begin{pmatrix} 0 \\ 0 \\ 0 \\ 0 \end{pmatrix} = \begin{pmatrix} 0 \\ 0 \end{pmatrix}; \quad P\begin{pmatrix} 1 \\ 0 \\ 0 \\ 0 \end{pmatrix} = \begin{pmatrix} 1 \\ 0 \end{pmatrix}; \quad P\begin{pmatrix} 0 \\ 1 \\ 0 \\ 0 \end{pmatrix} = \begin{pmatrix} 0 \\ 1 \end{pmatrix}; \quad P\begin{pmatrix} 0 \\ 0 \\ 0 \\ 1 \end{pmatrix} = \begin{pmatrix} 1 \\ 1 \end{pmatrix}.$$

Observe that to calculate the syndromes, we pick any complete set of coset representatives and multiply each by P.

The standard array above may now be simplified to

Table 15.2

Coset leader	Syndrome
0000	00
1000	10
0100	01
0001	11

This simplified array may be used to decode a received word β as follows:

(i) Compute $P\beta^T = \sigma^T$;

(ii) Find the coset associated with σ and suppose ε is the coset leader of this coset.

(iii) decode β as $\beta + \varepsilon$.

15.6.8 *Example*

We continue with Example 15.5.7. Suppose 1111 is received. The syndrome is $\begin{pmatrix} 1 & 0 \vdots 1 & 1 \\ 0 & 1 \vdots 0 & 1 \end{pmatrix}(1\,1\,1\,1)^T = (1\,0)^T$, and the corresponding coset leader is 1000. Therefore the word is decoded as $1111 + 1000 = 0111$. ∎

Remark. If we are interested only in error detection, we compute the syndrome of the received word and if it is zero, the word is accepted, otherwise a re-transmission is requested.

15.7 Hamming Codes

We shall construct a group code with r check digits, high information rate and single error-correcting capacity. To help us in the construction we use the following simple theorem.

15.7.1 *Theorem*

If $H = (\beta_1^T \ \beta_2^T ... \beta_n^T)$ is a $k \times n$ matrix and γ is a word of length n with 1 in the $a^{th}, b^{th}, c^{th} ... s^{th}$ coordinate positions and 0's elsewhere, then
$$H\gamma^T = \beta_a^T + \beta_b^T + ... + \beta_s^T.$$
Proof. Exercise. ∎

If we are to construct a single-error-correcting group code C, $\mu(C)$ must be at least 3, i.e, every nonzero word must have weight at least 3. This requirement imposes restrictions on P, the parity-check matrix.

15.7.2 *Theorem*

If P is the parity-check matrix of a single-error-correcting code C, then no column of P can consist entirely of 0's and no two columns can be equal.

Proof. Bearing in mind that the kernel of F_P is C, if the ath column consisted entirely of 0's, then, appealing to the previous theorem, the word $\varepsilon_a = 00...010...0$ with 0's everywhere except in the ath coordinate position where there is a 1, would be in C and so $\mu(C)$ would be 1.

Moreover, the columns of P must be distinct. For if $P = (\beta_1^T \ \beta_2^T ... \beta_n^T)$ with $\beta_a = \beta_b$, then, again by the preceding theorem, $\alpha = \varepsilon_a + \varepsilon_b$, a word of weight 2 would be in C. ∎

This theorem leads us to construct a matrix P with r rows and whose columns are all nonzero words of length r so arranged that the identity matrix appears as an $r \times r$ block to the left of all the other columns.

There are $2^r - 1$ nonzero words of length r and so P is an $r \times (2^r - 1)$ matrix. Therefore the code C associated with P is a $(2^r - 1, 2^r - 1 - r)$ group code.

15.7.3 *Definition*

A group code constructed as above is called a *Hamming code.*
Observe that the information rate

$$\frac{2^r - 1 - r}{2^r - 1} = 1 - \frac{r}{2^r - 1}$$

is close to 1 even when r is small

15.7.4 *Example*

We construct the $(2^3 - 1, 2^3 - 1 - 3)$ whose parity-check matrix is

$$P = \begin{pmatrix} 1 & 0 & 0 & \vdots & 1 & 1 & 0 & 1 \\ 0 & 1 & 0 & \vdots & 1 & 0 & 1 & 1 \\ 0 & 0 & 1 & \vdots & 0 & 1 & 1 & 1 \end{pmatrix}$$

and whose generator matrix is

$$M = \begin{pmatrix} 1 & 1 & 0 & 1 \\ 1 & 0 & 1 & 1 \\ 0 & 1 & 1 & 1 \\ 1 & 0 & 0 & 0 \\ 0 & 1 & 0 & 0 \\ 0 & 0 & 1 & 0 \\ 0 & 0 & 0 & 0 \end{pmatrix}.$$

The associated code is

$$C = \{0000000, 1101000, 1010100, 0110010, 1110001,$$
$$0111100, 1011010, 0011001, 1100110, 0100101,$$
$$1000011, 1001101, 1101011, 0001110, 0010111, 1111111\}.$$

Observe that the weight of each nonzero word is at least 3. The table below lists the coset leaders with their associated syndromes.

Table 15.3

Coset leader	Syndrome
0000000	000
1000000	100
0100000	010
0010000	001
0001000	110
0000100	101
0000010	011
0000001	111

Let us decode the received word 1001010. The syndrome is 001 and the coset leader 0010000. The word is decoded as $1001010 + 0010000 = 1011010$. ∎

Remark. For a $(2^r - 1, \ 2^r - 1 - r)$ Hamming code C there are $2^{2^r-1} / 2^{2^r-1-r} = 2^r$ cosets and 2^r words of weight at most 1, namely, $00...0, 10...0, 01...0, 00...1$. If P is the parity-check matrix of C, the images of these words under F_P are distinct since $F_P(\varepsilon_j)$ is the jth column of P and the columns of P are distinct. It follows that the syndromes of C are precisely the columns of P together with the zero word and that the words of weight at most 1 are the coset leaders.

15.8 Exercises

1) For $\alpha, \beta \in Z_2^n$, define $\alpha.\beta$ to be the word of length n such that
$$S(\alpha.\beta) = S(\alpha) \cap S(\beta).$$
Prove that
$$w(\alpha + \beta) = w(\alpha) + w(\beta) - 2w(\alpha.\beta).$$

2) Suppose a word of length n is transmitted across a noisy channel where the probability of a 1 (resp. 0) being received as a 0 (resp. 1) is p. Let P_s denote the probability that exactly s errors will occur. Prove that $P_{k+1} < P_k$ if, and only if, $p < {k+1} / {n+1}$.

3) Let $A(n,d)$ denote the number of code words in a code consisting of words of length n with minimum Hamming distance d. There is no known formula for $A(n,d)$ but some of its values can be computed. Find

   (i) $A(n,n)$;
   (ii) $A(n,1)$
   (iii) $A(n,2)$, $n \geq 2$.

4) If C is a code contained in $\mathbf{Z}_2^n$ and α a fixed word in $\mathbf{Z}_2^n$, prove that $\mu(C) = \mu(C+\alpha)$ where $C + \alpha = \{\gamma + \alpha \mid \gamma \in C\}$.

5) Either construct or prove that is impossible to construct a $(5,3)$ code which will detect two or fewer errors. Same question for a $(6,3)$ code.

6) For $\alpha \in \mathbf{Z}_2^n$ and any nonnegative real number r, define
$$S(\alpha;\ r) = \{\gamma \mid h(\alpha, \gamma) \leq r\}.$$
$S(\alpha;\ r)$ is the *sphere* with centre α and radius r.

   Prove that if $\alpha, \beta \in \mathbf{Z}_2^n$ and p and q are nonnegative integers, then
$$S(\alpha;\ p) \cap S(\beta;\ q) \neq \phi \text{ iff } h(\alpha, \beta) \leq p + q.$$

7) Let $\beta_0, \beta_1, ..., \beta_n$ be words of $\mathbf{Z}_2^n$. Prove by induction that
$$h(\beta_0, \beta_1) + h(\beta_1, \beta_2) + ... + h(\beta_{n-1}, \beta_n) \geq h(\beta_0, \beta_n).$$

8) If $\alpha, \beta, \gamma \in \mathbf{Z}_2^n$, we say that the word β is *between* α and γ if
$$h(\alpha, \beta) + h(\beta, \gamma) = h(\alpha, \gamma).$$

   (i) Prove that β lies between α and γ iff
$$S(\alpha) \cap S(\gamma) \subseteq S(\beta) \subseteq S(\alpha) \cup S(\beta);$$

   (ii) Prove or disprove: if β lies between α and γ and δ lies between α and β, then δ lies between α and γ.

9) If $\alpha, \beta \in \mathbf{Z}_2^n$ and r is a nonnegative real number, prove that
$$S(\alpha; r) + \beta = S(\alpha + \beta; r)$$
(see Problem 6 for the definition of $S(\alpha; r)$).

10) Given $\alpha \in \mathbf{Z}_2^n$ and a nonnegative integer m, find a formula for $|S(\alpha; r)|$. Use this formula to find $|S(0110101; 4)|$.

11) The subgroup of $\mathbf{Z}_2^n$ consisting of all words of even weight is called a parity-check code. Find the parity-check and generator matrices for these codes.

12) The *repetition code* of length n consists of the two words 00...0 and 11...1. Find its generator and parity-check matrices.

13) Show that in a group code, either all the words are of even weight or half are of even weight and half of odd weight.

14) If P is the parity-check matrix of a group code C, show that the coset of C whose syndrome is α contains a word of weight w iff some sum of w columns of P is equal to α.

15) Consider the parity-check matrix
$$P = \begin{pmatrix} 1 & 0 & 0 & 0 & 1 & 0 & 0 & 1 & 1 \\ 0 & 1 & 0 & 0 & 1 & 1 & 1 & 0 & 1 \\ 0 & 0 & 1 & 0 & 0 & 1 & 0 & 0 & 0 \\ 0 & 0 & 0 & 1 & 1 & 1 & 1 & 1 & 0 \end{pmatrix}$$
whose associated code is C.

(i) How many words are in C?

(ii) Find all the syndromes and choose coset leaders.

(iii) Draw up a table of syndromes and corresponding coset leaders and use it to decode the following:
$$110101011; 001101001; 110011001.$$

(iv) Find $\mu(C)$.

16) Suppose that the parity-check matrix P of a group code C is such that

(i) no column of P is the zero word;

(ii) no column of P is the sum of fewer than d other columns of P. Prove that $\mu(C) \geq d + 1$. Is the converse true?

17) Let $\alpha, \beta, \gamma \in \mathbf{Z}_2^n$.

(i) Prove that if $w(\alpha)$ is odd, and $\alpha = \beta + \gamma$, then one of β and γ is of odd weight and the other is of even weight.

(ii) Use the previous problem and part (i) of this problem to prove that the group code C whose parity-check matrix is

$$P = \begin{pmatrix} 1 & 0 & 0 & 0 & 1 & 1 & 1 & 0 \\ 0 & 1 & 0 & 0 & 1 & 1 & 0 & 1 \\ 0 & 0 & 1 & 0 & 1 & 0 & 1 & 1 \\ 0 & 0 & 0 & 1 & 0 & 1 & 1 & 1 \end{pmatrix}$$

has minimum Hamming distance 4.

(iii) How many cosets are there of C in $\mathbf{Z}_2^8$?

(iv) Find a full set of coset leaders and syndromes.
Draw up a table of cosets leaders and corresponding syndromes and use it to decode

$$11101100; \ 01110000; \ 11001100.$$

18) Let $\alpha = a_1 a_2 ... a_n$ and $\beta = b_1 b_2 ... b_n$ be words of $\mathbf{Z}_2^n$. Define $\alpha \bullet \beta$ to be $a_1 b_1 + a_2 b_2 + ... + a_n b_n$. (For example $110101 \bullet 101101 = 1$). Let C be an (n, k) code and define $C^\perp = \{\alpha \in \mathbf{Z}_2^n \mid \alpha \bullet \beta = 0 \text{ for all } \beta \in C\}$. This is the so-called *dual* of C.

(i) Prove $C^\perp$ is a subgroup of $\mathbf{Z}_2^n$.

(ii) Prove that if M is the generator matrix of C, then $C^\perp = \ker F_{M^T}$ where M^T is the transpose of M, i.e., the matrix obtained from M by interchanging rows and columns. Hence show that $|C^\perp| = 2^{n-k}$.

19) Let M be any $m \times n$ matrix over $\mathbf{Z}_2$. We write $M \to M_1$ if the $m \times n$ matrix M_1 can be obtained from M by either switching two rows of M or by adding one row of M to another row of M. We write

$M \sim M_s$ if there are matrices $M_1, M_2,..., M_s$ such that $M \rightarrow M_1 \rightarrow M_2$ $\rightarrow ... \rightarrow M_s$. In such a case we say that *M is row equivalent to M_s.*

(i) Prove that row equivalence is an equivalence relation.

(ii) If M is a matrix row equivalent to N, prove that

$$\ker F_M = \ker F_N.$$

(iii) If $P = (\alpha_1^T, \alpha_2^T,..., \alpha_n^T)$ is an $m \times n$ matrix and π a permutation of $\{1, 2,..., n\}$, define $\hat{\pi} : \mathbf{Z}_2^n \rightarrow \mathbf{Z}_2^n$ by

$$\hat{\pi}(a_1 a_2 ... a_n) = a_{\pi(1)} a_{\pi(2)} ... a_{\pi(n)}.$$

Let $Q = (\alpha_{\pi(1)}^T \alpha_{\pi(2)}^T ... \alpha_{\pi(n)}^T)$ and suppose $C = \ker F_P$ and $C' = \ker F_Q$. Prove $\hat{\pi}(C) = C'$. (Note that $|C| = |C'|$ and $\mu(C) = \mu(C')$.)

(iv) If

$$P = \begin{pmatrix} 0 & 1 & 0 & 1 & 0 & 1 & 0 & 1 & 0 & 1 \\ 1 & 1 & 0 & 0 & 1 & 1 & 0 & 0 & 1 & 1 \\ 1 & 0 & 1 & 0 & 0 & 0 & 1 & 0 & 0 & 0 \\ 0 & 0 & 1 & 1 & 1 & 0 & 1 & 1 & 1 & 1 \end{pmatrix}$$

and $C = \ker F_P$, find $|C|$ and $\mu(C)$. [Hint: Find a matrix N row equivalent to P with 1000, 0100, 0010, 0001 appearing as columns of N. Effect a permutation on the columns of N to obtain a matrix $Q = (I_4 : A)$ and use the preceding parts of this question.

20) Suppose $P = (I_{n-k} : A)$ is the $(n-k) \times n$ parity-check matrix associated with the code C and suppose $\mu(C) = d$ where d is odd. Let

$$Q = \begin{pmatrix} 1 & 1 & 1 & .... & 1 \\ 0 & & & \\ 0 & & P & \end{pmatrix}$$ (this is an $(n-k+1) \times n$ matrix) and let $C' = \ker F_Q$.

Prove that $\mu(C') = d + 1$.

Chapter 16

Polynomial Codes

16.1 Definitions and Elementary Results

With each binary word $\alpha = a_0 a_1 \ldots a_{n-1}$ we associate the polynomial

$$\pi(\alpha) = \sum_{j=0}^{n-1} a_j x^j.$$

If, for example $\alpha = 1011$, then the associated polynomial is

$$\pi(\alpha) = 1 + x^2 + x^3.$$

It is easy to check that the map π is an isomorphism of the group $(\mathbf{Z}_2^n, +)$ to the subgroup of the group $(\mathbf{Z}_2[x], +)$ consisting of all polynomials of degree less than n. If C is a subgroup of $\mathbf{Z}_2^n$, then clearly its image $\pi(C)$ is a subgroup of $\mathbf{Z}_2[x]$ isomorphic to C.

16.1.1 *Notation*

The set of polynomials in $\mathbf{Z}_2[x]$ of degree less than n will be denoted by $(\mathbf{Z}_2[x]; n)$.

Remark. The map $\pi: \mathbf{Z}_2^n \to (\mathbf{Z}_2[x]; n)$ is an isomorphism and so its inverse π^{-1} exists.

16.1.2 *Definition*

Let $p(x)$ be a fixed polynomial of degree $n - k$. The polynomial code $P_{n,k}$ *generated by* $p(x)$ consists of all words α of length n such that $\pi(\alpha)$ is divisible by $p(x)$. More explicitly,

$$P_{n,k} = \pi^{-1}\{c(x) \mid c(x) \in \mathbf{Z}_2[x] \text{ and } c(x) = p(x)q(x), \ q(x) \in (\mathbf{Z}_2[x]; k)\}.$$

Observe that there are 2^k polynomials over $\mathbf{Z}_2$ of degree at most k. Therefore $\mid P_{n,k} \mid = 2^k$.

16.1.3 *Example*

We construct the code $P_{5,3}$ using the *generator polynomial* $1 + x + x^2$. To obtain all polynomials in $(\mathbf{Z}_2[x]; 4)$ which are divisible by $1 + x + x^2$, we multiply $1 + x + x^2$ by all polynomials of degree at most two:

Table 16.1. A polynomial code

$(1 + x + x^2).0$	0	00000
$(1 + x + x^2).1$	$1 + x + x^2$	11100
$(1 + x + x^2).x$	$x + x^2 + x^3$	01110
$(1 + x + x^2).(1 + x)$	$1 + x^3$	10010
$(1 + x + x^2).x^2$	$x^2 + x^3 + x^4$	00111
$(1 + x + x^2).(1 + x^2)$	$1 + x + x^3 + x^4$	11011
$(1 + x + x^2).(x + x^2)$	$x + x^4$	01001
$(1 + x + x^2)^2$	$1 + x^2 + x^4$	10101

The code $P_{5,3}$ in the right-most column is a $(5,3)$ group code as can easily be checked. ∎

Remark. If the generator polynomial has zero constant term, then each code word will have zero in the first coordinate. This is clearly wasteful and so from now on we shall assume that the constant term is not zero.

In the last chapter, encoding was achieved by multiplying every word of length k by an $n \times k$ generator matrix M. This induced a monomorphism $F_M : \mathbf{Z}_2^k \to \mathbf{Z}_2^n$ with $\operatorname{Im} F_M$ the group of encoded words. In a similar manner we show in the case of a polynomial code $P_{n,k}$ with generator polynomial $p(x)$ that there is a monomorphism (or encoding) $E : \mathbf{Z}_2^k \to \mathbf{Z}_2^n$ so that

  (i) $\operatorname{Im} E = P_{n,k}$;
  (ii) each word of $\mathbf{Z}_2^k$ appears at the tail-end of some word of $P_{n,k}$.

16.1.4 *The encoding E*

Let $p(x)$ be the generator polynomial of degree $n-k$. Given $\alpha \in Z_2^k$, we want to end up with a word of the form $\beta\alpha$ of length n.

  Step 1. Form $\pi(\alpha)x^{n-k}$, a polynomial of degree n;

  Step 2. Apply the division algorithm to find the remainder when $\pi(\alpha)x^{n-k}$ is divided by $p(x)$:

$$\pi(\alpha)x^{n-k} = p(x)q(x) + r(x) \text{ where } \partial r(x) < n-k;$$

  Step 3. If $r(x) = r_0 + r_1x + ... + r_{n-k-1}x^{n-k-1}$, define
$$E(\alpha) = r_0 r_1 ... r_{n-k-1} : \alpha.$$

16.1.5 *Example*

Let $P_{5,3}$ be the polynomial code generated by $1+x+x^2$ in the example above and let us encode $\alpha = 101$. We have $\pi(101) = 1+x^2$. Therefore $\pi(\alpha)x^2 = x^2 + x^4$. Divide this by $1+x+x^2$ to obtain

$$\pi(\alpha)x^2 = (1+x+x^2)^2 + 1.$$

Therefore the remainder is $1 + 0x$ and so 101 is encoded as $10\dot{:}101$. We note that the polynomial corresponding to $10\dot{:}101$ is $1 + x^2 + x^4$ which is divisible by $1 + x + x^2$.

16.1.6   Theorem

The encoding map E defined above is an isomorphism from $\mathbf{Z}_2^k$ to $P_{n,k}$.

Proof. If $\alpha = a_0 a_1 ... a_{k-1}, \beta = b_0 b_1 ... b_{k-1} \in \mathbf{Z}_2^k$, then

$$\pi(\alpha) x^{n-k} = p(x)q_1(x) + r(x)$$

and

$$\pi(\beta) x^{n-k} = p(x)q_2(x) + s(x)$$

where

$$r(x) = \sum_{j=0}^{n-k-1} r_j x^j, \quad s(x) = \sum_{j=0}^{n-k-1} s_j x^j$$

and so

$$E(\alpha) = (r_0, r_1, ..., r_{n-k-1}, a_0, a_1, ..., a_{k-1})$$

and

$$E(\beta) = (s_0, s_1, ..., s_{n-k-1}, b_0, b_1, ..., b_{k-1}).$$

Then

$$\pi(\alpha) x^{n-k} + \pi(\beta) x^{n-k} = \pi(\alpha + \beta) x^{n-k} = p(x)[q_1(x) + q_2(x)] + r(x) + s(x)$$

and

$$\partial[r(x) + s(x)] < n - k = \partial(p(x)).$$

Therefore

$$E(\alpha + \beta) = (r_0 + s_0, r_1 + s_1, ..., r_{k-1} + s_{k-1}, a_0 + b_0, a_1 + b_1, ..., a_{n-k-1} + b_{n-k-1})$$
$$= E(\alpha) + E(\beta).$$

It remains to show that the image of E is $P_{n,k}$. First we must show that $p(x)$ divides $\pi(E(\alpha))$. We have

$$\pi(E(\alpha)) = r(x) + \pi(\alpha) x^{n-k} = r(x) + p(x)q_1(x) + r(x) = p(x)q_1(x)$$

since $r(x) + r(x) = 0$. Therefore $\mathrm{Im} E \subseteq P_{n,k}$.

Furthermore, E is clearly injective and so $|\operatorname{Im}E|=2^k$ and as we observed above, $|P_{n,k}|=2^k$. Hence $\operatorname{Im}E=P_{n,k}$ and so E is an isomorphism. ∎

Since $P_{n,k}$ is a group code, we should be able to construct the generator matrix M.

16.1.7 *Theorem*

The generator matrix M of the polynomial code $P_{n,k}$ is the matrix $M=\left(\beta_1^T\ \beta_2^T...\beta_k^T\right)$ where β_i is the encoding described above of ε_i, the word of length k with 1 in the ith coordinate position and 0's elsewhere.

Proof. From the encoding described above, $\beta_i=\gamma_i\varepsilon_i$, $i=1,2,...,k$. Therefore

$$M=\begin{pmatrix} \gamma_1^T & \gamma_2^T...\gamma_k^T \\ 1 & 0....\ 0 \\ 0 & 1....\ 0 \\ \vdots & \vdots ....\ 0 \\ 0 & 0....\ 1 \end{pmatrix}$$

which is of the required form. Moreover, $\operatorname{Im}F_M \subseteq P_{n,k}$ since $P_{n,k}$ is a group. But $|\operatorname{Im}F_M|=2^k=|P_{n,k}|$ and so $\operatorname{Im}F_M=P_{n,k}$. ∎

16.1.8 *Example*

We find the generator matrix of the code in Example 16.1.3:

$$\varepsilon_1=100\to1+0x+0x^2=\pi(\varepsilon_1);\ \pi(\varepsilon_1)x^2=x^2.$$

Using the division algorithm, we get

$$x^2=1.(1+x+x^2)+(1+x)$$

and so $\varepsilon_1\to11\dot{:}100$.

$$\varepsilon_2 \to \pi(\varepsilon_2) = 0 + x + 0x^2 = x; \ \pi(\varepsilon_2)x^2 = x^3.$$

An easy computation shows $\varepsilon_2 \to 10010$.

$$\varepsilon_3 \to \pi(\varepsilon_3) = 0 + 0x + x^2; \ \pi(\varepsilon_3)x^2 = x^4.$$

On division by $1 + x + x^2$ we are left with a remainder x. Therefore $\varepsilon_3 \to 01\vdots001$. The generator matrix is

$$M = \begin{pmatrix} 1 & 1 & 0 \\ 1 & 0 & 1 \\ \hline 1 & 0 & 0 \\ 0 & 1 & 0 \\ 0 & 0 & 1 \end{pmatrix}.$$

Using M we encode 101:

$$\begin{pmatrix} 1 & 1 & 0 \\ 1 & 0 & 1 \\ 1 & 0 & 0 \\ 0 & 1 & 0 \\ 0 & 0 & 1 \end{pmatrix} \begin{pmatrix} 1 \\ 0 \\ 1 \end{pmatrix} = \begin{pmatrix} 1 \\ 0 \\ 1 \\ 0 \\ 1 \end{pmatrix}.$$

(Compare with Example 16.1.5.)∎

We can also obtain the generator matrix directly from the generator polynomial $p(x)$:

For $i = 1, 2, \ldots k$ let $r_i(x)$ be the remainder when $x^{n-k-i+1}$ is divided by $p(x)$ and β be the word formed with the coefficients of $r_i(x)$ (coefficient of lowest term first). Then

$$M = \begin{pmatrix} \beta_1^T & \beta_2^T & \ldots \beta_k^T \\ \varepsilon_1^T & \varepsilon_2^T & \ldots \varepsilon_k^T \end{pmatrix}$$

and the parity-check matrix

$$P = \left(I_{n-k} \vdots \beta_1^T \ \ \beta_2^T \ldots \beta_k^T \right).$$

To prove these claims, the reader should follow the recipe for encoding ε_i, $i = 1, 2, \ldots, k$ outlined in 16.1.4.

16.1.9 *Examples*

(i) $P_{n,n-1}$ generated by $1+x$ is the $(n, n-1)$ parity-check code. By the preceding

$$P = (1\!:\!1\,1\,1....1).$$

(ii) $P_{n,1}$ generated by $1+x+x^2+...+x^{n-1}$ is the $(n, 1)$ repetition code with generator matrix

$$M = \begin{pmatrix} 1 \\ 1 \\ \vdots \\ 1 \\ 1 \end{pmatrix}. \quad \blacksquare$$

The following theorem gives us a condition on the generator polynomial which enables us to predict a lower bound for the minimum Hamming distance without actually having to compute the individual code words.

16.1.10 *Theorem*

If $p(x)$ is a generator polynomial of degree $n-k$ which does not divide x^s+1 for all s less than n, then $\mu(P_{n,k}) \geq 3$.

Proof. It is sufficient to show that $w(\alpha) \geq 3$ for all nonzero $\alpha \in P_{n,k}$. Let us follow the procedure of 16.1.4 to encode a word α. If $w(\alpha) \geq 3$, there is nothing to prove since the encoded word will have α at the tail-end. If the word is ε_i, following the procedure of 16.1.4. we get: $\varepsilon_i \to x^{n-k-1+i}$

$\to r_0 r_1...r_{n-k-1} \!:\! \varepsilon_i$ where $\sum_{j=0}^{n-k-1} r_j x^j$ is the remainder on dividing $x^{n-k-1+i}$ by

$p(x)$. But $p(x)$ does not divide $x^{n-k-1+i}$ since, by assumption, the constant term of $p(x)$ is nonzero. Hence the remainder is not zero and so $r_0 r_1...r_{n-k-1}$ is not the zero word. Therefore the encoded word has weight at least two. Suppose there is a word in $P_{n,k}$ of weight two. This corresponds to the polynomial $x^i+x^j, 1 \leq i < j < n$. Thus $p(x) | x^i$

$(1 + x^{j-i})$, a contradiction since $p(x)$ does not divide $1 + x^{j-i}$ and $\gcd(x^i, p(x)) = 1$. Hence $\mu(P_{n,k}) \geq 3$. ∎

16.1.11 *Example*

Let $p(x) = 1 + x + x^3$, $n = 5$. The generator matrix of this code is

$$
\begin{pmatrix}
1 & 0 \\
1 & 1 \\
0 & 1 \\
1 & 0 \\
0 & 1
\end{pmatrix}
$$

and the code words are $00000, 11010, 01101, 10111$. We note that $1 + x + x^3$ does not divide $1 + x^s$ for $s < 5$ and so, as predicted $\mu(P_{5,2}) \geq 3$.

16.2 BCH Codes

In this section we show how to construct double error-correcting codes. They are called Bose-Chaudhuri-Hocquenghern codes or BCH codes for short. To construct them we shall require much of the theory of finite fields which we have developed in Chapters 9 and 10. Any further results we require will be proved as we need them.

We restrict our discussion to fields of order 2^m. Each such field can be constructed by finding a polynomial $p(x)$ of degree m which is irreducible over $\mathbf{Z}_2$ and forming

$$
F = \mathbf{Z}_2[x] \Big/ (p(x))
$$

(see Remark following Theorem 9.6.3). We know (Theorem 9.6.10) that any two finite fields of the same order are isomorphic and so we could choose any other irreducible polynomial $q(x)$ of degree m and form $\mathbf{Z}_2[x]\big/_{q(x)}$ which would then be a field isomorphic to F.

Is there any advantage in choosing one polynomial over another? Since we are going to be making use of the multiplicative group of a finite field which we know is cyclic (Corollary 9.6.9), it simplifies matters to have a generator which is easy to calculate with and so it does make a difference what polynomial we choose. The simplest element we can hope for which will be a generator for the multiplicative group of $\mathbf{Z}_2[x]\big/_{(p(x))}$ is $\bar{x}$. It can be proved that polynomials $p(x)$ do exist such that $\bar{x}$ generates the multiplicative group of $\mathbf{Z}_2[x]\big/_{(p(x))}$ and there are methods for finding them. We shall confine ourselves to giving examples.

16.2.1 *Definition*

A polynomial $p(x)$, irreducible over $\mathbf{Z}_2$, is said to be *primitive* if $\bar{x} = x + (p(x))$ generates the multiplicative group of nonzero elements of $\mathbf{Z}_2[x]\big/_{(p(x))}$.

16.2.2 *Example*

We show that $x^2 + x + 1$ is a primitive polynomial. The elements of $\mathbf{Z}_2[x]\big/_{(x^2 + x + 1)}$ are

$$\bar{0}, \bar{1}, \bar{x}, \overline{1+x} \text{ and } <\bar{x}> = \{\bar{x}^0, \bar{x}, \bar{x}^2 = \overline{1+x}\} = \mathbf{Z}_2[x]\big/_{(x^2 + x + 1)} \setminus \{\bar{0}\}. \ \blacksquare$$

It is clear that any irreducible polynomial of degree n such that $2^n - 1$ is prime is primitive since every nonidentity element of a cyclic group of prime order is a generator. However, not all irreducible polynomials of a

given   degree   are   primitive.   For   example,   the   polynomial
$p(x) = x^4 + x^3 + x^2 + x + 1$ is irreduclble over $\mathbf{Z}_2$ but is not primitive since
$\bar{x}^5 = \bar{1}$ whereas the field $\mathbf{Z}_2[x]\big/(p(x))$ contains 15 nonzero elements.

16.2.3 *Example*

In   this   example   we   use   the   primitive   polynomial $x^3 + x + 1$ to
construct the field F of order 8. We also show how each element of F
can be represented in three different ways:

    (i) as a power of $\bar{x}$;

    (ii) as a coset with representative of degree less than 3;

    (iii) as an element of $\mathbf{Z}_2^3$.

For convenience we write b for $\bar{x}$.

Table 16.2

Elements of F as powers of b	Elements of F with representatives of degree <3	Elements of F as words of length 3
$\bar{0} = b^{-\infty}$	$\bar{0}$	000
$\bar{1} = b^0$	$\bar{1}$	$\alpha_1 = 100$
b	b	$\alpha_2 = 010$
b^2	b^2	$\alpha_3 = 001$
b^3	$1 + b$	$\alpha_4 = 110$
b^4	$b + b^2$	$\alpha_5 = 011$
b^5	$1 + b + b^2$	$\alpha_6 = 111$
b^6	$1 + b^2$	$\alpha_7 = 101$

Column 3 is obtained from column 2 by writing the coefficients of the
polynomials in column 2, coefficient of lowest term first. To add

elements of F it is most convenient to use column 3. For example, $b^3 + b^5 = 110 + 111 = 001 = b^2$. To multiply elements of F column 1 is best as the following example shows:

$$b^3 b^5 = (110).(111) = b^8 = b = 010 \ (b^8 = b \ \text{since} \ b^7 = \overline{1}).$$

We see that the table above may be used as a "dictionary" to translate from one notation to another. ∎

Let us now analyze how we might view the construction of a single-error-correcting code using the field F. Let $\alpha_1, \alpha_2, ..., \alpha_7$ be the nonzero elements of F (thought of as words of length 3 over $\{0,1\}$) and form the 3×7 matrix $P = \left(\alpha_1^T \ \alpha_2^T .... \alpha_7^T \right)$. If we use P as a parity-check matrix, we obtain a $(7,4)$ group code C with generator matrix

$$M = \begin{pmatrix} \alpha_4^T & \alpha_5^T & \alpha_6^T & \alpha_7^T \\ 1 & 0 & 0 & 0 \\ 0 & 1 & 0 & 0 \\ 0 & 0 & 1 & 0 \\ 0 & 0 & 0 & 1 \end{pmatrix}$$

Suppose that one error has occurred in the ith bit of a received word β. Then $\beta = \alpha + \varepsilon_i$ where $\alpha \in C$. Recall that $C = \ker F_p$ and so

$$P\beta^T = P(\alpha^T + \varepsilon_i^T) = P\alpha^T + P\varepsilon_i^T = P\varepsilon_i^T = \alpha_i^T,$$

the ith element of the field in our listing above. Thus the syndrome of β tells us what coordinate position the error has occurred in and we are able to correct it. Note that there are seven possible single-error words and so we need seven distinct elements of $F \setminus 0$ to distinguish between them.

Suppose now that we wish to be able to correct two or fewer errors. It seems reasonable to assume that, if three rows are required to pinpoint one error, it will take six rows to identify up to two errors. Let us therefore form the matrix

$$P = \begin{pmatrix} \alpha_1^T & \alpha_2^T .... \alpha_7^T \\ f(\alpha_1)^T & f(\alpha_2)^T .... f(\alpha_7)^T \end{pmatrix}$$

where f is some map from $F \setminus 0$ to $F \setminus 0$. In this case, if the received the word β has errors in the ith and j^{th} bits, the syndrome of β is

$$\begin{pmatrix} \alpha_i^T + \alpha_j^T \\ f(\alpha_i^T) + f(\alpha_j^T) \end{pmatrix} = \begin{pmatrix} \zeta_1 \\ \zeta_2 \end{pmatrix}.$$

Thus knowing ζ_1 and ζ_2 in F we must find α_i and α_j in F, that is, we must solve the equations

$$\alpha_i + \alpha_j = \zeta_1$$
$$f(\alpha_i) + f(\alpha_j) = \zeta_2$$

in the unknowns α_i and α_j. The key lies in finding a suitable map f. The simplest is $f(\alpha) = \alpha + \gamma$ where γ is a fixed nonzero element of F. Then the equations become

$$\alpha_i + \alpha_j = \zeta_1$$
$$\alpha_i + \alpha_j = \zeta_2.$$

These equations imply that $\zeta_1 = \zeta_2$ and since we anticipate ζ_1 and ζ_2 to be independent, this choice of f fails. In a similar manner we can show that if f is defined by $f(\alpha) = \gamma\alpha$ for some fixed γ or by $f(\alpha) = \alpha^2$, we are led to a dependence of ζ_1 and ζ_2. We next try $f(\alpha) = \alpha^3$. Then the equations become

$$\alpha_i + \alpha_j = \zeta_1$$
$$\alpha_i^3 + \alpha_j^3 = \zeta_2.$$

Now

$$\alpha_i^3 + \alpha_j^3 = (\alpha_i + \alpha_j)(\alpha_i^2 + \alpha_i\alpha_j + \alpha_j^2) = \zeta_2$$

and so

$$\zeta_1(\zeta_1^2 + \alpha_i\alpha_j) = \zeta_2.$$

Solving this for $\alpha_i\alpha_j$ we get

$$\alpha_i\alpha_j = \zeta_1^2 + \zeta_2 / \zeta_1.$$

Observe that we may assume that $\zeta_1 \neq 0$ since, if it were, ζ_2 would also be zero and this would imply that there are no errors. How do we disentangle α_i and α_j?

Consider the quadratic equation $(x + \alpha_i)(x + \alpha_j) = 0$ with roots α_i and α_j. Multiplying out we have

$$x^2 + (\alpha_i + \alpha_j)x + \alpha_i\alpha_j = 0, \text{ i.e.,}$$
$$x^2 + \zeta_1 x + (\zeta_1^2 + \zeta_2/\zeta_1) = 0.$$

Therefore, the error positions are the roots of the quadratic above, with the understanding that if there is exactly one solution, there is only one error.

16.2.4 *Summary for decoding β.*

(i) Calculate $P\beta^T = \begin{pmatrix} \zeta_1^T \\ \zeta_2^T \end{pmatrix}$.

(ii) If $\zeta_1 = 0$, decode β as β.

(iii) If $\zeta_1 \neq 0$, find the roots of

$$x^2 + \zeta_1 x + (\zeta_1^2 + \zeta_2/\zeta_1).$$

These roots are the error bits in β. Hence if the roots are α_i and α_j, then the ith and j^{th} bits in β are changed.

Although the above discussion applies to the field F, it is clear that we could use any field. As a matter of interest, the 6×7 matrix we get using the above construction with F is

$$P = \begin{pmatrix} \alpha_1^T & \alpha_2^T & \dots\dots & \alpha_7^T \\ (\alpha_1^3)^T & (\alpha_2^3)^T & \dots (\alpha_7^3)^T \end{pmatrix} = \begin{pmatrix} 1 & 0 & 0 & 1 & 0 & 1 & 1 \\ 0 & 1 & 0 & 1 & 1 & 1 & 0 \\ 0 & 0 & 1 & 0 & 1 & 1 & 1 \\ 1 & 1 & 1 & 0 & 1 & 0 & 0 \\ 0 & 1 & 0 & 0 & 1 & 1 & 1 \\ 0 & 0 & 1 & 1 & 1 & 0 & 1 \end{pmatrix}.$$

We shall not pursue this example any further since the code C with parity-check matrix P contains only two words. It was to be expected that we would have only a few code words in C since $\mu(C) \geq 5$ (because C is two-error-correcting) and the length of words in C is only 7.

For the convenience of the reader we list some primitive polynomials.

Table 16.3

Primitive polynomial	Degree m	Length of code word $= 2^m - 1$
$x + 1$	1	1
$x^2 + x + 1$	2	2
$x^3 + x + 1$	3	7
$x^4 + x + 1$	4	15
$x^5 + x^2 + 1$	5	31
$x^6 + x + 1$	6	63
$x^7 + x^3 + 1$	7	127

16.2.5 *Example*

We now construct a two-error-correcting code C applying the same method as above, but this time using the field K of order 16. We first construct the field, listing the elements as we did in the table on page 414.

Table 16.4

Elements of K as powers of $b = \bar{x}$	Elements of K using representatives of degree <4.	Elements of K as words of length 4
$\bar{0} = b^{-\infty}$	$\bar{0}$	0000
$\bar{1} = b^0$	$\bar{1}$	1000
b	b	0100
b^2	b^2	0010
b^3	b^3	0001
b^4	$1+b$	1100
b^5	$b+b^2$	0110
b^6	b^2+b^3	0011
b^7	$1+b+b^3$	1101
b^8	$1+b^2$	1010
b^9	$1+b^3$	0101
b^{10}	$1+b+b^2$	1110
b^{11}	$b+b^2+b^3$	0111
b^{12}	$1+b+b^2+b^3$	1111
b^{13}	$1+b^2+b^3$	1011
b^{14}	$1+b^3$	1001

We construct the 8×15 matrix as above:

$$P = \begin{pmatrix}
1 & 0 & 0 & 0 & 1 & 0 & 0 & 1 & 1 & 0 & 1 & 0 & 1 & 1 & 1 \\
0 & 1 & 0 & 0 & 1 & 1 & 0 & 1 & 0 & 1 & 1 & 1 & 1 & 0 & 0 \\
0 & 0 & 1 & 0 & 0 & 1 & 1 & 0 & 1 & 0 & 1 & 1 & 1 & 1 & 0 \\
0 & 0 & 0 & 1 & 0 & 0 & 1 & 1 & 0 & 1 & 0 & 1 & 1 & 1 & 1 \\
1 & 0 & 0 & 0 & 1 & 1 & 0 & 0 & 0 & 1 & 1 & 0 & 0 & 0 & 1 \\
0 & 0 & 0 & 1 & 1 & 0 & 0 & 0 & 1 & 1 & 0 & 0 & 0 & 1 & 1 \\
0 & 0 & 1 & 0 & 1 & 0 & 0 & 1 & 0 & 1 & 0 & 0 & 1 & 0 & 1 \\
0 & 1 & 1 & 1 & 1 & 0 & 1 & 1 & 1 & 1 & 0 & 1 & 1 & 1 & 1
\end{pmatrix}.$$

The code C with parity-check matrix P is a $(15,7)$ group code with information rate $\frac{7}{15}$. The word $\alpha = 11...11$ consisting of 15 1's is a word of C.

Suppose α is transmitted and errors in the third and eighth bits occur and let β be the received word. Applying the decoding algorithm given 16.2.4 we get:

$$P\beta^T = (111100000)^T = \begin{pmatrix} \zeta_1^T \\ \zeta_2^T \end{pmatrix}.$$

Thus $\zeta_1 = b^{12}$ and $\zeta_2 = 0$. We solve

$$x^2 + b^{12}x + b^9 = 0$$

Unfortunately the usual quadratic formula doesn't work for fields of characteristic 2 so we have to try various values of K. We find that $x = b^2$ or $x = b^7$. Using the table above, we see that b^2 is associated with the third coordinate position and b^7 with the eighth. ∎

16.3 Exercises

1) Find the generator and parity-check matrices for the following polynomial codes:

(i) the $(4,1)$ code generated by $1 + x + x^2 + x^3$;
(ii) the $(7,3)$ code generated by $(1+x)(1+x+x^3)$;
(iii) the $(9,4)$ code generated by $1 + x^2 + x^5$.

2) Construct a table of coset leaders and syndromes for each of the following codes:

(i) the $(3,1)$ code generated by $1 + x + x^2$;
(ii) the $(7,4)$ code generated by $1 + x + x^3$;
(iii) the $(9,3)$ code generated by $1 + x^3 + x^6$.

3) Consider the $(63,56)$ code generated by $(1+x)(1+x+x^6)$.

(i) What is the length of the code words before encoding?
(ii) What is the number of check-digits?
(iii) How many different syndromes are there?
(iv) What is the information rate?
(v) What sort of errors will it correct?
(vi) What sort of errors will it detect?

4) Using the primitive polynomial $x^5 + x^2 + 1$, construct a table of the field of order 32 similar to the one on page 9.

5) Let C be the BCH code whose parity-check matrix is given above. Decode the following words:

$$100010111000000$$
$$110011110100000$$
$$101110000001110$$
$$000100001100000$$
$$001111110000010$$
$$011000110010000$$

A *cyclic code* is a group code $C \subseteq \mathbf{Z}_2^n$ with the property that if $c_0 c_1 c_2 \ldots c_{n-1}$ is in C, so is $c_{n-1} c_0 c_1 \ldots c_{n-2}$. The following series of exercises aims to establish properties of these codes.

6) (i) Prove that the map $\varphi : (\mathbf{Z}_2^n, +) \to \left(\mathbf{Z}_2[x] \Big/ (x^n + 1), + \right)$ defined by

$$\varphi(c_0 c_1 \ldots c_{n-1}) = \sum_{i=0}^{n-1} \overline{c_i x^i} \text{ is an isomorphism.}$$

(ii) Let C be a cyclic code of length n and let $\bar{C} = \varphi(C)$. Prove that $\bar{C}$ is an ideal of $\mathbf{Z}_2[x] \Big/ (x^n + 1)$.

(iii) If J is an ideal of $\mathbf{Z}_2[x] \Big/ (x^n + 1)$ then $\varphi^{-1}(J)$ is a cyclic code.

(iv) Prove that every ideal of $\mathbf{Z}_2[x]\big/(x^n+1)$ is principal, generated by $\overline{g(x)}$ where $g(x)\,|\,x^n+1$.

Note that this means that once we know all divisors of x^n+1, we know all cyclic codes. The polynomial $g(x)$ above is called a *generator* of the cyclic code.

(v) Given that the factorization of x^6+1 as a product of irreducible is $(x+1)^2(x^2+x+1)^2$ calculate the ideal generated by $\overline{(x+1)(x^2+x+1)} =$ $\overline{x^3+1}$ in $\mathbf{Z}_2[x]\big/(x^6+1)$ and so calculate the cyclic code generated by x^3+1.

(vi) Prove that if $g(x)$ is a polynomial of degree $n-k$ dividing x^n+1, then the cyclic code with generator $g(x)$ is precisely the polynomial code $P_{n,k}$ with generator $g(x)$.

7) (i) Verify that over $\mathbf{Z}_2$,

$$x^9+1=(x+1)(x^2+x+1)(x^6+x^3+1).$$

Where each of the factors on the right-hand side is irreducible over $\mathbf{Z}_2$.

(ii) How many cyclic codes of length 9 are there?

8) (i) Factor x^7+1 into irreducible factors over $\mathbf{Z}_2$.

(ii) Construct two $(7,4)$ cyclic codes.

(iii) How many cyclic codes of length 7 are there? Find the minimum Hamming distance for each of these codes.

Appendix A

Rational, Real and Complex Numbers

A.1. Introduction

This appendix is included as a brief reminder to the reader of some of the elementary properties of the three systems in the title.

Given the integers $\mathbf{Z}$, we saw in Chapter 8 how we can extend $\mathbf{Z}$ to $\mathbf{Q} = \{ {}^{p}\!/\!_{q} \mid p, q \in \mathbf{Z}, q \neq 0 \}$, the field of quotients of the integral domain $\mathbf{Z}$ with addition and multiplication defined by

$$\frac{a}{b} + \frac{c}{d} = \frac{ad + bc}{bd}, \quad \frac{a}{b} \cdot \frac{c}{d} = \frac{ac}{bd},$$

respectively.

Within the rationals, $\mathbf{Q}$, we are able to solve equations of the form $ax = b$, where a and b are rationals, $a \neq 0$. Since simple equations like $x^2 = 2$ are unsolvable in $\mathbf{Q}$, we are led to extending the rationals to include irrational numbers so that equations of the form $x^m = a$ for nonnegative a are solvable. The collection of rational and irrational numbers form the real number system.

The equation $x^m = a$ with a negative and m even, however, is still unsolvable and leads us to a further extension of the number system to include numbers of the form $a + ib$ where a and b are real and $i = \sqrt{-1}$.

This last extension gives us the complex number system within which we are able to solve (in theory) any polynomial equation with complex coefficients.

A.2. The Real and Rational Number Systems

We shall assume that the reader has an intuitive grasp of the real number system and is comfortable dealing with equations and inequalities. Our presentation will not be rigorous and we shall not construct the real numbers from the rationals, a process which involves Dedekind cuts and is best left to a course in analysis.

We begin by introducing the so-called *least upper bound property*. which we adopt as an axiom. Heuristically, this axiom fills in the "holes" in the rational number line.

A.2.1. *Definition*

(i) A nonempty subset S of real numbers is *bounded above* if there is a real number M such that $s \le M$ for all $s \in S$. M is called an *upper bound* for S.

(ii) A nonempty subset S of the real numbers has a *least upper bound b* if:

(a) b is an upper bound for S;

(b) for any upper bound c for S, we have $b \le c$.

We remark that, if a least upper bound exists, then it is unique. This is easily seen since, if b and c are least upper bounds of some set, then, by the very definition, $b \le c$ by virtue of b being a least upper bound and c being an upper bound, and conversely, $c \le b$ by interchanging the roles of b and c.

We postulate:

A.2.2. *Axiom (The Least Upper Bound Property)*

Every nonempty subset of the reals which is bounded above has a least upper bound.

In view of the remark above, we may speak of *the* least upper bound of a set S and denote it by lub S. It is often called the *supremum* of the set and is denoted sup S.

If we restrict ourselves to **Q**, then, given a nonempty set S of rationals, it is not always the case that S has a rational least upper bound. We shall shortly give an example of a set of rational numbers having no rational supremum.

First we prove one of the consequences of the Least Upper Bound Property and use it to show that the rationals are "dense" in the reals, i.e., between any two real numbers there is a rational.

A.2.3. *Theorem*

Given any two positive real numbers a and b, then there exists a positive integer n such that $na > b$.

Proof. Suppose the contrary is true. Then for all positive integers n, $na \leq b$. Let $S = \{na \mid n \text{ is a positive integer}\}$.

Since S is bounded by b, it has a supremum, say l. Now since $l - a < l$, $l - a$ is not an upper bound for S. Therefore, there exists a positive integer m such that $ma > l - a$. But then $(m+1)a > l$, a contradiction since $(m+1)a \in S$. Hence there exists n such that $na > b$. ∎

Remark. The property proved above is called the *Archimedean* property and we say that **R** is an *Archimedean field*.

A.2.4. *Theorem*

Let a and b be real numbers with $a < b$. Then there exists a rational r such that $a < r < b$.

Proof. By the Archimedean property, there exists a positive integer m such that $n(b-a)>1$ and so $\frac{1}{n}<b-a$.

Case (i). $b>0$. By the Archimedean property and Well Ordering of the positive integers, there exists a least positive integer m such that $\frac{m}{n}\geq b$. Hence $\frac{(m-1)}{n}<b$ by the minimality of m. Also, $-\frac{1}{n}>a-b$ and so $a<b-\frac{1}{n}$. But since $b\leq\frac{m}{n}$, we have

$$a<b-\frac{1}{n}\leq\frac{m}{n}-\frac{1}{n}=\frac{(m-1)}{n}<b.$$

Case (ii). $b\leq 0$. Then $-b<-a$ and $-a>0$. By the first case, there exists a rational r such that $-b<r<-a$ and so $a<-r<b$. ∎

A.2.5. *Corollary*

Given any two positive real numbers, a and b with $a<b$ and a positive integer m, there exists a rational number r such that $a<r^m<b$.

Proof. Since $a<b$, we have $\sqrt[m]{a}<\sqrt[m]{b}$ and so, by the previous result, there exists a rational r such that $\sqrt[m]{a}<r<\sqrt[m]{b}$. Raising each member of the inequality to the mth power, we obtain the result. ∎

Remarks. It is an easy matter to extend the corollary above to include the cases where both or only one of a and b is negative and m is odd.

We now give the example promised above of a set of rationals bounded above but having no rational least upper bound.

A.2.6. *Example*

Let m be a fixed positive integer greater than 1 and p a prime. Consider $S=\{r\in\mathbf{Q}\mid r^m<p\}$. We claim that $\sup S$ is not rational.

Proof. Let $l = \sup S$ and suppose $l^m < p$. By Corollary A.2.5 there exists a rational k/n such that $l^m < \left(k/n\right)^m < p$. Hence $k/n \in S$, showing that l is not an upper bound of S since $k/n > l$.

Therefore, $\sup S = \sqrt[m]{p}$ which is irrational (apply Theorem 9.5.7). ∎

A.3. Decimal Representation of Rational Numbers

Given a decimal representation of a real number, we now show how to determine whether the number is rational or irrational. Many readers will be familiar with the topic from high school, but we include it for completeness sake. We note in passing that some real numbers have more than one representation as a decimal. For example, $1.\overline{9}$ (the bar over the 9 means that 9 is repeated ad infinitum) and 2.00 represent the same real number as we shall see.

A.3.1. *Definition*

We call a decimal representation of the form

$$m.a_1 a_2 ... a_k b_1 b_2 .... b_n b_1 b_2 ..... b_n b_1 ....$$

where m is an integer and the a's and b's are integers lying between 0 and 9 inclusive, a *repeating decimal* and write it as $m.a_1 a_2 ... a_k \overline{b_1 b_2 .... b_n}$ to denote that the block of b's is repeated ad infinitum.

A.3.2. *Theorem*

A real number is rational if, and only if, it is represented by a repeating decimal.

Proof. Suppose $r = m.a_1 a_2 ... a_k \overline{b_1 b_2 .... b_n}$. Then $10^k r = m a_1 a_2 ... a_k . \overline{b_1 b_2 ... b_n}$ and $10^{k+n} r = m a_1 a_2 ... a_k b_1 b_2 .... b_n . \overline{b_1 b_2 .... b_n}$.

Thus

$$10^{k+n} r - 10^k r = m a_1 a_2 ... a_k b_1 b_2 .... b_n . \overline{b_1 b_2 .... b_n} - m a_1 a_2 ... a_k . \overline{b_1 b_2 ... b_n}$$

and so

$$r = \frac{ma_1a_2...a_kb_1b_2...b_n - ma_1a_2...a_k}{10^{k+n} - 10^k},$$

a quotient of two integers.

The converse is left as an exercise for the reader. ∎

The proof of this theorem actually shows how to find the rational number corresponding to a given repeating decimal.

A.3.3. *Examples*

(i) Find the rational number corresponding to $1.\overline{345}$.

Solution. Let $r = 1.\overline{345}$. Then $10r = 13.\overline{45}$ and $10^3 r = 1345.\overline{45}$. Subtracting, we get $990r = 1332$ and so $r = \dfrac{1332}{990}$.

(ii) Find the rational number corresponding to $1.\overline{9}$

Solution. Let $r = 1.\overline{9}$. Then $10r = 19.\overline{9}$. Subtracting r from this, we get $9r = 18$ and so $r = 2$.

A.4. Complex Numbers

We define a complex number in what may appear to be an odd way. For readers who have some knowledge of matrices, however, it is probably the most satisfying way of doing things: if the reader believes in matrices, then (s)he must believe in complex numbers!

A.4.1. *Definition*

A complex number is a 2 by 2 matrix of the form

$$\begin{pmatrix} a & -b \\ b & a \end{pmatrix}$$

where a and b are real numbers. We denote the set of all such matrices by $\mathbf{C}$ and call this the *field of complex numbers*. That it is a field under the usual operations of matrix addition and multiplication is easy to check and is left as an exercise for the reader. Observe that it is not necessary to prove any properties which matrices possess under the usual operations, such as associativity.

We now introduce specific matrices in $\mathbf{C}$ in terms of which all other matrices of $\mathbf{C}$ can be expressed. These are:

(i) $\begin{pmatrix} a & 0 \\ 0 & a \end{pmatrix}$ for all $a \in \mathbf{R}$;

(ii) $\begin{pmatrix} 0 & -1 \\ 1 & 0 \end{pmatrix}$.

To simplify notation, we denote $\begin{pmatrix} a & 0 \\ 0 & a \end{pmatrix}$ by $\underline{a}$ and $\begin{pmatrix} 0 & -1 \\ 1 & 0 \end{pmatrix}$ by $\underline{i}$.

The reader is invited to verify the following:

(i) $\underline{a}.\underline{b} = \underline{b}.\underline{a} = \underline{ab}$;

(ii) $\underline{i}^2 = \underline{(-1)} = -\underline{1}$;

(iii) $\underline{a}.\underline{i} = \underline{i}.\underline{a} = \begin{pmatrix} 0 & -a \\ a & 0 \end{pmatrix}$;

(iv) $\begin{pmatrix} a & -b \\ b & a \end{pmatrix} = \underline{a} + \underline{i}.\underline{b}$;

If we simplify the notation further and dispense with the underlines, then our matrices in $\mathbf{C}$ can be expressed in the form $a + ib,\ a,b \in \mathbf{R}$. These matrices constitute the complex numbers and from now on we shall refer to them as such and call a the *real part* and b the *imaginary part* of the complex number. We denote the real part of the complex number z by $\operatorname{Re} z$ and the imaginary part by $\operatorname{Im} z$. If $\operatorname{Im} z = 0$, we say that z is *real* while if $\operatorname{Im} z \neq 0$ and $\operatorname{Re} z = 0$, we say that z is *pure imaginary*.

Bearing in mind that matrix addition is commutative and associative and that matrix multiplication is associative and distributive over matrix addition, we compute in $\mathbf{C}$ according to the formulas:

$$(a+ib)+(c+id)=(a+c)+i(b+d);$$
$$(a+ib)(c+id)=(ac-bd)+i(ad-bc).$$

In other words, we perform addition and multiplication of complex numbers as though we were dealing with ordinary numbers except that, whenever i^2 appears, we replace it with -1.

Using the well known properties of matrix addition and multiplication, it is now an easy matter to verify that $\mathbf{C}$ is a field. Furthermore, it is clear that the set of complex numbers whose imaginary part is zero is a field isomorphic to $\mathbf{R}$. We identify this subfield of $\mathbf{C}$ with $\mathbf{R}$ and think of $\mathbf{R}$. as actually contained in $\mathbf{C}$. In conformity with notation used in the reals, if w and $z(\neq 0)$ are complex numbers, we shall write $w\!/\!_z$ rather than wz^{-1}.

We can establish a one-to-one correspondence between the complex numbers and the points of the plane by associating with each complex number $x+iy$ the point (x,y) of the Cartesian plane. Often we think of a complex number as a vector and represent it in the usual way by an arrow emanating from the origin with arrowhead at the point (x,y). The x-axis is called the *real axis* and the y-axis the *imaginary axis*. The plane is referred to as the *complex plane* (also the *Argand* plane).

Note that, from a geometric standpoint, addition of complex numbers is the same as addition of vectors. Later, we shall give a geometric interpretation of multiplication of complex numbers.

A.4.2. *Definition*

If $z=a+ib$, then the *complex conjugate* of z is $a+i(-b)$ and we denote it by $\overline{z}$. $a+i(-b)$ is written as $a-ib$.
Regarded as a transformation of the complex plane, conjugation is reflection in the real axis.

A.4.3. *Theorem*

Let w and z be complex numbers. Then:

(i) $\overline{w+z} = \overline{w}+\overline{z}$;

(ii) $\overline{wz} = \overline{w}.\overline{z}$;

(iii) $\overline{\left(\dfrac{w}{z}\right)} = \dfrac{\overline{w}}{\overline{z}}$ provided that $z \neq 0$;

(iv) $z^{-1} = \dfrac{1}{z} = \dfrac{\overline{z}}{z\overline{z}}$;

(v) $\overline{\overline{w}} = w$;

(vi) $w+\overline{w} = 2\operatorname{Re} w$;

(vii) $w-\overline{w} = 2i \operatorname{Im} z$;

(viii) w is real iff $w = \overline{w}$.

Proof. We prove (ii) only, leaving the rest as an exercise for the reader. Let $w = a+ib$, $z = c+id$
Then

$$wz = ac-bd + i(ad \quad bc) \text{ and } \overline{wz} = (ac-bd) - i(ad-bc).$$

On the other hand,

$$\overline{w}.\overline{z} = (a-ib)(c-id) = (ac-bd) - i(ad-bc) = \overline{wz}. \blacksquare$$

A.4.4. *Definition*

The *modulus* of the complex number $z = a+ib$, denoted $|z|$, is $\sqrt{a^2+b^2}$. This is, of course, the length of the vector (a,b).

A.4.5. *Theorem*

Let z and w be complex numbers. Then

(i) $|z|^2 = z.\overline{z}$;

(ii) $|zw| = |z|.|w|$;

(iii) $\left|\dfrac{z}{w}\right| = \dfrac{|z|}{|w|}$ provided that $w \neq 0$.

Proof. We prove only (ii). Using (i),

$$|zw|^2 = zw\overline{zw} = zw\overline{z}.\overline{w} = z\overline{z}w\overline{w} = |z|^2|w|^2 = (|z|\,\|\,w|)^2.$$

Taking square roots, we obtain the result. ∎

A.4.6. *Theorem (A Fundamental Inequality)*

If $z = x + iy$ is a complex number, then

$$|\operatorname{Re} z| \leq |z| \leq |\operatorname{Re} z| + |\operatorname{Im} z|; \quad |\operatorname{Im} z| \leq |z| \leq |\operatorname{Re} z| + |\operatorname{Im} z|.$$

Proof. Let $z = x + iy$. Then

$$|\operatorname{Re} z|^2 = |x|^2 \leq x^2 + y^2 = |z|^2 \leq x^2 + 2|x|\,|y| + y^2 = (|x| + |y|)^2$$

and taking square roots, we obtain the result.

A similar proof works for the other inequality. ∎

A.4.7. *Theorem (The Triangle Inequality)*

If w and z are complex numbers, then

$$|w + z| \leq |w| + |z|.$$

Proof.
$$|w + z|^2 = (w + z)(\overline{w + z}) = (w + z)(\overline{w} + \overline{z}) = |w|^2 + (w\overline{z} + \overline{w}z) + |z|^2$$
$$= |w|^2 + 2\operatorname{Re} w\overline{z} + |z|^2.$$

But, by the inequality in A.4.6,

$$|w + z|^2 = |w|^2 + 2\operatorname{Re} w\overline{z} + |z|^2 \leq |w|^2 + 2|w|\,|z| + |z|^2 = (|w| + |z|)^2.$$

Taking square roots yields the inequality. ∎

A.5. Polar Form of a Complex Number

A.5.1. *Definition*

If $z = x + iy \neq 0$, then $z = r\cos\theta + ir\sin\theta$ where r is positive and (r, θ) are the polar coordinates of the point (x, y). The expression

$r\cos\theta + ir\sin\theta$ is called the *polar form* of the complex number z. Note that $r = \sqrt{x^2 + y^2}$, the modulus of z.

The angle θ is called the *argument* of z. It is determined up to multiples of 2π and is denoted by $\arg z$. If θ is a value of the argument, then $\theta + 2k\pi$, $k = 0, \pm 1, \pm 2, \ldots$ is the set of all values of the argument.

That value of the argument of z, denoted by $\operatorname{Arg} z$, such that $-\pi < \operatorname{Arg} z \leq \pi$ is called *the principal value* of the argument.

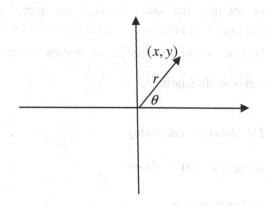

Fig. A.1 The complex plane

Notation. We write $\operatorname{cis}\theta$ for $\cos\theta + i\sin\theta$. In this notation, a complex number is of the form $r\operatorname{cis}\theta$.

A.5.2. *Theorem*

$$\operatorname{cis}\theta\operatorname{cis}\varphi = \operatorname{cis}(\theta + \varphi).$$

Proof.

$\operatorname{cis}\theta\operatorname{cis}\varphi = (\cos\theta\cos\varphi - \sin\theta\sin\varphi) + i(\cos\theta\sin\varphi + \sin\theta\cos\varphi)$.

By well known formulas of trigonometry,

$$\cos\theta\cos\varphi - \sin\theta\sin\varphi = \cos(\theta + \varphi),$$

$$\cos\theta\sin\varphi + \sin\theta\cos\varphi = \sin(\theta + \varphi).$$

Therefore $\operatorname{cis}\theta\operatorname{cis}\varphi = \operatorname{cis}(\theta+\varphi)$. ∎

Remark. Expressed in polar form, multiplication of complex numbers reads:

$$r\operatorname{cis}\theta.s\operatorname{cis}\varphi = rs.\operatorname{cis}(\theta+\varphi)..$$

The geometrical interpretation is that when a complex number z is multiplied by w, the vector representing wz is obtained from that representing z by rotating the latter through an angle $\arg w$ and magnifying (or shrinking) the modulus of z by a factor of $|w|$.

For example, multiplication by $i = \operatorname{cis}\frac{\pi}{2}$ represents a rotation through $\frac{\pi}{2}$ in a counter clockwise direction.

A.5.3. *Corollary (De Moivre's Theorem)*

If n is a positive integer, $(\operatorname{cis}\theta)^n = \operatorname{cis}n\theta$.

Proof. Use A.5.2 and induction. ∎

We can extend Corollary A.5.3 to negative integers.

A.5.4. *Theorem*

$(\operatorname{cis}\theta)^{-n} = \operatorname{cis}(-n\theta)$ where n is a positive integer.

Proof. By definition,

$$(\operatorname{cis}\theta)^{-n} = \frac{1}{(\operatorname{cis}\theta)^n} = \frac{1}{\operatorname{cis}n\theta}$$

(the last equality by Corollary A.5.3).
But $(\operatorname{cis}n\theta)(\operatorname{cis}(-n\theta)) = 1$ by Theorem 1.5.2, and so

$$\operatorname{cis}(-n\theta) = \frac{1}{\operatorname{cis}n\theta}.$$

Therefore, $(\operatorname{cis}\theta)^{-n} = \operatorname{cis}(-n\theta)$. ∎

The next result gives criteria for establishing whether two complex numbers given in polar form are equal.

A.5.5. *Theorem*

Suppose $r \neq 0$. Then
$$r.\operatorname{cis}\theta = s.\operatorname{cis}\varphi \text{ iff } r = s \text{ and } \theta = \varphi + 2k\pi, \text{ some } k \in \mathbf{Z}.$$

Proof. If two complex numbers are equal, then clearly their moduli are also equal. Hence, $|\,r\operatorname{cis}\theta\,| = |\,s\operatorname{cis}\varphi\,|$. But $|\,r\operatorname{cis}\theta\,| = r$ and $|\,s\operatorname{cis}\varphi| = s$ and so, $r = s \neq 0$.

Cancelling r, we get
$$\operatorname{cis}\theta = \operatorname{cis}\varphi \text{ and so } \operatorname{cis}\theta \big/ \operatorname{cis}\varphi = 1.$$
By A.5.4
$$\operatorname{cis}\theta.\operatorname{cis}(-\varphi) = 1$$
and by A.5.2,
$$\operatorname{cis}\theta.\operatorname{cis}(-\varphi) = \operatorname{cis}(\theta - \varphi) = 1.$$

Therefore, $\cos(\theta - \varphi) = 1$ and $\sin(\theta - \varphi) = 0$. It follows that $\theta - \varphi = 2k\pi$ for some integer k.

The converse is obvious since sine and cosine are periodic of period 2π. ∎

A.5.6. *Notation*

We denote the set of solutions of the equation $z^n = 1$ by $1^{\frac{1}{n}}$, n a positive integer and call any such solution *an nth root of unity*.

It is clear that 1 is a value of $1^{\frac{1}{n}}$. Are there any others? The next result establishes that there are, in fact, exactly n distinct complex n^{th} roots of unity.(See Corollary 9.5.11.)

A.5.7. *Theorem*

The equation $z^n = 1$ has exactly n distinct solutions, namely, $\text{cis}\, \frac{2k\pi}{n}, k = 0, 1, ..., n-1$.

Proof. Since $(\text{cis}\, \frac{2k\pi}{n})^n = \text{cis}\, 2k\pi = 1$, $\text{cis}\, \frac{2k\pi}{n}$, $k = 0, 1, ..., n-1$ are indeed solutions.

Suppose now that $z = r.\text{cis}\,\theta$ is a solution of $z^n = 1$. Then

$$r^n \text{cis}\, n\theta = 1$$

and by A.5.2,

$$r^n = 1 \text{ and } n\theta = 2k\pi, k = 0, \pm 1, \pm 2, ...$$

Therefore, $r = 1$ and $\theta = \frac{2k\pi}{n}$, $k = 0, \pm 1, \pm 2, ...$ Hence

$$\{\text{cis}\, \frac{2k\pi}{n} \mid k = 0, \pm 1, \pm 2, ...\}$$

contains all solutions. These solutions are not all distinct, however. Indeed

$$\text{cis}\, \frac{2k\pi}{n} = \text{cis}\, \frac{2l\pi}{n} \text{ iff cis}\, \frac{2(k-l)\pi}{n} = 1$$

$$\text{iff } \frac{2(k-l)\pi}{n} = 2m\pi, \text{some } m \in \mathbf{Z} \text{ iff } k \equiv l \bmod n.$$

Therefore, any n consecutive integral values of k will yield all solutions. We usually choose $k = 0, 1, 2, ..., n-1$. ∎

If we let $\omega = \text{cis}\, \frac{2\pi}{n}$, then any n^{th} root of unity can be written as ω^k, $k = 1, 2, ..., n-1$.

A.5.8. *Definition*

The particular root ω is called the *principal value of the nth root of unity*.

When plotted in the complex plane, the n n^{th} roots of unity form a regular n-gon with vertices on the unit circle. For example, the 6 sixth roots of unity are displayed in the diagram below.

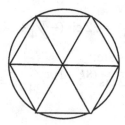

Fig. A.2   The sixth roots of unity

We now show that the equation $z^n = w$, w any nonzero complex number, has exactly n complex solutions.

We first write the equation in polar form with $z = r.\mathrm{cis}\,\theta$ and $w = s.\mathrm{cis}\,\alpha$. The equation then reads

$$r^n \, \mathrm{cis}\, n\theta = s.\mathrm{cis}\,\alpha.$$

By Theorem A.5.5,

$$r^n = s, \quad n\theta = \alpha + 2k\pi,$$

k an integer. Therefore

$$r = \sqrt[n]{s}, \quad \theta = \frac{(\alpha + 2k\pi)}{n}, \quad k \text{ an integer.}$$

As in Theorem A.5.5 above, the solutions of $z^n = w$ are

$$\sqrt[n]{s}\, \mathrm{cis}\, \frac{(\alpha + 2k\pi)}{n}, \quad k = 0, 1, 2, ..., n-1.$$

Remark. It is easily checked that these solutions can be written as

$$\sqrt[n]{s}\, \omega^k \, \mathrm{cis}\, \frac{\alpha}{n}, \quad k = 0, 1, 2, ..., n-1$$

where $\omega = \mathrm{cis}\left(\frac{2\pi}{n}\right)$ is the principal value of the nth root of unity.

A.5.9. *Examples*

(i) Find all sixth roots of $1 - i$.

Solution.

$$1 - i = \sqrt{2}\left(\frac{1}{\sqrt{2}} - \frac{i}{\sqrt{2}} \right) = \sqrt{2}\,\text{cis}\,(-\frac{\pi}{4} + 2k\pi),\ k = 0, \pm 1, \pm 2, \pm 3, \ldots.$$

Therefore, the 6 sixth roots are:

$$\sqrt[12]{2}\,\text{cis}\,\frac{(8k-1)\pi}{24},\ k = 0, 1, 2, 3, 4, 5.$$

(ii) Compute

$$\left(-\frac{1}{2} + \frac{\sqrt{3}}{2} \right)^{12}$$

Solution.

$$\left(-\frac{1}{2} + \frac{\sqrt{3}}{2} \right)^{12} = (\text{cis}\,\frac{2\pi}{3})^{12} = \text{cis}\,8\pi = 1.$$

In the second equality we have used DeMoivre's Theorem. ∎

The next example shows that for square roots we do not need to compute in polar form. The reader is invited to do the example below using the polar form.

A.5.10. *Example*

Show that the square roots of $a + ib,\ b \neq 0$ are

$$\pm \frac{1}{\sqrt{2}}\left(\sqrt{\sqrt{a^2 + b^2} + a} + \frac{b}{|b|}\left(\sqrt{\sqrt{a^2 + b^2} - a} \right)i \right).$$

Solution. Let $z = x + iy$ be a square root of $a + ib$. Then $(x + iy)^2 = a + ib$. Therefore,

$x^2 - y^2 + i(2xy) = a + ib$ and so $x^2 - y^2 = a,\ 2xy = b$, equating real and imaginary parts.

Since $b \neq 0$, it follows that $x \neq 0$ whence $y = \dfrac{b}{2x}$. Substituting in the other equation, we get

$$x^2 - \frac{b^2}{4x^2} = a .$$

Solving this quadratic in x^2, we find

$$4x^4 - 4ax^2 - b^2 = 0 \text{ and so } x^2 = \left(a \pm \sqrt{a^2 + b^2}\right)\Big/2 .$$

Since x is real and $a < \sqrt{a^2 + b^2}$, it follows that

$$x^2 = \left(a + \sqrt{a^2 + b^2}\right)\Big/2$$

and so,

$$x = \pm \frac{1}{\sqrt{2}}\left(\sqrt{\sqrt{a^2 + b^2} + a}\right).$$

Now

$$y^2 = x^2 - a = \left(a + \sqrt{a^2 + b^2}\right)\Big/2 - 2a\Big/2 = \left(\sqrt{a^2 + b^2} - a\right)\Big/2$$

and so

$$y = \pm \frac{1}{\sqrt{2}}\left(\sqrt{\sqrt{a^2 + b^2} - a}\right).$$

Case (i) $b > 0$. From $2xy = b$ we deduce that x and y have the same sign. Therefore, we obtain the two solutions

$$\pm \frac{1}{\sqrt{2}}\left(\sqrt{\sqrt{a^2 + b^2} + a} + i\sqrt{\sqrt{a^2 + b^2} - a}\right).$$

Case (ii). $b < 0$. In this case we see that x and y have opposite signs. This yields the two solutions

$$\pm \frac{1}{\sqrt{2}}\left(\sqrt{\sqrt{a^2 + b^2} + a} - i\sqrt{\sqrt{a^2 + b^2} - a}\right).$$

These two sets of solutions can be expressed in the single formula

$$\pm \frac{1}{\sqrt{2}}\left(\sqrt{\sqrt{a^2 + b^2} + a} + \frac{b}{|b|}\left(\sqrt{\sqrt{a^2 + b^2} - a}\right)i\right). \quad\blacksquare$$

A. 6   Exercises

1) Prove that if a is rational and b is irrational, then $a+b$, ab and a/b are all irrational.

2) Prove that between any two real numbers there is an irrational number. Note that we have proved that between any two real numbers there is a rational number.

3) Give examples to show that it is possible for the sum, product and quotient of two irrational numbers to be rational.

4) Find the decimal representation of each of the following rational numbers:

   (i) $\dfrac{22}{7}$;

   (ii) $\dfrac{5}{11}$;

   (iii) $\dfrac{7}{12}$;

   (iv) $\dfrac{5}{8}$;

   (v) $\dfrac{22}{9}$.

5) Find the rational equivalent of the following:

   (i) $1.2\overline{34}$;

   (ii) $0.0.12\overline{501}$;

   (iii) $2.2\overline{4}$;

   (iv) $0.\overline{9}$.

6) Prove that if $a+b$ is rational, and $a-b$ is irrational, then both a and b are irrational. Give an example of this.

7) Prove or supply a counterexample to the following assertions:

(i) If a and b are nonzero real numbers such that $a+b$ and ab are rational, then a and b are rational.

(ii) If a, b and $a+b$ are real nonzero numbers such that $a+b$ and $\dfrac{a}{b}$ are rational, then a and b are rational.

(iii) If ab and $\dfrac{a}{b}$ are rational, so are a and b.

(iv) If $a+b$ and ab are rational, so are a^2 and b^2.

8) Given any $\varepsilon > 0$, prove, using the Archimedean property, that there exists a positive integer n such that $\dfrac{1}{n} < \varepsilon$.

9) Prove that if w is the principal n^{th} root of 1, $\displaystyle\sum_{k=0}^{n-1} \omega^k = 0$.

10) If b is an imaginary root of $P(x) = 0$, prove that $x^2 + (2\,\mathrm{Re}\,b)x + |b|^2$ is a factor of $P(x)$.

11) Solve the following equations over the complex numbers:

(i) $z^3 + 1 = 0$;

(ii) $z^4 + 16 = 0$;

(iii) $z^3 + 1 + i = 0$;

(iv) $z^6 + 1 = 0$;

(v) $z^2 + z + 1 = 0$.

12) Express $z^4 + 16$ as a product of two quadratic polynomials with real coefficients.

13) Prove that the set of nth roots of 1 for a fixed n forms a cyclic group of order n under multiplication. Any generator of this group is called a *primitive root of unity*.

14) Prove that the set of all n^{th} roots of unity as n varies over the positive integers is a group under multiplication.

   If α is an nth root of unity and β is an mth root of unity, what root of unity is $\alpha\beta$?

15)   Use   De   Moivre's   Theorem   to   show   that

$$\cos n\theta = \sum_{k=0}^{m}(-1)^k \binom{n}{2k}\cos^{n-2k}\theta\sin^{2k}\theta \text{ where}$$

$$m = \begin{cases} n/2 & \text{if } n \text{ is even;} \\ (n-1)/2 & \text{if } n \text{ is odd.} \end{cases}$$

16) Using the preceding problem, express $\cos 4\theta$ in powers of $\cos\theta$.

17) Let a and b be fixed complex numbers, a not zero.

   (i) If t is a real variable, prove that the parametric equation $z = b + ta$ represents a straight line in the complex plane passing through b and parallel to a;

   (ii) Prove that a and ia represent two directions perpendicular to each other;

   (iii) Prove that a line perpendicular to the direction a can be written in the form

$$\bar{a}z + a\bar{z} = c, c \text{ real.}$$

18) Prove that the equation of a circle center $-a$ can be written in the form $z\bar{z} + \bar{a}z + a\bar{z} + b = 0$ where b is real and $b < |a|^2$.

19) Prove that if z and z' are reflections of each other in the line $\overline{a}z + a\overline{z} = c$, then $\overline{a}z' + a\overline{z} = c$. Using this, find the reflection of the point $(1,1)$ in the line $2x + y = 1$.

20) Formulate a definition for a greatest lower bound of a set S of real numbers and prove that it is unique and exists for all S which are bounded below.

Appendix B

Linear Algebra

B.1. Vector Spaces

It is assumed that the reader is familiar with the basic ideas of linear algebra and this appendix is provided, not as a detailed account of the subject, but rather as a resource for the reader to consult whenever (s)he feels unsure of some of the facts. The pace will be fast and some of the proofs may differ from the proofs encountered by the reader in more elementary courses. However, given that the reader will probably not have to consult this appendix until the later chapters, it is hoped that by that time (s)he will have acquired sufficient mathematical maturity to find the exposition digestible.

B.1.1. *Definition*

A *vector space* over a field consists of:

(i) An additive abelian group $(V, +)$ whose elements are called *vectors*;

(ii) A field F of *scalars;*

(iii) An action of F on V denoted $(a, \alpha) \rightarrow a\alpha$ satisfying:

(a) $(a + b)\alpha = a\alpha + b\alpha$ for all a and b in F and all α in V;

(b) $a(\alpha + \beta) = a\alpha + a\beta$ for all a in F and all α and β in V;

(c) $(ab)\alpha = a(b\alpha)$ for all a and b in F and all α in V;

(d) $1.\alpha = \alpha$ for all α in V.

445

B.1.2. *Example*

$V = F^n = \bigoplus\limits_{i=1}^{n} F_i = \{(b_1,b_2,b_3,...,b_n) \mid b_i \in F_i = F\}$, a direct sum of n copies
of the abelian group $(F,+)$. Multiplication by a scalar is defined by
$b(a_1,a_2,a_3,...,a_n) = (ba_1,ba_2,ba_3,...,ba_n)$. When $F = \mathbf{R}$, the reals, and
$n = 2$, this is the familiar Euclidean plane. ∎

B.1.3. *Definition*

(i) Let (F,V) be a vector space and $\{\alpha_1,\alpha_2,...,\alpha_k\}$ a subset of V. A
linear combination of the α's is an expression of the form $\sum\limits_{i=1}^{k} a_i\alpha_i$.

(ii) If T is a nonempty subset of V, then the *span* of T denoted sp(T)
is the set of all (finite) linear combinations of vectors from T. If
$T = \phi$, set sp(T) = $\{0\}$.

B.1.4. *Example*

If $T = \{(1,0,1),(1,1,0),(0,1,1)\}$ considered as a subset of $\mathbf{R}^3$, then

$$\text{sp}(T) = \{(a+b,b+c,a+c) \mid a,b,c \in \mathbf{R}\}. \blacksquare$$

B.1.5. *Definition*

A subspace W of the vector space (F,V) is a subset of V which is itself a
vector space under the operations inherited from (F,V).

B.1.6. *Example*

A line through the origin in $\mathbf{R}^3$ is a subspace as is a plane through the
origin.

B.1.7. *Theorem*

A subset W of the vector space (F,V) is a subspace iff:
(i) $(W,+)$ is a subgroup of $(V,+)$;
(ii) $s\alpha \in W$ for all $\alpha \in W$ and all $s \in F$.

Proof. Exercise (see 5.4.5). ■

B.1.8. *Theorem*

If T is a subset of the vector space (F,V), $\mathrm{sp}(T)$ is the smallest subspace containing T.

Proof. If $T = \phi$, $\mathrm{sp}(T) = \{0\}$ and clearly $\{0\}$ is the smallest subspace containing ϕ.

Suppose now that $T \neq \phi$. It is clear that the sum and difference of two finite linear combinations of elements of T is again a finite linear combination of elements of T, and so $(\mathrm{sp}(T),+)$ is a subgroup of $(V,+)$. Also, a scalar multiple of a linear combination of elements from T is again a linear combination of elements from T. By B.1.7, $\mathrm{sp}(T)$ is a subspace.

Obviously $T \subseteq \mathrm{sp}(T)$, and if W is a subspace containing T, then, since W is closed under the taking of linear combinations, we have $\mathrm{sp}(T) \subseteq W$, and so $\mathrm{sp}(T)$ is the smallest subspace containing T. ■

B.1.9. *Theorem*

(i) If $T \subseteq S$, $\mathrm{sp}(T) \subseteq \mathrm{sp}(S)$;
(ii) $\mathrm{sp}[\mathrm{sp}(T)] = \mathrm{sp}(T)$.

Proof. (i) Exercise (actually, it's obvious).
(ii) $\mathrm{sp}[\mathrm{sp}(T)]$ is the smallest subspace containing $\mathrm{sp}(T)$. On the other hand, since $\mathrm{sp}(T)$ is itself a subspace, it is the smallest subspace containing $\mathrm{sp}(T)$. Therefore, $\mathrm{sp}[\mathrm{sp}(T)] = \mathrm{sp}(T)$. ■

B.1.10. *Definition*

A nonempty subset B of a vector space is *linearly independent* if, for all $\beta \in B$, $\text{sp}(B \setminus \beta) \subset \text{sp}(B)$.

Note. (i) This says that there are no redundant vectors in B.

(ii) The negation of the condition for linear independence is: there exists $\beta \in B$ such that $\text{sp}(B \setminus \beta) = \text{sp}(B)$. In this case we say that B is linearly dependent.

B.1.11. *Theorem*

A nonempty subset of a vector space is linearly independent if, and only if, for all finite subsets $\{\beta_1, \beta_2, ..., \beta_k\}$ of B, $\sum_{i=1}^{k} a_i \beta_i = 0$ implies $a_i = 0$ for all i, $i = 1, 2, ..., k$.

Proof. Suppose B is linearly independent and assume that for some finite subset $\{\beta_1, \beta_2, ..., \beta_k\}$ of B we have $\sum_{i=1}^{k} b_i \beta_i = 0$ with $b_j \neq 0$ for some j. Then

$$\beta_j = -\sum_{i \neq j} b_j^{-1} b_i \beta_i.$$

Hence $\beta_j \in \text{sp}(B \setminus \beta_j)$ and so

$$B = (B \setminus \beta_j) \cup \beta_j \subseteq \text{sp}(B \setminus \beta_j).$$

Therefore

$$\text{sp}(B) \subseteq \text{sp}[\text{sp}(B \setminus \beta_j)] = \text{sp}(B \setminus \beta_j)$$

by Theorem B.1.9, proving that $\text{sp}(B) = \text{sp}(B \setminus \beta_j)$, a contradiction.

Conversely, suppose that the condition holds and assume that B is linearly dependent. Then there exists $\beta \in B$ such that $\text{sp}(B \setminus \beta) = \text{sp}(B)$. Then $\beta = \sum b_i \beta_i$ where $\beta_i \neq \beta$ for all i. Hence $\beta - \sum b_i \beta_i = 0$ but not all scalars are zero, a contradiction. ∎

B.1.12. *Definition*

A nonempty subset B of the vector space (F,V) is a *basis* of V if:
  (i) $\mathrm{sp}(B)=V$;
  (ii) B is linearly independent.

B.1.13. *Lemma (The Exchange Property)*

If $\alpha \notin \mathrm{sp}(B)$ but $\alpha \in \mathrm{sp}(B \cup \beta)$, then $\beta \in \mathrm{sp}(B \cup \alpha)$.

Proof. Since $\alpha \notin \mathrm{sp}(B)$ and $\alpha \in \mathrm{sp}(B \cup \beta)$, then $\alpha = b\beta + \sum b_i \beta_i$ where $\beta_i \in B$ for all i and $b \neq 0$. Solving for β we get

$$\beta = b^{-1}\alpha - \sum b^{-1}b_i \beta_i \in \mathrm{sp}(B \cup \alpha). \blacksquare$$

B.1.14. *Theorem*

If B is a finite subset of (F,V) such that $\mathrm{sp}(B)=V$, then B contains a subset which is a basis of V.

Proof. Let T be the smallest subset of B such that $\mathrm{sp}(T)=V$. If T is linearly dependent, then there exists $\beta \in T$ such that $\mathrm{sp}(T \setminus \beta) = \mathrm{sp}(T) = V$, contradicting the minimality of T. Therefore T is a basis. $\blacksquare$

B.1.15. *Theorem*

If B is linearly independent and $\alpha \notin \mathrm{sp}(B)$, then $B \cup \alpha$ is linearly independent.

Proof. Suppose $B \cup \alpha$ is linearly dependent. Then there exists $\beta \in B \cup \alpha$ such that

$$\mathrm{sp}((B \cup \alpha) \setminus \beta) = \mathrm{sp}(B \cup \alpha).$$

Clearly $\beta \neq \alpha$ since $\alpha \notin \mathrm{sp}(B)$, and so $\beta \in B$.

Since B is linearly independent $\beta \notin \mathrm{sp}(B \setminus \beta)$. However

$$\beta \in \mathrm{sp}((B \setminus \beta) \cup \alpha)$$

and so, by the Exchange Property, $\alpha \in \mathrm{sp}((B \setminus \beta) \cup \beta) = \mathrm{sp}(B)$, a contradiction. ∎

B.1.16. *Theorem*

If $B = \{\beta_1, \beta_2, ..., \beta_n\}$ is a basis of V and $A = \{\alpha_1, \alpha_2, ...\}$ another basis, then $|A| = n$.

Proof. Since A is linearly independent,

$$\mathrm{sp}(A \setminus \alpha_1) \subset \mathrm{sp}(B).$$

Hence there exists $\beta_1 \in B$ such that $\beta_1 \notin \mathrm{sp}(A \setminus \alpha_1)$. But

$$\beta_1 \in \mathrm{sp}((A \setminus \alpha_1) \cup \alpha_1).$$

By B.1.13, $\alpha_1 \in \mathrm{sp}((A \setminus \alpha_1) \cup \beta_1)$ and by the previous result, $(A \setminus \alpha_1) \cup \beta_1$ is a basis of V.

If $A_1 = A \setminus \alpha_1 \neq \phi$, pick $\alpha_2 \in A \setminus \alpha_1$. As before,

$$\mathrm{sp}(A_1 \setminus \alpha_2) \subset \mathrm{sp}(B)$$

and so $\beta_2 \notin \mathrm{sp}(A_1 \setminus \alpha_2)$. But $\beta_2 \in \mathrm{sp}((A_1 \setminus \alpha_2) \cup \alpha_2)$ and so, by B.1.13,

$$\alpha_2 \in \mathrm{sp}((A_1 \setminus \alpha_2) \cup \beta_2)$$

and by B.1.15, $A_2 = (A_1 \setminus \alpha_2) \cup \beta_2$. is a basis.

We can continue this procedure as long as $A \setminus \{\alpha_1, \alpha_2, ..., \alpha_k\} \neq \phi$. Since $\mathrm{sp}(B) = V$, it follows that $A \setminus \{\alpha_1, \alpha_2, ..., \alpha_k\} = \phi$ for some $k \leq n$. But then $\mathrm{sp}\{\beta_1, \beta_2, ..., \beta_k\} = V$ and since $\{\beta_1, \beta_2, ..., \beta_n\}$ is linearly independent it follows that $k = n$. Therefore $A = \{\alpha_1, \alpha_2, ..., \alpha_n\}$. ∎

B.1.17. *Theorem*

Suppose $B = \{\beta_1, \beta_2, ..., \beta_n\}$ is a basis of V and $A \neq \phi$ is a linearly independent set. Then A can be extended to a basis of V and so, in particular, $|A| \leq n$.

Proof. If $B \subseteq \text{sp}(A)$, then A is a basis of V and so, by B.1.16, $|A|=n$.
Suppose therefore that $B \not\subseteq \text{sp}(A)$. Then there exists a β, say β_1, (without loss of generality) such that $\beta_1 \notin \text{sp}(A)$. By B.1.15, $A \cup \beta_1$ is linearly independent.

If $B \subseteq \text{sp}(A \cup \beta_1)$, then $A \cup \beta_1$ is a basis and by B.1.16, $|A|+1=n$. If $B \not\subseteq \text{sp}(A \cup \beta_1)$, there exists a β, say $\beta_2 \notin \text{sp}(A \cup \beta_1)$. By B.1.15, $A \cup \beta_1 \cup \beta_2$ is linearly independent.

If $B \subseteq \text{sp}(A \cup \beta_1 \cup \beta_2)$, then $A \cup \beta_1 \cup \beta_2$ is a basis of V and by B.1.16, $|A|+2=n$. Continue in this fashion.

At the $(n-1)^{\text{st}}$ stage (if we need to go that far) we have a basis

$$A \cup \beta_1 \cup \beta_2 \cup ... \cup \beta_{n-1}$$

and so, by B.1.16, $|A|=1$ and A has been extended to a basis of V. ∎

B.1.18. *Definition*

The number of vectors in a basis of V is the *dimension of* V, denoted $\dim V$.

B.1.19. *Theorem*

If $\dim V = n$ and A is a subset of V with n vectors, then A is a basis of V iff either A is linearly independent or A spans V.

Proof. Exercise. ∎

B.1.20. *Theorem*

If $B = \{\beta_1, \beta_2, ..., \beta_n\}$ is a basis of V, then every element $\alpha \in V$ is uniquely expressible as a linear combination of $\beta_1, \beta_2, ..., \beta_n$.

Proof. Since $\text{sp}\{\beta_1, \beta_2, ..., \beta_n\} = V$, each element of V is a linear combination of the β's. Suppose

$$\alpha = \sum_{i=1}^{n} a_i \beta_i = \sum_{i=1}^{n} c_i \beta_i.$$

Then

$$\sum_{i=1}^{n} (a_i - c_i)\beta_i = 0$$

But B is a linearly independent set of vectors and so $a_i - c_i = 0$ for all i, $i = 1, 2, ..., n$. Therefore $a_i = c_i$ for all i, $i = 1, 2, ..., n$. ∎

B.2. Linear Transformations

B.2.1. *Definition*

If (F,V) and (F,W) are vector spaces, a map $T : V \to W$ is a *linear transformation* if

$$T(r\alpha + s\beta) = rT(\alpha) + sT(\beta)$$

for all r and s in F and all α and β in V. If T is bijective we call T an isomorphism, in which case we say that *V is isomorphic to W*.

If $V = W$, then T is a *linear operator on V*.

Remark. If A is an $m \times n$ matrix, then the map $T_A : F^n \to F^m$ defined by $T_A(\xi) = \eta$ where $\eta^T = A\xi^T$ is a linear transformation.

Most elementary texts on linear algebra emphasize matrices and, as an afterthought, introduce the concept of a linear transformation. We have chosen to emphasize linear transformations because they have a more geometric flavour which often leads to a more intuitive grasp of the subject. However, both matrices and linear transformations have their uses, the former when we need to compute and the latter when we want to develop the theoretical aspects of the subject.

The following theorems are well known and their method of proof follows almost verbatim the corresponding proofs for groups and homomorphisms in Chapter 6.

B.2.2. Theorem

Let $T : V \to W$ be a linear transformation. Then:

(i) $T(0_V) = 0_W$;

(ii) if S is a subspace of V, then $T(S)$ is a subspace of W;

(iii) if U is a subspace of W, then $T^{-1}(U)$ is a subspace of V.

Proof. Exercise. ■

B.2.3. Definition

The null space of a linear transformation T from V to W is the subspace $N(T) = T^{-1}(0_W)$. (Compare this with the definition of the kernel of a homomorphism in Chapter 6.)

As with groups, we have:

B.2.4. Theorem

T is injective iff $N(T) = \{0_V\}$.

Proof. Exercise. ■

Since the additive structure of an F-vector space is that of an abelian group, any subspace is a normal subgroup of $(V, +)$. Therefore, for a given subspace W, we can form the additive group V / W. We can endow this with the structure of a vector space by defining

$$s.\bar{\alpha} = s.(\alpha + W) = \overline{s\alpha}.$$

B.2.5. Theorem

$(F, V / W)$ is a vector space.

Proof. We prove that the action of F on V/W is well defined and leave the reader to verify that the resulting algebraic system is a vector space. As usual, $\bar{\alpha}$ denotes the coset $\alpha + W$.

Suppose $\bar{\alpha} = \bar{\beta}$. We need to show that for all $s \in F$, $\overline{s\alpha} = \overline{s\beta}$. From $\bar{\alpha} = \bar{\beta}$ we get $\alpha - \beta \in W$. Since W is a subspace, $s(\alpha - \beta) \in W$ and so $s\alpha - s\beta \in W$, proving that $\overline{s\alpha} = \overline{s\beta}$. ■

B.2.6. *Theorem*

If W is a subspace of dimension m of a vector space V of dimension n, then V/W is a vector space of dimension $n - m$.

Proof. Let $\alpha_1, \alpha_2, ..., \alpha_m$ be a basis of W and extend it to a basis $\alpha_1, \alpha_2, ...,$ $\alpha_m, \alpha_{m+1}, ..., \alpha_n$ of V. We show that $\overline{\alpha_{m+1}}, \overline{\alpha_{m+2}}, ..., \overline{\alpha_n}$ is a basis of V/W.

Suppose

$$c_{m+1}\overline{\alpha_{m+1}} + c_{m+2}\overline{\alpha_{m+2}} + ... + c_n\overline{\alpha_n} = \bar{0}.$$

Then

$$c_{m+1}\alpha_{m+1} + c_{m+2}\alpha_{m+2} + ... + c_n\alpha_n \in W$$

and so

$$c_{m+1}\alpha_{m+1} + c_{m+2}\alpha_{m+2} + ... + c_n\alpha_n = c_1\alpha_1 + c_2\alpha_2 + ... + c_m\alpha_m.$$

Therefore

$c_1\alpha_1 + c_2\alpha_2 + ... + c_m\alpha_m - c_{m+1}\alpha_{m+1} - c_{m+2}\alpha_{m+2} - ... - c_n\alpha_n = 0$ and since $\alpha_1, \alpha_2, ..., \alpha_m, \alpha_{m+1}, ..., \alpha_n$ is a basis of $V, c_i = 0$ for all $i, i = 1, 2, ..., n$ In particular, $c_{m+1} = c_{m+2} = ... = c_n = 0$.

Moreover, if $\bar{\beta} \in V/W$, then

$$\beta = b_1\alpha_1 + b_2\alpha_2 + ... + b_m\alpha_m + b_{m+1}\alpha_{m+1} + ... + b_n\alpha_n$$

and

$$\bar{\beta} = b_1\bar{\alpha}_1 + b_2\bar{\alpha}_2 + ... + b_m\bar{\alpha}_m + b_{m+1}\bar{\alpha}_{m+1} + ... + b_n\bar{\alpha}_n.$$

But $\bar{\alpha}_i = \bar{0}$ for all $i, i = 1, 2, ..., m$ and so $\bar{\beta} = b_{m+1}\bar{\alpha}_{m+1} + ... + b_n\bar{\alpha}_n$. ■

Just as with groups, we have the various isomorphism theorems. The reader should bear in mind that everything is in additive notation and the maps are linear transformations rather than group homomorphisms. Other than that, the theorems are proved in exactly the same as they were for groups.

B.2.7. *Theorem (First Isomorphism Theorem)*

If $T:V \to W$ is a linear transformation then

$$V/N(T) \cong T(V).$$

Proof. Since T is a homomorphism from the additive abelian group $(V,+)$ to the additive abelian group $(W,+)$, by the First Isomorphism Theorem for groups, $V/N(T)$ and $T(V)$ are isomorphic as groups by means of the map

$$\hat{T}:V/N(T) \to T(V)$$

defined by

$$\hat{T}(\beta + N(T)) = T(\beta).$$

But since T is a linear transformation, $T(s\beta) = sT(\beta)$ and so

$$\hat{T}[s(\beta + N(T))] = \hat{T}(s\beta + N(T)) = T(s\beta) = sT(\beta) = s\hat{T}(\beta + N(T)). \blacksquare$$

B.2.8. *Definition*

The *rank* of a linear transformation T, denoted $\rho(T)$, is the dimension of ImT and the *nullity*, denoted $\nu(T)$, is the dimension of the null space.

B.2.9. *Corollary*

If V is a vector space of dimension n and $T:V \to W$ is a linear transformation, then

$$\rho(T) + \nu(T) = n.$$

Proof. By the First Isomorphism Theorem,

$$\frac{V}{N(T)} \cong T(V)$$

and by B.2.6,

$$\dim T(V) = \dim V - \dim N(T).$$

Therefore

$$\dim T(V) + \dim N(T) = \dim V$$

and so $\rho(T) + \nu(T) = n.$ ∎

B.2.10. *Corollary*

Let $T : V \to W$ be a linear transformation and suppose $\dim V = \dim W = n$. Then T is surjective iff T is injective.

Proof. Since $\rho(T) + \nu(T) = n$, it follows that T is surjective iff $\rho(T) = n$ iff $\nu(T) = 0$ iff T is injective. ∎

B.2.11. *Definition*

A linear transformation is said to be *non-singular* if it is injective. If it is both injective and surjective, the it is said to be *invertible*.

Remark. If a linear transformation is bijective, then the inverse exists as a map. However, just as with groups, it is easy to show that the map is actually linear.

B.2.12. *Theorem*

If $\alpha_1, \alpha_2, ..., \alpha_n$ is a basis of V and $\beta_1, \beta_2, ..., \beta_n$ are any n vectors in W, then there is a unique linear transformation $T : V \to W$ such that $T(\alpha_j) = \beta_j$ for all j, $j = 1, 2, ..., n$.

Proof. Since each vector in V is uniquely a linear combination of the basis vectors, we can extend T linearly as follows: if

$$\xi \in V \text{ and } \xi = \sum_{i=1}^{n} x_i \alpha_i, \text{ define } T \text{ by } T(\xi) = \sum_{i=1}^{n} x_i \beta_i.$$

It is an easy matter to show that T is a linear transformation.

The uniqueness follows from the fact that every vector in V is a linear combination of the α's. ■

Note. The theorem just proved shows that any two linear transformations which agree on a basis must be equal.

We now show how to associate a matrix with a linear transformation T from V to W. We require the following ingredients:

(i) an ordered basis $(\alpha_1, \alpha_2, ..., \alpha_n)$ of V;

(ii) an ordered basis $(\beta_1, \beta_2, ..., \beta_m)$ of W;

(iii) the images $T(\alpha_i)$ of the chosen basis of V expressed as a linear combinations of the chosen basis of W.

B.2.13. *Definitions and notation*

(i) Let $A = (\alpha_1, \alpha_2, ..., \alpha_n)$ and $B = (\beta_1, \beta_2, ..., \beta_m)$ be ordered bases of V and W respectively and let $T : V \to W$ be a linear transformation. Suppose

$$T(\alpha_j) = \sum_{i=1}^{m} a_{ij} \beta_i.$$

Then the matrix of T relative to the bases A and B is the $m \times n$ matrix $[a_{ij}]$ whose entry in the ith row j^{th} column is a_{ij}. We denote this matrix by $[T]_A^B$.

(ii) If $\xi \in V$ and $\xi = \sum_{i=1}^{n} x_i \alpha_i$ we denote the n-tuple of coefficients $(x_1, x_2, ..., x_n)$ by $[\xi]_A$ and call it *the (row) coordinate vector of ξ relative to the A basis*.

(iii) The *transpose* of a given matrix M is the matrix M^T, obtained by interchanging rows and columns. Thus the *column coordinate vector of ξ relative to the A basis is $([\xi]_A)^T$*.

B.2.14. *Theorem*

If V is an n-dimensional F-vector space, then V is isomorphic to F^n, the vector space of n-tuples of elements of F.

Proof. Let $A = (\alpha_1, \alpha_2, ..., \alpha_n)$ be an ordered basis of V and define

$$\varphi : V \to F^n \text{ by } \varphi(\xi) = [\xi]_A.$$

We show φ is an isomorphism.

(i) Let $\xi = \sum_{i=1}^{n} x_i \alpha_i$ and $\eta = \sum y_i \alpha_i$. Then

$$r\xi + s\eta = r\sum_{i=1}^{n} x_i \alpha_i + s\sum_{i=1}^{n} y_i \alpha_i = \sum_{i=1}^{n} (rx_i + sy_i)\alpha_i.$$

Therefore

$$[r\xi + s\eta]_A = (rx_1 + sy_1, rx_2 + sy_2, ..., rx_n + sy_n) = r[\xi]_A + s[\eta]_A$$

and so

$$\varphi(r\xi + s\eta) = r\varphi(\xi) + s\varphi(\eta).$$

(ii) If $\xi = \sum_{i=1}^{n} x_i \alpha_i$ and $\varphi(\xi) = 0$, then

$$[\xi]_A = (x_1, x_2, ..., x_n) = (0, 0, ..., 0)$$

and so $x_i = 0$ for all $i, i = 1, 2, ..., n$. Therefore $\xi = 0$ and φ is injective.

(iii) Given $(x_1, x_2, ..., x_n) \in F^n$, $\varphi(\sum_{i=1}^{n} x_i \alpha_i) = (x_1, x_2, ..., x_n)$ and so φ is

onto. ∎

B.2.15. *Theorem*

If $T : V \to W$ is a linear transformation and $(\alpha_1, \alpha_2, ..., \alpha_n)$, $(\beta_1, \beta_2, ..., \beta_m)$, are ordered bases of V and W, respectively, then

$$[T]_A^B ([\xi]_A)^T = [T(\xi)]_B^T.$$

Proof. The expression above is somewhat indigestible, so here is what it says: to find the (column) coordinate vector of $T(\xi)$ relative to the B basis, multiply the (column) vector of ξ relative to the A basis by the matrix of T relative to the two bases $(\alpha_1, \alpha_2, ..., \alpha_n)$ and $(\beta_1, \beta_2, ..., \beta_m)$.

Letting $[T]_A^B = [a_{ij}]$, we compute: If

$$\xi = \sum_{j=1}^n x_j \alpha_j,$$

then

$$T(\xi) = \sum_{j=1}^n x_j T(\alpha_j) = \sum_{j=1}^n x_j \sum_{i=1}^m a_{ij}\beta_j = \sum_{i=1}^m (\sum_{j=1}^n a_{ij}x_j)\beta_i.$$

But

$$[a_{ij}].(x_1,x_2,...,x_n)^T = (\sum_{j=1}^n a_{1j}x_j, \sum_{j=1}^n a_{2j}x_j,..., \sum_{j=1}^n a_{mj}x_j),^T$$

thus proving the result. ∎

B.2.16. *Theorem (Change of Bases)*

If A and B are ordered bases as above and $C = (\gamma_1,\gamma_2,...,\gamma_n)$ and $D = (\delta_1,\delta_2,...,\delta_m)$ are another pair of ordered bases of V and W, respectively, then:

$$[T]_A^B[P]_A = [Q]_B[T]_C^D$$

where P is the linear transformation from V to W which takes the "old" basis $(\alpha_1,\alpha_2,...,\alpha_n)$ to the "new" basis $(\gamma_1,\gamma_2,...,\gamma_n)$; that is, $P(\alpha_i) = \gamma_i$ for all i, $i = 1,2,...,n$ and $[P]_A$ is the matrix of this linear transformation relative to the A basis, considered as a basis of both the domain and codomain. To conform with the notation used when the bases for the domain and codomain are different, we could write $[P]_A^A$. Similar remarks apply to Q and $[Q]_B$.

Proof. Let $[T]_A^B = [a_{ij}]$, $[T]_C^D = [c_{ij}]$, $[P]_A = [p_{ij}]$ and $[Q]_B = [q_{ij}]$. Then

$$T(\gamma_j) = T(\sum_{i=1}^n p_{ij}\alpha_i) = \sum_{i=1}^n p_{ij}T(\alpha_i)$$

$$= \sum_{i=1}^n p_{ij}\sum_{k=1}^m a_{ki}\beta_k = \sum_{k=1}^m (\sum_{i=1}^n a_{ki}p_{ij})\beta_k.$$

On the other hand

$$T(\gamma_j) = \sum_{i=1}^{m} c_{ij}\delta_i = \sum_{i=1}^{m} c_{ij} \sum_{k=1}^{m} q_{ki}\beta_k = \sum_{k=1}^{m} (\sum_{i=1}^{m} q_{ki}c_{ij})\beta_k.$$

It follows that

$$\sum_{i=1}^{n} a_{ki}p_{ij} = \sum_{i=1}^{m} q_{ki}c_{ij} \text{ for all } k, \ k=1,2,...,m \text{ and for all } j, \ j=1,2,...,n. \text{ This}$$

says precisely that

$$[T]_A^B[P]_A = [Q]_B[T]_C^D. \ \blacksquare$$

B.2.17. *Corollary*

Let T be a linear operator on V and $A = (\alpha_1, \alpha_2, ..., \alpha_n)$ and $B = (\beta_1, \beta_2, ..., \beta_n)$ bases of V. Then

$$[T]_A[P]_A = [P]_A[T]_B$$

where $P : V \to V$ is the linear transformation taking the "old" basis to the "new" basis, i.e., $P(\alpha_i) = \beta_i$, $i = 1,2,...,n$. Thus $[T]_B = [P]_A^{-1}[T]_A[P]_A$.

Proof. In the previous theorem let $B = A$, $C = B$ and $D = B$. $\blacksquare$

B.2.18. *Theorem*

Let V, W and X be vector spaces with ordered bases $A = (\alpha_1, \alpha_2, ..., \alpha_n)$, $B = (\beta_1, \beta_2, ..., \beta_m)$ and $C = (\gamma_1, \gamma_2, ..., \gamma_p)$, respectively and suppose $T : V \to W$ and $S : W \to X$ are linear transformations. Then

$$[S]_B^C[T]_A^B = [ST]_A^C.$$

Proof. Let $[T]_A^B = [a_{ij}]$ and $[S]_B^C = [b_{ij}]$. Then

$$T(\alpha_j) = \sum_{i=1}^{m} a_{ij}\beta_i$$

and

$$S(T(\alpha_j)) = \sum_{i=1}^{m} a_{ij}S(\beta_i) = \sum_{-=1}^{m}\sum_{k=1}^{p} b_{ki}\gamma_k = \sum_{k=1}^{p} (\sum_{i=1}^{m} b_{ki}a_{ij})\gamma_k.$$

But $\displaystyle\sum_{i=1}^{m} b_{ki} a_{ij}$ is the (k, j) entry of $[S]_B^C [T]_A^B$. Therefore

$$[S]_B^C [T]_A^B = [ST]_A^C. \quad \blacksquare$$

B.2.19. *Corollary*

Let V and W be F-vector spaces of the same dimension n and $T : V \to W$ a nonsingular linear transformation. If A and B are bases of V, then

$$[T]_A^B [T^{-1}]_B^A = I_n.$$

Therefore the matrix of a nonsingular transformation from a vector space to another of the same dimension is invertible and $[T^{-1}]_B^A = ([T]_A^B)^{-1}$.

Proof. Apply the previous theorem. $\blacksquare$

The last theorem of this section shows how we can endow the set of all linear transformation from (F, V) to (F, V) with the structure of a ring. We show further that this ring is isomorphic to the ring of matrices under the usual matrix operations.

B.2.20. *Theorem*

Let $\Lambda(V, V)$ denote the set of linear operators on V where $\dim V = n$ and define on $\Lambda(V, V)$ two operations, $+$ and $\cdot$ by:

(i) $(S + T)(\alpha) = S(\alpha) + T(\alpha)$ for all S and T in $\Lambda(V, V)$ and all α in V;

(ii) $(ST)(\alpha) = S(T(\alpha))$ for all S and T in $\Lambda(V, V)$ and all α in V. Then $(\Lambda(V, V), +, .)$ is a ring with unity. Moreover, this ring is isomorphic to the ring of $n \times n$ matrices over F.

Proof. It is an easy matter to prove that $(\Lambda(V, V), +, .)$ is a ring and we leave most of the details to the reader.

To give a flavour of the proof, we show that multiplication is distributive over addition.

Let P, S and T be elements of $\Lambda(V, V)$ and consider $P(S+T)$. By definition,

$$[P(S+T)](\alpha) = P((S+T)(\alpha)) = P(S(\alpha) + T(\alpha)) = P(S(\alpha)) + P(T(\alpha)).$$

The last equality is valid since P is a linear transformation. Again by definition,

$$P(S(\alpha)) + P(T(\alpha)) = (PS)(\alpha) + (PT)(\alpha) = (PS + PT)(\alpha)$$

for all α in V. Therefore $P(S+T) = PS + PT$.

To prove that $\Lambda(V, V)$ and $M^{n \times n}$ (the ring of $n \times n$ matrices over F) are isomorphic as rings, chose a basis $A = (\alpha_1, \alpha_2, ..., \alpha_n)$ of V and define a map

$$\Pi : \Lambda(V, V) \to M^{n \times n}$$

by

$$\Pi(T) = [T]_A.$$

We check a few of the properties of an isomorphism and leave the others to the reader.

(1) $\Pi(I_V) = [I_V]_A = I_n$ where I_V and I_n are the identities of the respective rings since $I_V(\alpha_i) = \alpha_i$.

(2) $\Pi(ST) = [ST]_A = [S]_A[T]_A = \Pi(S)\Pi(T)$ by Theorem B.2.17.

Note that when T is invertible, so is $[T]_A$ by the Corollary above. Therefore units of $\Lambda(V, V)$ map to units of $M^{n \times n}$. ∎

B.3. Inner Product Spaces

So far metric notions (distance and angle) have not entered our discussions. We now specialize to vector spaces over the reals **R** and define an *inner product* which will allow us to introduce the concepts of length, distance and angle.

B.3.1. *Definition*

An inner product on a real vector space V is a function $< , >: V \times V \to \mathbf{R}$ satisfying:

(i) $<\alpha, \alpha> > 0$ for all $\alpha \neq 0$;

(ii) $<\alpha, \beta> = <\beta, \alpha>$ for all α and β in V;

(iii) $<\alpha+\beta,\gamma>=<\alpha,\gamma>+<\beta,\gamma>$ for all α,β and γ in V;

(iv) $<r\alpha,\beta>=r<\alpha,\beta>$ for all α and β in V and for all real numbers r.

Remarks. Property (i) is known as *positive definiteness*, property (ii) is *symmetry* and properties (ii), (iii) and (iv) together imply that the inner product is linear in the two variables. Such a function is said to be *bilinear*.

A vector space endowed with an inner product is called an *inner product space*. We write $(V,<,>)$ to convey that $<,>$ is an inner product on V.

B.3.2. *Example*

On the vector space of $\mathbf{R}^n$, of n-tuples of real numbers, we have the usual inner product which is known as the *dot* product, defined by

$$(x_1,x_2,...,x_n)\bullet(y_1,y_2,...,y_n)=\sum_{i=1}^{n}x_iy_i.$$

B.3.3. *Definition*

Let $(V,<,>)$ be an inner product space. The *length* of a vector α is defined to be

$$\sqrt{<\alpha,\alpha>}=\|\alpha\|$$

and the distance between α and β

$$\|\alpha-\beta\|.$$

To give a definition of angle, we first need to prove a famous inequality.

B.3.4. *Theorem (The Cauchy-Schwartz inequality)*

If $(V,<,>)$ is an inner product space, then

$$|<\alpha,\beta>|\leq\|\alpha\|.\|\beta\|.$$

Proof. Consider

$$<x\alpha-\beta,x\alpha-\beta>=\|\alpha\|^2 x^2-2x<\alpha,\beta>+\|\beta\|^2$$

where x is a real variable. Since the inner product is positive definite, the quadratic equation either has no roots or a repeated root. By the familiar formula for the roots of a quadratic,

$$x = \frac{<\alpha,\beta> \pm \sqrt{<\alpha,\beta>^2 - \|\alpha\|^2\|\beta\|^2}}{\|\alpha\|^2}$$

and so

$$<\alpha,\beta>^2 - \|\alpha\|^2\|\beta\|^2 \le 0.$$

Therefore

$$<\alpha,\beta>^2 \le \|\alpha\|^2\|\beta\|^2$$

and taking square roots, we get the result. ∎

Writing the inequality as $-1 \le \dfrac{<\alpha,\beta>}{\|\alpha\|.\|\beta\|} \le 1$, we see that there is a unique angle θ, $0 \le \theta \le \pi$ such that $\cos\theta = \dfrac{<\alpha,\beta>}{\|\alpha\|.\|\beta\|}$ and we call this the angle between α and β. Beyond 3-dimensional space we are unable to visualize the angle, but, as we shall see in a moment, in 2- and 3-dimensional space with the dot product as the inner product, θ is indeed the angle of elementary geometry between the vectors.

B.3.5. *Theorem*

If $\mathbf{R}^2$ or $\mathbf{R}^3$ is endowed with the usual dot product, then

$$\frac{\alpha.\beta}{\|\alpha\|.\|\beta\|} = \cos\theta$$

where θ is the actual geometric angle between α and β.

Proof. Consider the diagram below which can be thought of as being in $\mathbf{R}^2$ or $\mathbf{R}^3$.

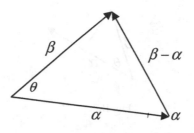

Fig. B.1

By a well known formula from high school days,

$$\| \beta - \alpha \|^2 = \| \alpha \|^2 + \| \beta \|^2 - 2 \| \alpha \| . \| \beta \| \cos \theta.$$

But

$$\| \beta - \alpha \|^2 = (\beta - \alpha).(\beta - \alpha) = \| \alpha \|^2 + \| \beta \|^2 - 2\alpha.\beta.$$

Equating the right-hand side of these two equations and cancelling $\| \alpha \|^2$ and $\| \beta \|^2$ from both sides, we obtain the result. ∎

B.3.6. *Definition*

If $< \alpha, \beta > = 0$, we say that α and β are *orthogonal*. ("Orthogonal" is just a fancy word for "perpendicular").

In the 2- and 3-dimensional cases with the dot product as inner product, $\alpha.\beta = 0$, means that if the vectors are drawn, they will be at right angles to each other since $\cos \theta = 0$ and so $\theta = \pi/2$.

Suppose that $(V, <, >)$ is an inner product space and suppose we are given two nonzero vectors α and β. We can express β as the sum of two vectors, one along α, the other perpendicular to α. The vector along α is the *orthogonal projection of β onto α.* This is precisely what we do in physics when we resolve forces or velocities. The procedure for doing this in the case of two vectors leads to the general situation with any number of linearly independent vectors. We tackle the case of two vectors first.

Consider the diagram below.

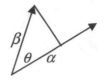

Fig. B.2

The orthogonal projection of β onto α is

$$\left(\frac{\|\beta\|\cos\theta}{\|\alpha\|}\right)\alpha = \left(\frac{\|\beta\|<\alpha,\beta>}{\|\alpha\|^2\|\beta\|}\right)\alpha = \left(\frac{<\alpha,\beta>}{<\alpha,\alpha>}\right)\alpha.$$

Set

$$\gamma = \beta - \left(\frac{<\alpha,\beta>}{<\alpha,\alpha>}\right)\alpha.$$

Then

$$\beta = \gamma + \left(\frac{<\alpha,\beta>}{<\alpha,\alpha>}\right)\alpha$$

and $<\alpha,\gamma> = 0$. Note that $\mathrm{sp}\{\alpha,\beta\} = \mathrm{sp}\{\beta,\gamma\}$.

We apply this procedure to the general case.

B.3.7. *Theorem*

Let $\alpha_1,\alpha_2,...,\alpha_k$ be a set of linearly independent vectors. Then we can find vectors $\beta_1,\beta_2,...,\beta_k$ with $\beta_1 = \alpha_1$ such that $<\beta_i,\beta_j> = 0$ if $i \neq j$. Moreover, $\mathrm{sp}\{\alpha_1,\alpha_2,...,\alpha_s\} = \mathrm{sp}\{\beta_1,\beta_2...,\beta_s\}$ for all $s, 1 \leq s \leq k$.

Proof. We have shown above the case where $k = 2$. We shall prove the general case by induction on s.

Assume therefore that the theorem is true for some $2 \leq s < k$. Then $\{\beta_1,\beta_2,...,\beta_s\}$ is a set of mutually orthogonal vectors such that

$$\mathrm{sp}\{\beta_1,\beta_2...,\beta_s\} = \mathrm{sp}\{\alpha_1,\alpha_2,...,\alpha_s\}.$$

Note that since $\{\alpha_1, \alpha_2, ..., \alpha_s\}$ is a linearly independent set, so is $\{\beta_1, \beta_2, ..., \beta_s\}$. Therefore $<\beta_i, \beta_i> \neq 0$ for all i, $i = 1, 2, ..., s$. Let

$$\beta_{s+1} = \alpha_{s+1} - \left(\frac{<\alpha_{s+1}, \beta_1>}{<\beta_1, \beta_1>}\right)\beta_1 - \left(\frac{<\alpha_{s+1}, \beta_2>}{<\beta_2, \beta_2>}\right)\beta_2 - ... - \left(\frac{<\alpha_{s+1}, \beta_s>}{<\beta_s, \beta_s>}\right)\beta_s.$$

If $\beta_{s+1} = 0$, then $\alpha_{s+1} \in \text{sp}\{\beta_1, \beta_2, ..., \beta_s\} = \text{sp}\{\alpha_1, \alpha_2, ..., \alpha_s\}$ contradicting the linear independence of $\{\alpha_1, \alpha_2, ..., \alpha_k\}$. Solving for α_{s+1} we see that $\alpha_{s+1} \in \text{sp}\{\beta_1, \beta_2, ... \beta_{s+1}\}$ and so

$$\text{sp}\{\alpha_1, \alpha_2, ..., \alpha_{s+1}\} = \text{sp}\{\beta_1, \beta_2, ..., \beta_{s+1}\}.$$

Moreover, $<\beta_i, \beta_j> = 0$ for all i and j, $i \neq j$ whenever $1 \leq i, j \leq s$ by the induction hypothesis and $<\beta_i, \beta_{s+1}> = 0$ for all $i, 1 \leq i \leq s$ since

$$<\beta_i, \beta_{s+1}> = <\beta_i, \alpha_{s+1}> - <\beta_i, \left(\frac{<\alpha_{s+1}, \beta_i>}{<\beta_i, \beta_i>}\right)\beta_i>$$

$$= <\beta_i, \alpha_{s+1}> - \left(\frac{<\alpha_{s+1}, \beta_i>}{<\beta_i, \beta_i>}\right)<\beta_i, \beta_i> = <\beta_i, \alpha_{s+1}> - <\alpha_{s+1}, \beta_i> = 0.$$

The result follows by induction. ∎

Remark. (i) The procedure which orthogonalizes a set of vectors is called the Gramm-Schmidt Orthogonalization Process.

(ii) If for a set of vectors it is impossible to complete the orthogonalization process, then the set of vectors is not independent. This therefore affords us a method for determining whether a set of vectors is independent (in general not the easiest method).

B.3.8. *Definition*

If each vector of a set of vectors is of unit length and every pair of distinct vectors in the set is orthogonal, the set is said to be *orthonormal*.

B.3.9. *Example*

In $\mathbf{R}^n$ with the usual dot product, let $\varepsilon_i = (0, 0, ..., 1, 0, 0..., 0)$ where all coordinate positions are zero except the ith where there is a 1. Then

$\{\varepsilon_1, \varepsilon_2, ..., \varepsilon_n\}$ forms an orthonormal basis of $\mathbf{R}^n$. We call this the *standard basis of* $\mathbf{R}^n$.

B.3.10. *Corollary*

Given any set A of linearly independent vectors of the inner product space V, we can construct an orthogonal basis $C \cup D$ of V such that $\operatorname{sp}(A) = \operatorname{sp}(C)$. Moreover, by dividing each vector in $C \cup D$ by its length, we obtain an orthonormal basis $C' \cup D'$ with $\operatorname{sp}(C') = \operatorname{sp}(A)$.

Proof. Exercise. ∎

B.4. Orthogonal Linear Transformations and Orthogonal Matrices

In this section we focus our attention on those linear transformations which preserve distance. This will lay the foundation for our investigation of groups of isometries of $\mathbf{R}^n$ endowed with the usual dot product, with particular emphasis on groups of isometries of $\mathbf{R}^2$ and $\mathbf{R}^3$ in Chapter 13.

B.4.1. *Definition*

A linear operator on an inner product space $(\mathbf{R}, V)$ is orthogonal if for all α and β in $V, \| \alpha - \beta \| = \| T(\alpha) - T(\beta) \|$.

Note that an orthogonal linear transformation is injective and so invertible by Corollary B.2.10.

Remark. Since $T(0) = 0, \| \alpha \| = \| \alpha - 0 \| = \| T(\alpha) - T(0) \| = \| T(\alpha) \|$; that is to say, T preserves lengths.

B.4.2. *Example*

Rotations and reflections in a line in $\mathbf{R}^2$; rotations about an axis and reflections in a plane in $\mathbf{R}^3$.

B.4.3. *Theorem*

If T is an orthogonal linear operator on V, then
$$<T(\alpha), T(\beta)> = <\alpha, \beta>$$
for all α and β in V.

Proof. Since
$$\|\alpha - \beta\|^2 = \|T(\alpha) - T(\beta)\|^2$$
we have
$$<\alpha - \beta, \alpha - \beta> = <T(\alpha) - T(\beta), T(\alpha) - T(\beta)>.$$
Therefore
$$\|\alpha\|^2 + \|\beta\|^2 - 2<\alpha, \beta> = \|T(\alpha)\|^2 + \|T(\beta)\|^2 - 2<T(\alpha, \beta>.$$
By the remark above, $\|\alpha\|^2 = \|T(\alpha)\|^2$ and $\|\beta\|^2 = \|T(\beta)\|^2$. Hence
$$<\alpha, \beta> = <T(\alpha), T(\beta)>. \blacksquare$$

Another way of saying this is that a linear transformation which preserves distances also preserves angles.

B.4.4. *Lemma*

A set of nonzero mutually orthogonal vectors is linearly independent.

Proof. If $\{\alpha_1, \alpha_2, ..., \alpha_k\}$ is a set of mutually orthogonal vectors and $\sum_{i=1}^{k} a_i \alpha_i = 0$, then
$$<\alpha_j, \sum_{i=1}^{k} a_i \alpha_i> = a_j <\alpha_j, \alpha_j> = 0.$$
Since $<\alpha_j, \alpha_j> \neq 0$, it follows that $a_j = 0$ for all $j, j = 1, 2, ..., k$. $\blacksquare$

B.4.5. *Theorem*

A linear operator T on $(V,< >)$ is orthogonal if, and only if, it takes an orthonormal basis to an orthonormal basis.

Proof. Suppose T is orthogonal and let $\{\alpha_1 \alpha_2,...,\alpha_n\}$ be an orthonormal basis. Since $\| \alpha_i \|=\|T(\alpha_i)\|$ for all α_i, the images of the α_i are all unit vectors. Also, because of B.4.3, $0 = <\alpha_i,\alpha_j> = <T(\alpha_i),T(\alpha_j)>$ for all $i \neq j$ and so the $T(\alpha_i)$ form an orthonormal set of vectors and so, by the preceding lemma, it is a basis.

Conversely, suppose $\{T(\alpha_1),T(\alpha_2),...,T(\alpha_n)\}$ is an orthonormal basis. Let $\alpha = \sum_{i=1}^{n} a_i\alpha_i$ and $\beta = \sum_{i=1}^{n} b_i\alpha_i$. Then

$$\| \alpha - \beta \|^2 = < \sum_{i=1}^{n}(a_i - b_i)\alpha_i, \sum_{i=1}^{n}(a_i - b_i)\alpha_i > = \sum_{i=1}^{n}(a_i - b_i)^2$$

and

$$\| T(\alpha) - T(\beta) \|^2 = \| T(\alpha - \beta) \|^2 = < \sum_{i=1}^{n}(a_i - b_i)T(\alpha_i), \sum_{i=1}^{n}(a_i - b_i)T(\alpha_i) >$$

$$= \sum_{i=1}^{n}(a_i - b_i)^2$$

since $\{T(\alpha_1),T(\alpha_2),...,T(\alpha_n)\}$ is an orthonormal basis. ∎

To find the inverse of a matrix is often a daunting task. For one type of matrix, however, all we need to do is to switch columns and rows. Such matrices are closely related to orthogonal linear transformations as we shall see.

B.4.6. *Definition*

The *transpose* of a matrix $A=[a_{ij}]$ is the matrix $A^T =[a_{ij}^*]$ where $a_{ij}^* = a_{ji}$. The matrix A is said to be *orthogonal* if $A^{-1} = A^T$.

B.4.7. Theorem

Let $A = (\alpha_1, \alpha_2, ..., \alpha_n)$ be an orthonormal basis of V and T a linear operator on V. Then T is an orthogonal linear operator if, and only if, $[T]_A$ is an orthogonal matrix.

Proof. Suppose T is orthogonal. By the previous theorem, $\{T(\alpha_1), T(\alpha_2), ..., T(\alpha_n)\}$ is an orthonormal basis of V. If $[T]_A = [a_{ij}]$, then

$$< \sum_{i=1}^{n} a_{is}\alpha_i, \sum_{i=1}^{n} a_{it}\alpha_i >= \delta_{st}$$ where $\delta_{st} = 0$ is $s \neq t$ and $\delta_{ss} = 1$ for all s and t.

But

$$< \sum_{i=1}^{n} a_{is}\alpha_i, \sum_{i=1}^{n} a_{it}\alpha_i >= \sum_{i=1}^{n} a_{is}a_{it} = \delta_{st}$$

using the bilinearity of the inner product and the fact that A is an orthonormal basis. Therefore $[T]_A [T]_A^T = I_n$, proving that $[T]_A$ is orthogonal.

Conversely, assume that $[T]_A = [a_{ij}]$ is orthogonal. Then

$$\sum_{i=1}^{n} a_{is}a_{it} = \delta_{st}$$ since $[a_{ij}].[a_{ij}]^T = I_n$

and so

$$< \sum_{i=1}^{n} a_{is}\alpha_i, \sum_{i=1}^{n} a_{it}\alpha_i >= \sum_{i=1}^{n} a_{is}a_{it} = \delta_{st}$$

since A is an orthonormal basis. But $T(\alpha_s) = \sum_{i=1}^{n} a_{is}\alpha_i$. Therefore

$$< T(\alpha_s), T(\alpha_t) >= \delta_{st}$$

showing that $\{T(\alpha_1), T(\alpha_2), ..., T(\alpha_n)\}$ is an orthonormal basis. By the previous theorem, T is orthogonal. ∎

B.5. Determinants

Although determinants arose naturally in the solution of systems of n linear equations in n unknowns, they are very rarely used for that purpose since the computations very quickly boil over for even fairly small values of n. Nevertheless, from a theoretical standpoint, they are indispensible. We give a brief and somewhat abstract treatment of them to derive their most fundamental properties.

In what follows, $\alpha_j = (a_{1j}, a_{2j}, ..., a_{ij}, ..., a_{nj})$, an n-dimensional row vector and α_j^T the corresponding n-dimensional column vector.

B.5.1. *Definition*

A map "det" from the set of all $n \times n$ matrices over a field F to F is a *determinant* provided that det satisfies the following three conditions:
   (i) For all k, $k=1, 2, ..., n$ and for all b and c in F,

$$\det(\alpha_1^T, \alpha_2^T, ..., b\beta_k^T + c\gamma_k^T, ..., \alpha_n^T)$$
$$= b\det(\alpha_1^T, \alpha_2^T, ..., \beta_k^T, ..., \alpha_n^T) + c\det(\alpha_1^T, \alpha_2^T, ..., \gamma_k^T, ..., \alpha_n^T)$$

   (ii) If two adjacent columns of a matrix are equal, then its determinant is 0.

   (iii) The determinant of the identity matrix is the unity of the field.

For the moment, we shall set aside the question of the existence and uniqueness of such a map and derive some of its properties.

B.5.2. *Theorem*

If a map det has properties (i), (ii) and (iii) above, then:
   (1)
$$\det(\alpha_1^T, \alpha_2^T, ..., \alpha_k^T + a\alpha_{k+1}^T, \alpha_{k+1}^T, ..., \alpha_n^T) = \det(\alpha_1^T, \alpha_2^T, ..., \alpha_k^T, \alpha_{k+1}^T, ..., \alpha_n^T);$$
(The value of a determinant is unchanged if a constant multiple of one column is added to an adjacent column.)
   (2) $\det(\alpha_1^T, \alpha_2^T, ..., \alpha_k^T, \alpha_{k+1}^T, ..., \alpha_n^T) = -\det(\alpha_1^T, \alpha_2^T, ..., \alpha_{k+1}^T, \alpha_k^T, ..., \alpha_n^T);$
(If two adjacent columns of a determinant are interchanged the value of the determinant changes sign.)
   (3) if any two columns of A are equal, then $\det A = 0$;
   (4)
$$\det(\alpha_1^T, \alpha_2^T, ..., \alpha_s^T, ..., \alpha_t^T, ..., \alpha_n^T) = -\det(\alpha_1^T, \alpha_2^T, ..., \alpha_t^T, ..., \alpha_s^T, ..., \alpha_n^T).$$
(If any two columns are interchanged the value of the determinant changes sign.)

(5)
$$\det(\alpha_1^T, \alpha_2^T, ..., \alpha_s^T, ..., \alpha_t^T, ..., \alpha_n^T) = \det(\alpha_1^T, \alpha_2^T, ..., \alpha_s^T + a\alpha_t^T, ..., \alpha_t^T, ..., \alpha_n^T).$$
(The value of a determinant is unchanged if a constant multiple of *any* column is added to *another*)

Proof. (1) By (i) of the definition with $\beta_k = \alpha_k$, $b = 1$, $c = a$ and $\gamma_k = \alpha_{k+1}$ we have
$$\det(\alpha_1^T, \alpha_2^T, ..., \alpha_k^T + a\alpha_{k+1}^T, \alpha_{k+1}^T, ..., \alpha_n^T)$$

$$= \det(\alpha_1^T, \alpha_2^T, ..., \alpha_k^T, \alpha_{k+1}^T, ..., \alpha_n^T) + a\det(\alpha_1^T, \alpha_2^T, ..., \alpha_{k+1}^T, \alpha_{k+1}^T, ..., \alpha_n^T).$$
The last determinant has two adjacent columns equal and so its value is 0 by (ii). Hence
$$\det(\alpha_1^T, \alpha_2^T, ..., \alpha_k^T + a\alpha_{k+1}^T, \alpha_{k+1}^T, ..., \alpha_n^T) = \det(\alpha_1^T, \alpha_2^T, ..., \alpha_k^T, \alpha_{k+1}^T, ..., \alpha_n^T).$$

(2) Adding the $(k+1)^{\text{st}}$ column to the kth column and appealing to (i), we get
$$\det(\alpha_1^T, \alpha_2^T, ..., \alpha_k^T, \alpha_{k+1}^T, ..., \alpha_n^T) = \det(\alpha_1^T, \alpha_2^T, ..., \alpha_k^T + \alpha_{k+1}^T, \alpha_{k+1}^T, ..., \alpha_n^T).$$

Appealing to (i) again, we obtain
$$\det(\alpha_1^T, ..., \alpha_k^T + \alpha_{k+1}^T, \alpha_{k+1}^T, ..., \alpha_n^T) = \det(\alpha_1^T, ..., \alpha_k^T + \alpha_{k+1}^T, -\alpha_k^T, ..., \alpha_n^T)$$
by subtracting the kth column from the $(k+1)^{\text{st}}$. But
$$\det(\alpha_1^T, \alpha_2^T, ..., \alpha_k^T + \alpha_{k+1}^T, -\alpha_k^T, ..., \alpha_n^T) = \det(\alpha_1^T, \alpha_2^T, ..., \alpha_{k+1}^T, -\alpha_k^T, ..., \alpha_n^T)$$
by adding the $(k+1)^{\text{st}}$ column to the kth. Finally, applying (i), we get
$$\det(\alpha_1^T, \alpha_2^T, ..., \alpha_{k+1}^T, -\alpha_k^T, ..., \alpha_n^T) = -\det(\alpha_1^T, \alpha_2^T, ..., \alpha_{k+1}^T, \alpha_k^T, ..., \alpha_n^T)$$
and so
$$\det(\alpha_1^T, \alpha_2^T, ..., \alpha_k^T, \alpha_{k+1}^T, ..., \alpha_n^T) = -\det(\alpha_1^T, \alpha_2^T, ..., \alpha_{k+1}^T, \alpha_k^T, ..., \alpha_n^T).$$

(3) If the jth column is the same as the kth column where $j < k$, then by a series of $k - j - 1$ adjacent column switches we can move the kth column adjacent to the jth one. The value of this last determinant is 0 by (2).

In passing from the original determinant to the final one we have changed signs $k - j - 1$ times while leaving invariant the absolute value of the determinant. Therefore the value of the original determinant is 0.

(4) To switch the positions of the sth column with the tth column requires $2(t - s) - 1$ adjacent column interchanges and so the value of the

last determinant is $(-1)^{2(t-s)-1} = -1$ times the value of the original determinant by (ii).

(5) This follows immediately by applying (1) and (3). ∎

B.5.3. *Notation*

If $\pi \in S_n$, we set $\operatorname{sgn} \pi = 1$ if π is an even permutation and $\operatorname{sgn} \pi = -1$ if π is odd.

B.5.4. *Corollary*

If $A = (\alpha_1^T, \alpha_2^T, ..., \alpha_n^T)$ and $\pi \in S_n$, then

$$\det A = \operatorname{sgn} \pi \det(\alpha_{\pi(1)}^T, \alpha_{\pi(2)}^T, ..., \alpha_{\pi(n)}^T).$$

Proof. Every permutation is the product of transpositions and, by (iv) of the previous theorem, every transposition of columns multiplies the determinant by −1. ∎

So far we are unable to compute the value of even the simplest determinant. The next theorem and corollary give an explicit formula which we can use to compute. Note, however, that what we shall have proved is that *if a function det exists possessing the properties in Definition B.5.1, then the function is of the given form.* Uniqueness is thus established but the question of existence is still unresolved.

B.5.5. *Theorem*

Suppose A, B and C are $n \times n$ matrices with $B = (\beta_1^T, \beta_2^T, ..., \beta_n^T)$, $A = (\alpha_1^T, \alpha_2^T, ..., \alpha_n^T)$ and $C = BA$. Then

$$\det(C) = \det B \sum_{\pi \in S_n} \operatorname{sgn} \pi a_{\pi(1)1} a_{\pi(2)2} ... a_{\pi(n)n}.$$

Proof. We compute using the bilinearity in the columns.

$$C = BA = (\beta_1^T, \beta_2^T, ..., \beta_n^T) \begin{pmatrix} a_{11} & a_{12} & a_{13} .....a_{1n} \\ a_{21} & a_{22} & a_{23} .....a_{2n} \\ \vdots & \vdots & \vdots & \vdots \\ a_{n1} & a_{n2} & a_{n3} .....a_{nn} \end{pmatrix}$$

and so

$$\det(C) = \det(\sum_{j=1}^{n} a_{j1}\beta_j^T, \sum_{j=1}^{n} a_{j2}\beta_j^T, ..., \sum_{j=1}^{n} a_{jn}\beta_j^T)$$
$$= \sum_{k_1,k_2,...,k_n} a_{k_1 1} a_{k_2 2} ... a_{k_n n} \det(\beta_{k_1}^T, \beta_{k_2}^T, ..., \beta_{k_n}^T)$$

where the sum is taken over all possible sequences of length n over $\{1,2,3,...,n\}$ (there are n^n of them).

The determinant $\det(\beta_{k_1}^T, \beta_{k_2}^T, ..., \beta_{k_n}^T) = 0$ whenever two of the k's are equal by (3) of the previous theorem. Therefore the only (possibly) nonzero contribution from $\det(\beta_{k_1}^T, \beta_{k_2}^T, ..., \beta_{k_n}^T)$ will occur when the k's are distinct, that is, when $k_1, k_2, ..., k_n$ is a permutation of $\{1,2,...,n\}$. Moreover, $\det(\beta_{\pi(1)}^T, \beta_{\pi(2)}^T, ..., \beta_{\pi(n)}^T) = \text{sgn } \pi \det B$ by the corollary above. Hence

$$\det(C) = \det B \sum_{\pi \in S_n} \text{sgn } \pi a_{\pi(1)1} a_{\pi(2)2} ... a_{\pi(n)n}. \blacksquare$$

B.5.6. Corollary

$$\det A = \sum_{\pi \in S_n} \text{sgn } \pi a_{\pi(1)1} a_{\pi(2)2} ... a_{\pi(n)n}.$$

Proof. Set $B = I_n$ and use (iii) of the definition. $\blacksquare$

Observe that our discussion above gives us the following theorem:

B.5.7. *Theorem*

$$\det AB = \det A.\det B.$$

Proof. The theorem follows immediately from Theorem B.5.5 and Corollary B.5.6. $\blacksquare$

If T is a linear operator on V whose matrix relative to a basis A is $[T]_A$, we define the determinant of T to be $\det[T]_A$. If we choose another basis B, then $[T]_B = [P]_A^{-1}[T]_A[P]_A$ by Theorem B.2.16 and so $\det[T]_B = \det[P]_A^{-1}[T]_A[P]_A = \det([P]_A)^{-1}\det[T]_A\det[P]_A$ by B.2.18 and the previous theorem. Therefore $\det[T]_A = \det[T]_B$ and so the determinant of a linear transformation is independent of the matrix representing it.

B.5.8. *Lemma*

The columns of an $n \times n$ matrix $A = (\alpha_1^T, \alpha_2^T, ..., \alpha_n^T)$ over the field F are linearly dependent if, and only if, $\det A = 0$.

Proof. Suppose the columns are linearly dependent. Then for some j,
$$\alpha_j = a_1\alpha_1 + a_2\alpha_2 + ... + a_{j-1}\alpha_{j-1} + a_{j+1}\alpha_{j+1} + ... + a_n\alpha_n.$$
Subtracting a_i times the ith column, $i = 1, 2, ..., j-1, j+1, ..., n$ from the jth column, we get a matrix A' whose determinant is the same as $\det A$ and whose j^{th} column consists entirely of 0's By the linearity in the columns, we conclude that $\det A' = 0$ and so $\det A = 0$.

Conversely, suppose the columns are linearly independent. Then $\text{sp}\{\alpha_1, \alpha_2, ..., \alpha_n\} = F^n$ and so there exist $b_{ij} \in F$ such that
$$\sum_{i=1}^n b_{ij}\alpha_i = \varepsilon_j, j = 1, 2, ..., n.$$
Setting $B = (b_{ij})$, we get $AB = I_n$, the $n \times n$ identity matrix. Hence
$$1 = \det I_n = \det AB = \det A.\det B$$
and so $\det A \neq 0$. ∎

B.5.9. *Theorem*

A linear operator $T : V \to V$ is invertible if, and only if, $\det T \neq 0$.

Proof. If T is invertible, then
$$1 = \det I = \det TT^{-1} = \det T.\det T^{-1}$$
and so $\det \neq 0$.

Conversely, suppose $\det T \neq 0$ and let A be a basis of V. By the lemma, the columns of $[T]_A$ are linearly independent. Therefore $[T]_A[\alpha]_A^T = \mathbf{0}$ implies $[\alpha]_A^T = \mathbf{0}$ where $\mathbf{0}$ is the n-dimensional zero column vector. But

$$[T]_A[\alpha]_A^T = [T(\alpha)]_A^T$$

by B.2.14. and so the null space of T consists only of the zero vector. By Theorem B.2.4, T is injective and so bijective by B.2.10. Therefore T is invertible. ∎

We have still to prove that a map det possessing the properties listed in Definition B.5.1 exists. There are essentially two ways this can be done:

    (i) If $A = (a_{ij})$, we can define $\det A = \sum_{\pi \in \Sigma_n} \operatorname{sgn} \pi a_{\pi(1)1} a_{\pi(2)2} \ldots a_{\pi(n)n}$ and show that this map satisfies the conditions in Definition B.5.1;

    (ii) (a) We define a map for 1×1 matrices by $\det(a) = a$, show quite trivially that this satisfies the required properties;

        (b) We assume inductively that a determinant map exists for $(n-1) \times (n-1)$ matrices and show that, based on this, we can construct a determinant map for $n \times n$ matrices.

We choose the first method and outline the second.

B.5.10. *Theorem*

If $A = (a_{ij})$, the map $\det A = \sum_{\pi \in S_n} \operatorname{sgn} \pi a_{\pi(1)1} a_{\pi(2)2} \ldots a_{\pi(n)n}$ has the properties listed in Definition B.5.1.

Proof. Property (i): Let

$$A = (\alpha_1^T, \alpha_2^T, \ldots, \alpha_k^T, \ldots, \alpha_n^T) \text{ where } \alpha_k^T = b\beta_k^T + c\gamma_k^T,$$
$$B = (\alpha_1^T, \alpha_2^T, \ldots, \beta_k^T, \ldots, \alpha_n^T) \text{ and } C = (\alpha_1^T, \alpha_2^T, \ldots, \gamma_k^T, \ldots, \alpha_n^T).$$

Then

$$\det A = \sum_{\pi \in S_n} \operatorname{sgn} \pi a_{\pi(1)1} a_{\pi(2)2} \ldots (bb_{\pi(k)k} + cc_{\pi(k)k}) \ldots a_{\pi(n)n}$$

$$= b \sum_{\pi \in S_n} \operatorname{sgn} \pi a_{\pi(1)1} a_{\pi(2)2} \dots b_{\pi(k)k} \dots a_{\pi(n)n} + c \sum_{\pi \in S_n} \operatorname{sgn} \pi a_{\pi(1)1} a_{\pi(2)2} \dots c_{\pi(k)k} \dots a_{\pi(n)n}$$

$$= b \det B + c \det C.$$

Property (ii): Suppose $A = (a_{ij})$ and $a_{ik} = a_{i(k+1)}$ for all $i, i = 1, 2, \dots, n$. For a given permutation π consider the terms

$$\operatorname{sgn} \pi a_{\pi(1)1} a_{\pi(2)2} \dots a_{\pi(k)k} a_{\pi(k+1)(k+1)} \dots a_{\pi(n)n}$$

and

$$\operatorname{sgn} \pi \sigma a_{\pi\sigma(1)1} a_{\pi\sigma(2)2} \dots a_{\pi\sigma(k)k} a_{\pi\sigma(k+1)(k+1)}$$

where σ is the transposition $(k, k+1)$. The second term is

$$\operatorname{sgn} \pi \sigma a_{\pi(1)1} a_{\pi(2)2} \dots a_{\pi(k+1)k} a_{\pi(k)(k+1)} \dots a_{\pi(n)n}.$$

But

$$a_{\pi(k+1)k} = a_{\pi(k+1)(k+1)} \text{ and } a_{\pi(k)(k+1)} = a_{\pi(k)k}$$

whence the two terms cancel since $\operatorname{sgn} \pi \sigma = -\operatorname{sgn} \pi$. Therefore $\det A = 0$.

Property (iii): Let $I_n = (\delta_{ij})$ where $\delta_{ij} = 0$ if $i \neq j$ and $\delta_{ii} = 1$. Then $\det I_n = \sum_{\pi \in S_n} \operatorname{sgn} \pi \delta_{\pi(1)1} \delta_{\pi(2)2} \dots \delta_{\pi(n)n}$ and every term in the sum is zero except for the term $\delta_{11} \delta_{22} \dots \delta_{nn}$ which contributes 1 to the sum. Hence $\det I_n = 1$. ■

B.5.11. *Corollary*

$$\det A = \det A^T.$$

Proof. Let $A = (a_{ij})$, $A^T = (a_{ij}^*)$ where $a_{ij} = a_{ji}^*$. Then

$$\det A = \sum_{\pi \in S_n} \operatorname{sgn} \pi a_{\pi(1)1} a_{\pi(2)2} \dots a_{\pi(n)n}$$

and

$$\det A^T = \sum_{\pi \in S_n} \operatorname{sgn} \pi a_{\pi(1)1}^* a_{\pi(2)2}^* \dots a_{\pi(n)n}^*$$

$$= \sum_{\pi \in S_n} \operatorname{sgn} \pi a_{1\pi(1)} a_{2\pi(2)} \dots a_{n\pi(n)}$$

$$= \sum_{\pi \in S_n} \operatorname{sgn} \pi a_{\pi^{-1}(1)1} a_{\pi^{-1}(2)2} \dots a_{\pi^{-1}(n)n}.$$

But as π varies over S_n, so does π^{-1} and $\operatorname{sgn}\pi = \operatorname{sgn}\pi^{-1}$. Therefore

$$\det(A^T) = \sum_{\pi^{-1}\in S_n} \operatorname{sgn}\pi^{-1} a_{\pi^{-1}(1)1} a_{\pi^{-1}(2)2}\cdots a_{\pi^{-1}(n)n}$$

$$= \sum_{\pi\in S_n} \operatorname{sgn}\pi\, a_{\pi(1)1} a_{\pi(2)2}\cdots a_{\pi(n)n}$$

$$= \det A. \blacksquare$$

The reader is encouraged to fill in the gaps in the outline given below of the second method of proving the existence of determinants. This method has the added bonus of supplying us with the well known process of evaluating determinants. It is, however, a considerably longer proof.

B.5.12. *Theorem*

The determinant map exists.

Proof. By induction on n, the dimension of the matrix.
If $n=1$, define $\det(a) = a$. It is clear that this has the required properties
Assume that det exists for $(n-1)\times(n-1)$ matrices and, if A is an $n\times n$ matrix, define

$$\det A = \sum_{j=1}^{n} a_{ij} co(A_{ij})$$

where A_{ij} is the $(n-1)\times(n-1)$ matrix obtained from A by deleting the ith row and j^{th} column of A and

$$co(A_{ij}) = (-1)^{i+j}\det A_{ij}.$$

($co(A_{ij})$ is the (i,j) *cofactor* of A.)

Based on the inductive hypothesis that $\det(A_{ij})$ satisfies the properties listed in Definition B.5.1, it is not too hard to prove that det as defined above also possesses the same properties. $\blacksquare$

B.6. Eigenvalues and Eigenvectors

If T is a linear operator on V, it is natural to ask if there are any vectors which are fixed by T and if so, how to find them. For example, if T is a rotation in $\mathbf{R}^3$, how do we find the axis of rotation?

More generally, can we find vectors whose direction is fixed under T? If there is such a vector, say α, then there is a scalar s such that $T(\alpha) = s\alpha$. In this last section we show how to find these vectors and their associated scalars.

B.6.1. *Definition*

An *eigenvector* of a linear operator T on V is a nonzero vector α such that $T(\alpha) = s\alpha$ for some scalar $s \in F$. The scalar s is the *eigenvalue belonging to* α and the vector α is said to *belong* to s. If A is an $n \times n$ matrix, then α is an eigenvector belonging to an eigenvalue s if $A\alpha^T = s\alpha^T$.

B.6.2. *Examples*

(i) Every nonzero vector of the null space $N(T)$ of T (if there is one) is an eigenvector belonging to 0.

(ii) If T is a reflection in a plane in $\mathbf{R}^3$ passing through the origin, every nonzero vector in the plane is an eigenvector belonging to the eigenvalue 1 and any nonzero vector orthogonal to the plane is an eigenvector belonging to the eigenvalue -1.

B.6.3. *Theorem*

If T is a linear operator on V, then the scalar s is an eigenvalue of T if, and only if, $\det(T - sI) = 0$.

Proof. Suppose s is an eigenvalue of T. Then there exists a nonzero vector α such that $T(\alpha) = s\alpha$. Therefore $(T - sI)(\alpha) = 0$ and so the null space of the linear operator $T - sI$ is not trivial. Hence $T - sI$ is not invertible and so, by Theorem B.5.7, $\det(T - sI) = 0$.

Conversely, if $\det(T - sI) = 0$, then, by Theorem B.5.8, T is not invertible and so the null space of T is not trivial. Therefore there exists $\alpha \neq 0$ such that $(T - sI)(\alpha) = 0$. Hence $T(\alpha) = s\alpha$. ∎

B.6.4. Definition

The polynomial $\det(T - xI)$ is the *characteristic polynomial* of the linear operator T.

To compute the characteristic polynomial of T we need to find $[T]_A$ for some basis A. It is left to the reader to prove that no matter what matrix we use to represent T, we obtain the same characteristic polynomial (see Problem 21).

B.6.5. Example

Consider the linear operator whose matrix relative to the standard basis of $\mathbf{R}^2$ is

$$M = \begin{pmatrix} \cos\theta & \sin\theta \\ \sin\theta & -\cos\theta \end{pmatrix}.$$

To find the eigenvalues, we solve

$$\det \begin{pmatrix} \cos\theta - x & \sin\theta \\ \sin\theta & -\cos\theta - x \end{pmatrix} = 0 \text{ or } -\cos^2\theta + x^2 - \sin^2\theta = 0.$$

We get $x = \pm 1$.

To find the eigenvector belonging to 1, we solve

$$\begin{pmatrix} \cos\theta - 1 & \sin\theta \\ \sin\theta & -\cos\theta - 1 \end{pmatrix}\begin{pmatrix} x \\ y \end{pmatrix} = \begin{pmatrix} 0 \\ 0 \end{pmatrix}.$$

Multiplying we obtain the two equations

$$(\cos\theta - 1)x + \sin\theta\, y = 0$$
$$\sin\theta\, x - (\cos\theta + 1)y = 0.$$

These two equations are not independent and so, solving the first one, we obtain the line spanned by $(-\sin\theta, \cos\theta - 1)$. Using some trigonometric identities, this gives $(\cos\frac{\theta}{2}, \sin\frac{\theta}{2})$ provided $\sin\frac{\theta}{2} \neq 0$. If

$\sin \theta/2 = 0$, then $\theta/2 = 0$ or π and so the matrix $M = \begin{pmatrix} 1 & 0 \\ 0 & -1 \end{pmatrix}$.   This

represents a reflection in the x-axis.

For the eigenvector belonging to -1, we solve

$$\begin{pmatrix} \cos\theta+1 & \sin\theta \\ \sin\theta & -\cos\theta+1 \end{pmatrix}\begin{pmatrix} x \\ y \end{pmatrix} = \begin{pmatrix} 0 \\ 0 \end{pmatrix}.$$

An easy computation yields $(-\sin\theta/2, \cos\theta/2)$. Hence M is a reflection

in the line determined by $(\cos\theta/2, \sin\theta/2)$. Note that the eigenvector

belonging to -1 is orthogonal to $(\cos\theta/2, \sin\theta/2)$. ■

B. 7   Exercises

1) If $A = \begin{pmatrix} 1 & 2 \\ 3 & 1 \\ 2 & 1 \end{pmatrix}$, find infinitely many matrices B such that $BA = I_2$.

Also prove that there is no matrix C such that $AC = I_3$.

2) If $A = (a_{ij})$ is an $n \times n$ matrix, define the *trace* of A to be $\text{tr}(A) = \sum_{i=1}^{n} a_{ii}$.

Prove

(i) $\text{tr}(A+B) = \text{tr}(A) + \text{tr}(B)$;

(ii) $\text{tr}(AB) = \text{tr}(BA)$.

3) Prove that over a field of characteristic 0, there is no solution in $n \times n$ matrices of the equation $AB - BA = I_n$.

4) Prove the associativity of matrix multiplication by using the fact that composition of maps is associative and Theorem B.2.18.

5) If A and B are bases of a vector space V and $\xi \in V$, find a relationship between $[\xi]_A$, $[\xi]_B$ and $[P]_A$ where P is the linear operator taking the basis A to the basis B.

6) (i) Prove that if X and Y are subspaces of a vector space V, then $X + Y$ and $X \cap Y$ are subspaces of V.
    (ii) Show that $\dim(X + Y) = \dim X + \dim Y - \dim X \cap Y$.

7) Prove that the ordered set of vectors $A = ((1,1,1),(0,1,1),(0,0,1))$ is a basis of $\mathbf{R}^3$ and find $[(1,2,-3)]_A$.

8) Prove that if V, W and X are vector spaces over F and $T:V \rightarrow W, S:W \rightarrow X$, then $\rho(ST) \leq \min\{\rho(S), \rho(T)\}$. When is $\rho(ST) = \rho(T)$?

9) If A is an $m \times n$ matrix over $F (n > m)$, prove that the null space of the linear transformation whose matrix relative to the standard bases of F^n and F^m is A, has at least dimension $n - m$. Note that this says that the system of linear equations $AX = 0$ has a solution space of dimension at least $n - m$.

10) Find all linear operators on $\mathbf{R}^3$ which carry the plane $x + y + z = 0$ to the plane $2x - y - 2z = 0$.

11) If A is an $m \times n$ matrix and B an $n \times m$ matrix, prove that $AB - I_m$ is invertible iff $BA - I_n$ is invertible.
[Hint: Prove that $\det(AB - I_m) = 0$ iff $\det(BA - I_n) = 0$ and think of eigenvectors.]

12) (Vandermonde determinant) Prove that

$$\det \begin{pmatrix} 1 & 1 & 1 \\ a_1 & a_2 & a_3 \\ a_1^2 & a_2^2 & a_3^2 \end{pmatrix} = (a_2 - a_1)(a_3 - a_1)(a_3 - a_2).$$

Generalize to an $n \times n$ matrix.

13) If T is a linear operator on V and s an eigenvalue of T, prove that
$$E_s = \{\alpha \in V \mid T(\alpha) = s\alpha\}$$
is a subspace of V. (E_s is the *eigenspace* belonging to s.)

14) Let $(V, <\ >)$ be an inner product space and T an orthogonal linear operator on V.

   (i) Prove that the only possible eigenvalues of T are ± 1.

   (ii) If both 1 and -1 are eigenvalues of T, prove that E_1 is orthogonal to E_{-1}, i.e., every vector in E_1 is orthogonal to every vector in E_{-1}. (See previous problem for the definition of E_1 and E_{-1}.)

15) A linear operator T on V is a *projection* of V if $T^2 = T$.

   (i) Prove that if T is a projection of V, then $V = T(V) \oplus N(T)$ where $N(T)$ is the null space of T.

   (ii) Prove that if T is a projection of V, then there exists a basis A of V such that $[T]_A = \begin{pmatrix} I_k & \mathbf{0} \\ \mathbf{0} & \mathbf{0} \end{pmatrix}$ where k is $\dim T(V)$ and the $\mathbf{0}$'s are zero matrices of the appropriate dimensions.

16) An $n \times n$ matrix A is said to be *symmetric* if $A^T = A$. If A is a 2×2 real symmetric matrix, prove that its eigenvalues are real.

17) If A is a symmetric $n \times n$ matrix over $\mathbf{R}$, prove that eigenvectors belonging to different eigenvalues are orthogonal relative to the ordinary dot product.

18) Determine the characteristic polynomial of the matrix
$$\begin{pmatrix} 0 & 1 & 0 & 0 & 0 \\ 1 & 0 & 1 & 0 & 0 \\ 0 & 1 & 0 & 1 & 0 \\ 0 & 0 & 1 & 0 & 1 \\ 0 & 0 & 0 & 1 & 0 \end{pmatrix}.$$
Generalize to an $n \times n$ matrix of the same type.

19) Let V be a vector space with a basis $A = (\alpha_1, \alpha_2, ..., \alpha_n)$ over a field F and let $a_1, a_2, ..., a_{n-1}$ be elements of F. Define a linear operator on V

by $T(\alpha_i) = \alpha_{i+1}$ if $i < n$ and $T(\alpha_n) = \sum_{i=1}^{n-1} a_i \alpha_i$. Determine $[T]_A$ and the characteristic polynomial of T.

20) Let T be a linear operator on the finite-dimensional vector space V. Prove that for some positive integer n, $\operatorname{Im} T^n \cap \ker T^n = (0)$. Hence show that $V = \operatorname{Im} T^n \oplus \ker T^n$.

21) Prove that the characteristic polynomial of a linear operator is independent of the matrix chosen to represent the operator.

22) Do A and A^T have the same eigenvalues? The same eigenvectors?

23) If A is a 3×3 invertible matrix over $\mathbf{R}$, show that the volume of the image of the unit cube with edges parallel to the three coordinate axes and a vertex at $(0,0,0)$ is given by $\det(A\varepsilon_1, A\varepsilon_2, A\varepsilon_3)$.

24) Let T be a linear operator on an n-dimensional vector space with eigenvalues $s_1, s_2, ..., s_n$.
Prove that $\operatorname{tr} T = \sum_{i=1}^{n} s_i$ and that $\det T = s_1 s_2 ... s_n$.

25) Let P be a real matrix such that $P^T = P^2$. What are the possible eigenvalues of P?

26) If A is an $n \times n$ matrix with n linearly independent eigenvectors, prove that there exists a matrix P such that $P^{-1}AP$ is a diagonal matrix.

Index